PESTICIDE DERMATOSES

DERMATOLOGY: CLINICAL AND BASIC SCIENCE SERIES

Series Editor Howard I. Maibach, M.D.

Published Titles:

Bioengineering of the Skin: Cutaneous Blood Flow and Erythema
Enzo Berardesca, Peter Elsner, and Howard I. Maibach

Bioengineering of the Skin: Water and the Stratum Corneum
Peter Elsner, Enzo Berardesca, and Howard I. Maibach

Bioengineering of the Skin: Methods and Instrumentation
Enzo Berardesca, Peter Elsner, Klaus P. Wilhelm, and Howard I. Maibach

Bioengineering of the Skin: Skin Surface, Imaging, and Analysis
Klaus P. Wilhelm, Peter Elsner, Enzo Berardesca, and Howard I. Maibach

Dermatologic Research Techniques
Howard I. Maibach

Pigmentation and Pigmentary Disorders
Norman Levine

Protective Gloves for Occupational Use
Gunh Mellström, J.E. Walhberg, and Howard I. Maibach

Skin Cancer: Mechanisms and Human Relevance
Hasan Mukhtar

Human Papillomavirus Infections in Dermatovenereology
Gerd Gross and Geo von Krogh

Contact Urticaria Syndrome
Smita Amin, Arto Lahti, and Howard I. Maibach

Skin Reactions to Drugs
Kirsti Kauppinen, Kristiina Alanko, Matti Hannuksela, and Howard I. Maibach

Dermatologic Botany
Javier Avalos and Howard I. Maibach

Hand Eczema, Second Edition
Torkil Menné and Howard I. Maibach

Dry Skin and Moisturizers: Chemistry and Function
Marie Loden and Howard I. Maibach

The Irritant Contact Dermatitis Syndrome
Pieter van der Valk, Pieter Coenrads, and Howard I. Maibach

Nickel and The Skin: Immunology and Toxiology
Howard I. Maibach and Torkel Menne

DERMATOLOGY: CLINICAL & BASIC SCIENCE SERIES

PESTICIDE DERMATOSES

Edited by
Homero Penagos, M.D.
Dermatologist
Occupational Dermatology Service
Social Security Bureau of Panama
David Chiriqui, Panama
Michael O'Malley, M.D.
Health Service
University of California, Davis
Associate Clinical Professor
UC Davis School of Medicine, Preventive Medicine and Epidemiology
Howard Maibach, M.D.
Professor of Dermatology
Department of Dermatology
University of California Medical Center
San Francisco

CRC Press
Boca Raton London New York Washington, D.C.

Library of Congress Cataloging-in-Publication Data

Pesticide dermatoses / [edited by] Homero Penagos, Michael O'Malley, Howard I. Maibach.
p. ; cm. -- (Dermatology)
Includes bibliographical references and index.
ISBN 1-56670-293-3 (alk. paper)
1. Dermatotoxicology. 2. Pesticides--Toxicology. 3. Contact dermatitis. I. Penagos, Homero. II. O'Malley, Michael, 1952- III. Maibach, Howard I. IV. Title. V. CRC series in dermatology.
[DNLM: 1. Dermatitis, Contact--etiology. 2. Agrochemicals--adverse effects. 3. Dermatitis, Contract--diagnosis. 4. Pesticides--adverse effects. WR 175 P476 2000]
RL803 .P47 2000
616.5′1--dc21 00-057951
CIP

This book contains information obtained from authentic and highly regarded sources. Reprinted material is quoted with permission, and sources are indicated. A wide variety of references are listed. Reasonable efforts have been made to publish reliable data and information, but the author and the publisher cannot assume responsibility for the validity of all materials or for the consequences of their use.

Direct all inquiries to CRC Press LLC, 2000 N.W. Corporate Blvd., Boca Raton, Florida 33431.

Trademark Notice: Product or corporate names may be trademarks or registered trademarks, and are used only for identification and explanation, without intent to infringe.

International Standard Book Number 1-56670-293-3
Library of Congress Card Number 00-057951
Printed in the United States of America 1 2 3 4 5 6 7 8 9 0
Printed on acid-free paper

Series Preface

Our goal in creating the *Dermatology: Clinical & Basic Science Series* is to present the insights of experts on emerging applied and experimental techniques and theoretical concepts that are, or will be, at the vanguard of dermatology. Books will cover new and exciting multidisciplinary areas of cutaneous research. We want these to be the books every physician will use in order to get acquainted with new methodologies in skin research. These books can be given to graduate students and postdoctoral fellows when they are looking for guidance to start a new line of research.

The series will consist of books that will be edited by an expert and will consist of chapters written by the leaders in a particular field. Books will be richly illustrated and will contain comprehensive bibliographies. Each chapter will provide substantial background material relevant to the particular subject. These books will contain detailed "tricks of the trade" and information as to where the methods presented can be safely applied. In addition, information on where to buy equipment and web sites that will be helpful in solving both practical and theoretical problems will be included.

We are working with these goals in mind and hope that as the books become available, the effort put in by the publisher, the book editors, and the individual authors will contribute to the further development of dermatology research and clinical practice. The extent that we achieve this goal will be determined by the utility of these books.

Howard I. Maibach, M.D.

Preface

Skin reactions account for about one third of the acute illnesses and injuries caused by pesticides. Nevertheless, they have received comparatively scant attention and many researchers find it particularly difficult to obtain basic information about results of animal tests, cases reported to the existing pesticide illness registries, and cases reported in the relevant scientific literature. This work attempts to simplify the task of understanding skin reactions to the bewilderingly diverse group of compounds sold as pesticides.

The initial sections of the book review scientific issues related to the skin effects of pesticides, including basic animal tests used to predict the occurrence of skin irritation and sensitization, related material on the basic elements of quantitative structure activity relationships (QSAR), and basic concepts in clinical dermatology. The chapter on diagnostic methods emphasizes the proper method of doing provocation testing for both delayed and immediate dermal allergic reactions. The section on treating skin reactions describes managing work exposures as well as basic treatment for irritant and allergic contact dermatitis. Chapter 5 explores evaluation of exposures, with emphasis on understanding the patterns of pesticide contact associated with particular work activities. Current means of regulating skin reactions to pesticides, and requirements for skin testing are discussed in Chapter 6. Basic concepts of epidemiology are described in Chapter 7, with an emphasis on understanding how differing study designs contribute to the understanding of cause and effect relationships. This material is supplemented with a brief field questionnaire and examination form for use in outbreak investigations (Appendix IA). The Tuohilampi questionnaire, a more extensive survey instrument comparable to those available for respiratory disease research from the British Medical Research Council and the American Thoracic Society, is also included in Appendix IB.

The chapter describing the reported skin effects of pesticides by use and structural category makes extensive use of available information on animal data testing and reported illness information for pesticide handlers from the California pesticide illness registry, as well as data reported in the public scientific literature. The animal and human data are integrated with QSAR principles where possible to emphasize the common effects of related compounds. This material is supplemented with a detailed review of available information on appropriate patch testing concentrations (Appendix II) and summary statistics for illnesses in handlers of individual pesticides (Appendix III).

The final, lengthy section of the book reviews the international experience with the skin effects of pesticides, relying upon pesticide registry data where available. The California experience with varying rates of pesticide dermatitis in different types of agricultural production is noted and considered in evaluating case reports, outbreak investigations, and comparative agricultural production, pesticide trade, and pesticide use data elsewhere. Relevant to the epidemiologic information about the skin effects of pesticides, prevalence data on common skin conditions from the U.S. population is given in Appendix IV.

Dedication

This work is dedicated to the many individual clinicians and scientists who have diligently documented the research that is reviewed here, especially Carola Liden, Toshio Matsushita, and Paolo Lisi.

Acknowledgments

The authors acknowledge the contributions of many colleagues who have worked for years in evaluating the skin effects of pesticides on both animals and humans. At the California Department of Pesticide Regulation, this especially includes Joshua Johnson, Peter Leung, Thomas Moore, and the late Rhoda Wang. Contributions from the existing pesticide illness registries in Washington, Arizona, Oregon, California, Texas, Nicaragua, and Costa Rica were crucial for the section reviewing international experience with the pesticide skin effects. These include especially Catherine Thomsen, Barbara Morrissey, Lynden Baum, Ernest Arvisu, and Roberto Castro.

The Editors

Homero Penagos, M.D., is a dermatologist with the Bureau of Social Security in the Chiriqui region of northern Panama. He has been active in the investigation of skin reactions to pesticides among agricultural workers and Panama and throughout Central America. The collaboration that resulted in this book began during the course of his Fulbright research with the University of California, San Francisco, and the California Department of Pesticide Regulation. His special interests include pigmentary skin reactions to pesticides, and skin reactions to paraquat.

Michael O'Malley, M.D., is employed at the Employee Health Service of the University of California, Davis, and is an Associate Clinical Professor at the UC Davis School of Medicine in Preventive Medicine and Epidemiology. He works as a medical consultant to the California Department of Pesticide Regulation, and has worked with the California Pesticide Illness Registry since 1987. He has written extensively on the health effects of pesticides, and has had a long-standing interest in their effects on the skin. His special interests include the skin effects of propargite, and the evaluation of nursery workers' skin reactions to plants and pesticides.

Howard Maibach, M.D., is a Professor of Dermatology at the University of California, San Francisco, and has been a leading contributor to experimental research in dermatopharmacology, and to clinical research on contact dermatitis, contact uticaria, and other skin conditions. His work on pesticides includes clinical research on glyphosate, chlorothalonil, sodium hypochlorite, norflurazon, diethyl toluamide, and isothiazolin compounds. His experimental work include research on the local lymph node assay, and the evaluation of the percutaneous absorption of atrazine, boron-containing pesticides, phenoxy herbicides, acetochlor, glyphosate, and many other compounds.

Contributors

Homero Penagos, M.D.
Division of Dermatology
Social Security Bureau of Panama

Howard I. Maibach, M.D.
Department of Dermatology
University of California
San Francisco

Michael O'Malley, M.D., M.P.H.
Employee Health Services
University of California
Davis Medical Consultant
California Department of Pesticide Regulation

Saqib T. Bashir, M.D.
Department of Dermatology
University of California
San Francisco

Ai-Lean Chew,
Department of Dermatology
University of California
San Francisco

Alicia Beatriz Cannavó, M.D.
Buenos Aires, Argentina

Luis Mirande, M.D.
Buenos Aires, Argentina

Luis Conde Salazar Gomez, M. D.,
Head, Occupational Dermatology Department
National Institute of Health
National Institute of Hygiene and Working Safety
Madrid, Spain

Paivikki Susitaival, M.D., Ph.D.
National Institute of Occupational Health, Finland
Kuopio Regional Institute of Occupational Health
Kuopio, Finland

Woraphong Manuskatti
Departament of Dermatology
University of California
San Francisco

Kenneth Abrams
Departament of Dermatology
University of California
San Francisco

Daniel J. Hogan
Departament of Dermatology
University of California
San Francisco

Thomas J. Ferguson, M.D.
Employee Health Services
University of California

Table of Contents

1 Methods for Testing the Irritation and Sensitization Potential of Agricultural Chemicals

Saqib J. Bashir and Howard I. Maibach

CONTENTS

1-56670-293-3/00/$0.00+$.50

This chapter considers the rationale and methodology for testing first, the irritation potential, and second, the sensitization potential of topically applied substances. Animal and human models are discussed, providing perspectives on their relative strengths and weaknesses in various settings.

1.1 THE NEED FOR MODELS

Accurate prediction of the irritation potential of agricultural materials is necessary to advise suitable health and safety precautions. Presently, animal models fulfill licensing criteria for regulatory bodies. However, a search for alternative methods is stimulated by difficulties in relating animal data to human settings, and also by humane motives.

Many aspects of irritation have been described, ranging from the visible erythema and edema to molecular mediators such as interleukins and prostaglandins. Therefore, a variety of *in vivo* and *in vitro* approaches to experimental assay are possible. However, no model assays inflammation in its entirety. Each model is limited by our ability to interpret and extrapolate the features of inflammation to the desired context. Therefore, predicting human responses based on data from non-human models requires particular care.

Various human experimental models have also been proposed, providing irritant data for the relevant species. However, these studies are also limited by pitfalls in interpretation and, of course, by the fear of applying new substances to human skin before their irritant potential has been evaluated.

1.2 ANIMAL AND HUMAN MODELS FOR TESTING IRRITATION

1.2.1 Draize Rabbit Models

The Draize model[1] and its modifications are commonly used to assay skin irritation using albino rabbits. Various governmental agencies have adopted these methods as standard test procedure. The procedure adopted in the U.S. Federal Hazardous Substance Act (FHSA) is described in Tables 1.1 and 1.2.[2-4] Table 1.3 compares this method with some other modifications of the Draize model.

TABLE 1.1
The Draize-FHSA Model

Number of animals:	6 albino rabbits (clipped)
Test sites:	2 x 1 in.^{2} sites on dorsum
	One site intact, the other abraded, e.g., with hypodermic needle
Test materials:	Applied undiluted to both test sites
	Liquids: 0.5 ml
	Solids/semisolids: 0.5 g
Occlusion:	1 in.^{2} surgical gauze over each test site
	Rubberized cloth over entire trunk
Occlusion period:	24 h
Assessment:	24 and 72 h
	Visual scoring system

Draize utilized the above scoring system to calculate the Primary Irritation Index (PII) — averaging the erythema scores and also averaging the edema scores of all sites (abraded and non-abraded). The two averages are added together to give the PII value. A value of <2 is considered non-irritating, 2-5 mildly irritating, and >5 severely irritating. A value of 5 defines an irritant by Consumer Product Safety Commission (CPSC) standards. Subsequent laboratory and clinical experience has demonstrated that the value judgments (i.e. non-, mild, severely irritating) proposed in 1944 require clinical judgment and perspective and should not be viewed in an absolute sense. Many materia0ls irritating to the rabbit are tolerated by human skin.

Although the Draize scoring system does not include vesiculation, ulceration, and severe eschar formation, all of the Draize-type tests are used to evaluate corrosion as well as irritation. When severe and potentially irreversible reactions occur, the test sites are further observed on days 7 and 14, or later if necessary.

Modifications to the Draize assay have attempted to improve its prediction of human experience. The model is criticized for inadequately differentiating between mild and moderate irritants. However, it serves well in hazard identification, often over-predicting the severity of human skin reactions.[4] Therefore, Draize assays continue to be recommended by regulatory bodies.

1.2.2 Cumulative Irritation Assays

Several assays study the effects of cumulative exposure to a potential irritant. Justice et al.[5] administered 7 applications of surfactant solutions at 10-min intervals to the clipped dorsum of albino mice. The test site was occluded with a rubber dam to prevent evaporation and the skin was examined microscopically for epidermal erosion.

Frosch et al.[6,7] described the guinea pig repeat irritation test (RIT) to evaluate protective creams against the chemical irritants sodium lauryl sulfate (SLS), sodium hydroxide (NaOH), and toluene. The irritants were applied daily for 2 weeks to shaved back skin of young guinea pigs. Barrier creams were applied to the test animals 2 h prior to and immediately after exposure to the irritant. Control animals were treated with the irritant only. Erythema was measured visually and by bioengineering methods using laser doppler flowmetry and transepidermal water loss. One barrier cream was effective against SLS and toluene, while the other tested was not.

TABLE 1.2
The Draize-FHSA Scoring System[2]

Erythema and Eschar Formation	Score
No erythema	0
Very slight erythema (barely perceptible)	1
Well-defined erythema	2
Moderate to severe erythema	3
Severe erythema (beet redness) to slight eschar formation (injuries in depth)	4
Edema Formation	
No edema	0
Very slight edema (barely perceptible)	1
Slight edema (edges of area well defined by definite raising)	2
Moderate edema (raised >1 mm)	3
Severe edema (raised >1 mm and extending beyond the area of exposure)	4

In a follow-up study, another allegedly protective cream failed to inhibit irritation caused by SLS and toluene and exaggerated irritation to NaOH, contrary to its recommended use.[6] The RIT is proposed as an animal model to test the efficacy of barrier creams, and further propose as a human version, described below.

TABLE 1.3
Examples of Modified Draize Irritation Method

	Draize	FHSA	DOT	FIFRA	OECD
No. of animals	3	6	6	6	6
Abrasion/intact	Both	Both	Intact	2 of each	Intact
Dose liquids	0.5 ml undiluted	0.5 ml undiluted	0.5 ml	0.5 ml undiluted	0.5 ml
Dose solids in solvent	0.5 g	0.5 g moistened	0.5 g moistened	0.5 g	0.5 g
Exposure period (h)	24	24	4	4	4
Examination (h)	24, 72	24, 72	4, 48	0.5, 1, 24, 48, 72	0.5, 1, 24, 48, 72
Removal of test materials	Not specified	Not specified	Skin washed	Skin wiped	Skin washed
Excluded from testing	—	—	—	Toxic materials pH <2 or >11.5	Toxic materials pH <2 or >11.5

Note: FHSA: Federal Hazardous Substance Act; DOT: Department of Transportation; FIFRA: Federal Insecticide, Fungicide, and Rodenticide Act; OECD: Organization for Economic Cooperation and Development.

Source: From Esther and Maibach

Repeat application patch tests have been developed to rank the irritant potential of products. Putative irritants are applied to the same site for 3-21 days, under occlusion. The degree of occlusion influences percutaneous penetration, which may in turn influence the sensitivity of the test. Patches used vary from Draize-type gauze dressings to metal chambers. Therefore, a reference irritant material is often included in the test to facilitate interpretation of the results. Various animal species have also been used, such as the guinea pig and the rabbit.[8,9] Wahlberg[9] measured skin fold thickness with Harpenden calipers to assess the edema-producing capacity of chemicals in guinea pigs. This model demonstrated clear dose-response relationships and discriminating power, except for acids and alkalis, where no change in skin fold thickness was found.

Open application assays are also used for RIT. Marzulli and Maibach[10] described a cumulative irritation assay in rabbits, which utilizes open applications and control reference compounds. The test substances are applied 16 times over a 3-week period and the results are measured with a visual score for erythema and skin thickness measurements. These two parameters correlated highly. A significant correlation was also demonstrated between the scores of 60 test substances in the rabbit and in man, suggesting that the rabbit assay is a powerful predictive model.

Anderson et al.[11] utilized an open application procedure in guinea pigs to rank weak irritants. A baseline response to SLS solution was obtained after 3 applications per day for 3 days to a 1-cm^2 test area. This baseline is used to compare other irritants, of which trichloroethane was the most irritating, similar to 2% SLS. Histology demonstrated a mononuclear dermal inflammatory response.

1.2.3 Immersion Assay

The guinea pig immersion assay was developed to assess the irritant potential of aqueous surfactant solutions, but might also be useful for evaluating aqueous dilutions of agricultural chemicals. Restrained guinea pigs are immersed in the test solution while maintaining their head above water. The possibility of systemic absorption of a lethal dose restricts the study to products of limited toxic potential and the test concentration to a maximum of 10%.

Ten guinea pigs are immersed in a 40°C solution 4 h daily for 3 days. A comparison group is immersed in a reference solution. Then 24 h after the final immersion, the animals' flanks are shaved and evaluated for erythema, edema, and fissures.[12-15] Gupta et al.[16] concomitantly tested the dermatotoxic effects of detergents in guinea pigs and humans, utilizing the immersion test, and the patch test, respectively. Epidermal erosion and a 40-60% increase in the histamine content of the guinea pig skin was found, in addition to a positive patch test reaction in 7 of 8 subjects.

1.2.4 Mouse Ear Model

Uttley and Van Abbe[17] applied undiluted shampoos to one ear of mice daily for four days, visually quantifying the degree of inflammation as vessel dilatation, erythema, and edema. Patrick and Maibach[18] measured ear thickness to quantify the inflammatory response to surfactant-based products and other chemicals. This allowed quantification of dose-response relationships and comparison of chemicals. Inoue et al.[19] used this model to compare the mechanism of mustard oil induced skin inflammation to the mechanism of capsaicin induced inflammation. Mice were pretreated with various receptor antagonists, such as 5-HT_2, H_1, and tachykinin antagonists, demonstrating that the tachykinin NK1 receptor was an important mediator of inflammation induced by mustard oil. The mouse models provide simplicity and objective measurements. Relevance for man requires elucidation.

1.2.5 Other Methods

Several other assays of skin irritation have been suggested. Humphrey,[20] quantified the amount of Evans blue dye recovered from rat skin following exposure to skin irritants. Trush et al.[21] utilized myleoperoxidase in polymorphonuclear leukocytes as a biomarker for cutaneous inflammation.

1.2.6 Human Models

Human models for skin imitation testing are species relevant, thereby eliminating the precarious extrapolation of animal and *in vitro* data to the human setting. As the required test area is small, several products or concentrations can be tested simultaneously and compared. Inclusion of a reference irritant substance facilitates interpretation of the irritant potential of the test substances. Prior animal studies can be utilized to exclude particularly toxic substances or concentrations before human exposure.

1.2.6.1 Single Application Patch Testing

The National Academy of Sciences (NAS)[22] outlined a single application patch test procedure determining skin irritation in humans. Occlusive patches may be applied to the intrascapular region of the back or the dorsal surface of the forearms, utilizing a relatively non-occlusive tape for new or volatile materials. More occlusive tapes or chambers generally increase the severity of the responses. A reference material is included in each battery of patches.

The exposure time may vary to suit the study. NAS suggests a 4-h exposure period, although it may be desirable to test new or volatile materials for 30 to 60 min. Studies greater than 24 h have been performed. Skin responses are evaluated 30 to 60 min after removal of the patch, using the animal Draize scale (Table 1.1) or similar. Kligman and Wooding[23] described statistical analysis on test data to calculate the IT_{50} (time to produce imitation in 50% of the subjects) and the ID_{50} (dose required to produce irritation in 50% of the subjects after a 24-h exposure).

Robinson et al.[24] suggested a 4-h patch test as an alternative to animal testing. Assessing erythema by visual scoring, they tested a variety of irritants on Caucasians and Asians. A relative ranking of irritancy was obtained utilizing 20% SLS as a benchmark. Taking this model further, McFadden et al.[25] investigated the threshold of skin irritation in the six different skin types. Again using SLS as a benchmark, they defined the skin irritant threshold as the lowest concentration of SLS that would produce skin irritation under the 4 h occluded patch conditions. They found no significant difference in irritation between the skin types.

1.2.6.2 Cumulative Irritation Testing

Lanman et al.[26] and Phillips et al.[8] described a cumulative irritation assay, which has become known as the "21 day" cumulative irritation assay. The purpose of the test was to screen new formulas prior to marketing. A l-in. square of Webril was saturated with liquid or 0.5 g of viscous substances and applied to the surface of the pad to be applied to the skin. The patch was applied to the upper back and sealed with occlusive tape. The patch is removed after 24 h, and then reapplied after examination of the test site. This is repeated for 21 days and the IT_{50} can then be calculated.

Modifications have been made to this method. The chamber scarification test (see below) was developed to predict the effect of repeated applications of a potential irritant to damaged skin, rather than healthy skin. The cumulative patch test described above had failed to predict adverse reactions to skin damaged by acne or shaving, or sensitive areas such as the face.[27]

Wigger-Alberti et al.[28] compared two cumulative models, testing skin reaction to metalworking fluids (MWF). Irritation was assessed by visual scoring, transepidermal water loss, and chromametry. In the first method, MWFs were applied with Finn chambers on the volunteers' mid-back, removed after 1 day of exposure, and reapplied for a further 2 days. In the second method, cumulative irritant contact dermatitis was induced using an RIT for 2 weeks (omitting weekends) 6 h per day. The 3-day model was preferred because of its shorter duration and better discrimination of irritancy. For low irritancy materials in which discrimination is not defined with visual and palpatory scores, bioengineering methods (i.e., transepidermal water loss) may be helpful.

1.2.6.3 The Chamber Scarification Test

This test was developed [29,30] to test the irritant potential of products on damaged skin. Six to eight l-mm sites on the volar forearm were scratched 8 times with a 30-gauge needle, without causing bleeding. Four scratches were parallel and the other 4 were perpendicular to these. Duhring chambers, containing 0.1 g of test material (ointments, creams or powders), were then placed over the test sites. For liquids, a fitted pad saturated (0. l ml) may be used. Chambers containing fresh materials are reapplied daily for 3 days. The sites are evaluated by visual scoring 30 min after removal of the final set of chambers. A scarification index may be calculated if both normal and scarified skin is tested to reflect the relative degree of irritation between compromised and intact skin — this is the score of scarified sites divided by the score of intact sites. However, the relationship of this assay to routine use of substances on damaged skin remains to be established. Another compromised skin model, the arm immersion model of compromised skin, is described in the immersion tests section below.

1.2.6.4 Immersion Tests

These tests of aqueous solutions were developed in order to improve irritancy prediction by mimicking consumer use. Kooyman and Snyder[31] describe a method in which soap solutions of up to 3% are prepared in troughs. The temperature was maintained at 105°F while subjects immersed one hand and forearm in each trough, comparing different products (or concentrations). The exposure period ranged from 10 to 15 min, 3 times daily for 5 days, or until irritation was observed in both arms. The antecubital fossa was the first site to demonstrate irritation, followed by the hands.[5, 31] Therefore, antecubital wash tests (see below) and hand immersion assays were developed.[4]

Clarys et al.[32] used a 30-min/4-day immersion protocol to investigate the effects of temperature and also anionic character on the degree of irritation caused by detergents. The irritation was quantified by assessment of the stratum corneum barrier function (transepidermal water loss), skin redness (a* color parameter), and skin dryness (capacitance method). While both detergents tested significantly affected the integrity of the skin, higher anionic content and temperature, respectively, increased the irritant response.

Allenby et al.[33,34] describe the arm immersion model of compromised skin, which is designed to test the irritant or allergic potential of substances on damaged skin. Such skin may demonstrate an increased response, which may be negligible or undetectable in normal skin. The test subject immersed one forearm in a solution of 0.5% sodium dodecyl sulfate for 10 min, twice daily until the degree of erythema reached 1 to 1+ on visual scale. This degree of damage corresponded to a morning's wet domestic work. Patch tests of various irritants were applied to the dorsal and volar aspects of both the pretreated and untreated forearms, and also to the back. Each irritant produced a greater degree of reaction on the compromised skin.

1.2.6.5 Assessing Protective Barriers

Zhai et al.[35] proposed a model to evaluate skin protective materials. Ten subjects were exposed to the irritants SLS and ammonium hydroxide (in urea) and Rhus allergen. The occluded test sites were on each forearm, with one control site on each. The irritant response was assessed visually using a 10-point scale, which included vesiculation and maceration, unlike standard Draize scales. The scores were statistically analyzed for non-parametric data. Of the barrier creams studied, paraffin wax in cetyl alcohol was found to be the most effective in preventing irritation.

Wigger-Alberti and Elsner[28] investigated the potential of petrolatum to prevent epidermal barrier disruption induced by various irritants in an RIT assessing its potential as a standard reference product. White petrolatum was applied to the backs of 20 human subjects who were exposed to SLS, NaOH, toluene and lactic acid. Irritation was assessed by transepidermal water loss and colorimetry in addition to visual scoring. It was concluded that petrolatum was an effective barrier cream against SLS, NaOH, and lactic acid, and moderately effective against toluene.

Frosch et al.[6] adapted the guinea pig RIT described above for use in humans. Two barrier creams were evaluated for their ability to prevent irritation to SLS. In this repetitive model, the irritant was applied to the ventral forearm, using a glass cup, for 30 min daily for 2 weeks. One arm of each subject was pretreated with a barrier cream. As in the animal model, erythema was assessed by visual scoring, laser doppler flow, and transepidermal water loss. Skin color was also measured by colorimetry (La* value). One barrier cream decreased skin irritation to SLS, the most differentiating parameter being transepidermal water loss and the least differentiating being colorimetry. Zhai[36] (1998) has detailed the subject.

1.2.6.6 Bioengineering Methods in Model Development

Many of the models described above do not employ the modern bioengineering techniques available, and therefore data based on these models may be imprecise. Despite the investigator skill, subjective assessment of erythema, edema and other visual parameters may lead to confounding by inter- and intra- observer variation. Although the eye may be more sensitive than current spectroscopy and chromametric techniques, the reproducibility and increased statistical power of such data may provide greater benefit. A combination of techniques, such as transepidermal water loss, capacitance, ultrasound, laser doppler flowmetry, spectroscopy, and chromametric analysis, in addition to skilled observation, may increase the precision of the test. Andersen and Maibach[37] compared various bioengineering techniques, finding that clinically indistinguishable reactions induced significantly different changes in barrier function and vascular status. An outline of many of these techniques is provided by Patil et al.[4]

1.3 METHODS FOR TESTING THE SENSITIZATION POTENTIAL

Prediction of skin sensitization is important to both the consumer and to industry. The consumer is protected by a wealth of data allowing commercial or pharmaceutical substances to be licensed for use. Industry, applying predictive data, can confidently launch products in the marketplace, knowing that they are unlikely to cause significant skin allergy.

Currently, predictive methods consist of a variety of guinea pig assays and some human assays. These can be used to test substances under laboratory conditions, which aim to reflect "in use" conditions. Attempts to replace these varied methods, with their varying strengths and weaknesses, have led to the local lymph node assay and also mathematical modeling of a substance's likelihood of causing allergy, based on its physico-chemical properties, discussed below.

1.3.1 Guinea Pig Models

The guinea pig assay depends on the expertise of the planner and interpreter. Properly executed and interpreted, the assay provides hazard identification and, with assays such as the open epicutaneous test (OET) risk assessment — the dose that will induce (threshold inducing concentration, TIC) and elicit (threshold eliciting concentration, TEC) allergic contact dermatitis. Conversely, in less sophisticated milieus, sensitivity and specificity may be compromised. Anderson (1985) provides a rich background, not only of methodology but also of reliability.

Several guinea pig models exist to predict sensitization. Common features include exposure to a test substance to induce sensitization, a rest period of several days, and re-exposure to the same substance at a virgin skin site (challenge exposure). This can be followed by a series of rechallenges over a period of several weeks to confirm allergy. Alternatively, the sensitized guinea pig can be exposed to putative cross-allergens.

Many variations exist around this core process. Some tests favor intradermal injections to induce allergy, others favor epicutaneous administration. Some tests combine both. Similarly, the challenge exposure may be by these methods. The duration of the initiation also varies from model to model, as does the method of skin irritation. Each model, therefore, should be selected with care, paying particular regard to how that model fits with the purpose of the test, and the nature of the test substance.

Irritation Testing — Initially, a group of four to six animals is used to identify a suitable concentration of the test substance for the induction treatment, which should at most be moderately irritant. The highest non-irritant concentration is used for the challenge dose.
Controls — The control animals are usually treated exactly as the test group, except the test substance is omitted during the induction phase of the experiment. They are then challenged in exactly the same manner as the test animals and the reactions compared.
Evaluation — Visual scoring of the erythema and edema can be used to evaluate the skin reaction to the test substance. The frequency, intensity, and duration of reactions elicited in test and control animals allow determination of whether sensitization had occurred or not. Several models below attempt to quantify the reaction, by measuring the size of the reaction, and performing statistical analysis on this data.
Classification — A potential contact sensitizer is considered to be a substance which produces allergic contact dermatitis in 15% of the test animals in a non-adjuvant assay and 8% in an adjuvant assay.[38] A negative result is when no allergic reaction is elicited in any of the animals. If less than 15% of the animals demonstrate a reaction, the result is defined as questionable and rechallenging or study repetition is necessary.

1.3.1.1 Guinea Pig Maximization Test[39,40]

In this assay, 20 test and 10 to 20 control guinea pigs are used. The induction process is in two phases: intradermal injections on day 0 and a 48-h occlusive patch on day 7 (boosting). This is followed by a challenge test on day 21.

1.3.1.1.1 Method

Day 0: Paired intradermal injections (0.1 ml each) into the clipped and shaved shoulder region.
Freund's Complete Adjuvant (FCA).
Test article in a vehicle (water, paraffin oil, propylene glycol injections)
mixture of dissolved or suspended test article with FCA (1:1).
Day 7: Topical occlusive patch applied for 48 h over shoulder (clipped on day 6); concentration applied should be moderately irritating; if substance is nonirritating, then pretreat region with 10% SLS on day 6 freshly after clipping.
Day 21: Challenge test is performed on a 4-cm^2 area on the left flank.
Shave test region.
Apply nonirritating concentrations, e.g., via Finn chamber.
Apply occluded vehicle for control.
Day 28: Rechallenge/cross test at contralateral flank weekly.

Challenge reactions are examined at 24 and 48 h after removal of the patch and scored according to a standard rating scale. Control animals are treated similarly to the test animals, except no test substance is applied. Kligman and Basketter[41] recommend in their critique of the test that one control group should receive only FCA as this substance lowers the skin's irritant threshold. Without this control, irritant reactions may be interpreted erroneously as allergic reactions, when the skin is challenged with the "non-irritating" concentration identified in preliminary irritation studies. Further, the control group should be exposed to an irritant stimulus comparable in intensity to that of the test substance (e.g., SLS). The purpose of this is to control for excited skin syndrome, which may cause a false-positive result. Rechallenge reactions should also be performed to rule out false-positive reactions. Truly allergic reactions last for several weeks, whereas non-specific irritation diminishes after 2-3 weeks. Re-challenging is performed weekly on the contralateral flank.

1.3.1.2 Split Adjuvant Technique

In this assay, involving 2 groups of 10-20 guinea pigs each, the test substance is administered separately from the FCA.[42-44]

1.3.1.2.1 Method

Day 0: Induction: Window dressing fixed on shaved suprascapular region
2-cm^2 induction site is exposed to "dry ice" for 5 s.
Apply topically 0.2 ml semisolid or 0.1 ml liquid test sample.
Occlude under filter paper for 48 h.
Day 2: Induction: Open dressing, reapply test substance, and reocclude.
Day 4: Induction: Prior to induction application, inject 2 ∞ 0.1 ml FCA into test site.
Apply test substance as before; reocclude.
Day 7: Reapply and reocclude as before.
Day 9: Remove entire dressing.
Day 22: Challenge test over normal appearing skin on mid-back
0.5 ml of semisolid or 0.2 ml of liquid is applied to a clipped area and occluded for 24 h. Challenge readings are taken on days 23, 24, and 25.

Control animals are treated similarly except no test substance is added.

This test differs from the guinea pig maximization test as it uses exclusively topical application of the test substance to elicit sensitization, although FCA is still injected intradermally to enhance sensitivity. Also, dry ice is used to irritate the test site.

In a modified version of this test, the test site is excoriated with sandpaper or tape and is left open rather than occluded. The aim is to achieve a test close to "in use" conditions.[38]

1.3.1.3 Optimization Test

Maurer[45] models a test of 20 guinea pigs per test substance and 20 per control group. This test induces sensitization by intradermal injection. The test sites are depilated both mechanically and chemically. The challenge is by intradermal injection and by topical occlusion. Injection solutions are made up in physiological saline solution (0.1%). The reactions were assessed by visual scoring and also quantitatively by measuring skin fold thickness ∞ greatest diameter of erythema to give a "reaction volume."

1.3.1.3.1 Method

Wk 1: Induction
0.1 ml injections are made into the right flank (0.05 ml) and back (0.1 ml) three times/week (on day 0, two injections to each site).

Wk 2: Induction: test compound suspended in FCA, is injected into the nuchal skin three times/week.
Wk 3: Induction: As for week 2.
Wk 4: Rest.
Wk 5: Rest.
Wk 6: Challenge 1: Intradermal injection into contralateral flank.
Same solution used as week 1 induction (0.05 ml of 0.1% solution).
Reactions measured at 24 h.
Wk7: Challenge 2: Epidermal occlusive challenge
Results read at 24, 48, and 72 h.

Control animals are treated similarly to test animals, except the test substance is omitted during induction.

The reaction volume is used to calculate the skin irritant threshold (mean reaction volume + 1 SD). If the challenge reaction volume post injection is greater than the threshold, the reaction is considered to be allergic. Visual scoring is used to evaluate the occlusion challenge. The exact Fisher test is used to determine statistical significance.

1.3.1.4 Freund's Complete Adjuvant Test

For the control and treatment groups 10 to 20 guinea pigs each are required.[38, 46] This test was devised to test weaker allergens as well as strong/moderate allergens. The test substance is mixed with FCA in a 1:1 ratio, at a concentration of less than 5%. The induction phase involves intradermal application, and the challenge phase is an open application of the test substance. The challenge dose is the minimal irritating dose and the 1:3 non-irritating dilutions. This allows a threshold level to be estimated.

1.3.1.4.1 Method

Day 0: Induction suprascapular intradermal injection of 0.1 ml test substance into shaved test site.
Day 4: Repeat above.
Day 8: Repeat above.
Day 21: Challenge on clipped flank skin.
Apply up to 6 concentrations (minimal irritating and subsequent dilutions) in vehicle.
Results interpreted at 24, 48, and 72 h and visually scored.
Re-challenge/cross-test at 10–14 day intervals on contralateral flank.

Control animals are treated similarly, except the test article is omitted during induction.

1.3.1.5 OET

In this assay, the test substance is applied epicutaneously to a shaved flank, using constant volumes per square centimeter to standard areas.[38,46] Six guinea pigs are used for each test concentration group, while end-use products are tested on 20 animals. Ten guinea pigs are used in the control group.

The induction period requires daily application over 5 consecutive days per week in the 4-week study, or daily in the 3-week study, until a total of 20 applications.

1.3.1.5.1 Method

Wk 1: Induction application over 5 or 7 consecutive days.
Apply 0.1 ml of neat test article, or progressive (1:3) dilutions to same 8-cm^2 site
Wk 2: Induction as above
Wk 3: Induction as above.Total of 20 applications over the induction period
Wk 4: Induction as above.
Day 21/29: Challenge application of test substance/dilutions at 2-cm^2 test site.
Results interpreted at 24, 48, and 72 h post challenge
Re-challenge/cross-test at 10-14 day intervals on contralateral flank

Control animals are treated similarly, except the test article is omitted during induction.

Klecak[46] suggests that a combination of OET and the FCA test is useful in defining conditions under which borderline substances may be allergenic.

1.3.1.6 Modified Draize Test

The aim of the modifications to the Draize test[47] was to increase its sensitivity to weaker allergens. The test consists of two parts, involving 10 guinea pigs in the treatment group and in the control group. Skin reactions are assessed for erythema and edema, which are graded, and longitudinal and lateral diameters of reaction are measured to determine reaction size.

The control animals are treated with FCA only during the induction. During the challenge and rechallenge, they are treated as the test animals.

1.3.1.6.1 Method

Part 1.
Day 0: Induction intradermal injections of 0.1 ml each onto four sites into clipped skin.
Test sites are over inguinal and axillary lymph nodes.
Test dose is 2.5 times the challenge dose.
Reactions measured and scored at 24 h.
Day 14: Challenge test.
0.1 ml of test substance at nonirritant or slightly irritant concentration injected intradermally to one clipped flank.
0.1 ml of test substance applied epicutaneously onto 8 cm^2 clipped contralateral flank, results measured and scored at 24 h.
Day 21: Confirmation rechallenge.
Day 28: Confirmation rechallenge.
Part 2: If the challenge tests are negative in Part 1, a second set of intradermal injections is administered on day 35. The challenge procedure is as above, however, rechallenge is performed intradermally in weekly intervals. New control animals are necessary for Part 2 of the test.

1.3.1.7 Buehler Test

This test involves three groups of guinea pigs: (1) the test group, (2) the vehicle control, which is treated in the same way as the test group except for the omission of the test substance, and (3) the negative control, which is challenged only.[38,48]

The induction phase consists of 1 or 3 weekly occlusive applications of the test substance to an 8-cm^2 clipped area of skin over the left shoulder. A 4-cm^2 occlusive patch/chamber containing the highest possible (moderately irritating) concentration is applied for 6 h, during which time the animals are placed in restrainers. The induction lasts between 3 and 9 weeks.

The challenge procedure occurs 14 days after the final induction. About 0.2 – 0.5 ml of the test substance is applied under an occlusive patch on naïve back skin for 6 h. Non-irritating concentrations are used. The guinea pigs are restrained for induction. Ten control guinea pigs are also challenged. Grading of observed skin reactions occurs at 24 and 48 h post challenge. Re-challenges/cross-tests follow at intervals of 1–2 weeks at virgin sites on the back. If indicated, additional control animals may be used.

Assessment is by a calculation of an incidence index, which is the number of responding animals out of the test group, and a severity index, which is a sum of the reaction grades divided by the number of animals used.

Other models include the modified guinea pig maximization test, the single injection adjuvant test, the teirexperimentaller nachweis (TINA) test, the cumulative contact enhancement test, the epicutaneous maximization test, the ear/flank test, the footpad test, and the guinea pig test adapted to cosmetic ingredients. These are reviewed by Klecak[38] and extensively discussed by Anderson and Maibach.[37]

1.3.1.8 The Local Lymph Node Assay

The murine local lymph node assay (LLNA) measures the activity of injected [^{3}H]-methylthymidine in the lymph nodes draining a topical test site test.[49] This is thought to reflect the clonal expansion of T-lymphocytes in these nodes during induction sensitization.

The test substance is applied topically to the dorsum of both ears in a vehicle, which is determined by the solubility of the test material. After five days, [^{3}H]-methylthymidine is injected intravenously into test and control mice, which are sacrificed five hours later. The auricular lymph nodes are then excised and the [^{3}H]-methylthymidine count is measured by β-scintillation. A count three times greater than the control group is consistent with sensitization.

This method has the advantage of objectivity — counting radioactivity. As a relatively new assay, we are just learning to deal with its imprecision in terms of specificity and sensitivity. Note that it assays for hazard identification and not risk assessment.[50]

1.3.2 Human Allergy Tests

Currently, the modified maximization technique[51] and the modified Draize procedure[52,53] are the methods of choice for testing sensitization in humans.[54] A large pool of subjects is necessary to ensure statistical validity.

In these studies, the test material is applied to the skin under an occlusive patch. This induction sensitization is repeated 7–10 times over a 3 week induction period. This is followed by a rest period of 10–14 days. The challenge is performed on virgin skin, under occlusion using a non-irritating dose. Enhancing techniques such as repetition of the chemical insult,[55] using a high concentration of test material,[52,53] treating the skin with SLS,[56] skin stripping,[57] or freezing were used.[58] Freund's adjuvant[59] was among the earliest techniques employed to enhance sensitization.

1.4 QUANTITATIVE STRUCTURE ANALYSIS RELATIONSHIPS (QSAR)

The premise behind QSAR is that physicochemical data available on previously identified sensitizers can be used to predict the likelihood of the test substance being a sensitizer. As many sensitizers have been reported, their properties can be cataloged to form a vast database, upon which correlations between their structures and sensitizing ability can be drawn. The larger this database, the more likely statistically significant mathematical relationships can be identified between structural elements (independent variables) and biological activity (dependent variables). The structures that are significantly associated with allergy in the database may also be associated with allergy *in vivo*. If QSAR models can accurately predict sensitization, then expensive, time-consuming animal and human studies can be avoided. Potential new compounds can be analyzed for sensitivity prior to development, allowing manufacturers to modify products at a conceptual stage.

1.4.1 Classification Model

This model has been developed to discriminate between allergens and non-allergens,[60-65] utilizing a database of structures that had been reported to cause allergic contact dermatitis over a 20-year period. These structures were analyzed for size, polarity, and protein reactive functions such as amino and alcohol groups, aldehydes, and esters. Similar data were obtained for structures that are non-allergenic, thereby creating a pool of characteristics responsible for sensitization and the degree of reactivity.

Importantly, this model is based on both features that contribute to allergenicity: a combination of the *diffusion ability* of the compound across the stratum corneum, and *protein reactivity* of the compound. The former is reflected by transport/binding descriptors and the latter by a count of dichotomous descriptors (1/0) of toxophores (haptens and prohaptens). These descriptors are demonstrated in Table 1.1.

Each descriptor has a coefficient value, which is used in an algorithm. This expresses the *Y*-estimate, which scores allergens and non-allergens above and below 0.5, respectively. Values in the range 0.4–0.6 are considered indeterminate. Hostýnek et al.[60] modeled six randomly selected chemicals not previously investigated for allergenicity, validating the results *in vivo*. Full concordance was found between the model analysis and the murine local lymph node assay.

Enslein et al.[66] tested the model on 74 known allergenic and non-allergenic compounds. Twelve compounds fell in the indeterminate range, 57 were correctly predicted, and 5 were incorrectly predicted. These 5 compounds were reported allergens in literature, but falsely identified as negative by the QSAR model. No false-positives were predicted, suggesting that the model more accurately identifies non-allergens (100% specificity, 90% sensitivity).

1.4.2 The Rank Model

In contrast to the classification model above, this model attempts to rank the allergic response to a substance as non-allergenic, weak, moderate, and strong, on a scale of 1–4, respectively. The descriptors are identical in both the rank and discriminatory models, although the coefficient values assigned to each descriptor vary. Rank 1 (non) is clearly distinct from rank 4 (strong) in this model; however, there is considerable overlap between the intermediate ranks.

Testing this model on the same 74 compounds above, Enslein et al.[66] demonstrated model sensitivity of 95%, allowing an error of plus or minus one rank in prediction. This error margin was considered equivalent to experimental assays (Gad, 1988). The model specificity was 60%.

Combining the two models provided an overall concordance of 93%, or one model able to identify some of the "false-negatives" predicted in the other.

1.4.3 The DEREK System

The QSAR knowledge-based computer system DEREK was developed under the collaboration of a multinational group of expert toxicologists. In addition to skin sensitization prediction, it is also used to predict genotoxicity.

Test structures can be input into the computer graphically, and analysis is carried out by screening each structure against the structural alerts in the DEREK database. Toxophores are identified, and the computer provides information on the toxic effect and also lists literature references. User interpretation is required to assess the risk posed if several toxophores are identified, and also if no toxophores are identified. The latter does not simply reflect the compound's safety, but also reflects the depth of the database.

The database consists of separate rule sets, which can be chosen according to the user's needs. The standard rule set, developed originally by Sanderson and Earnshaw[67] contains 52 rules, with end-points such as carcinogenicity, neurotoxicity, and some adverse reproductive effects. This has been expanded on by the developmental rule set, which resulted from collaborative work. Utilizing a Unilever database from approximately 300 guinea pig maximization tests, 40 structural alerts to skin sensitizers were derived. Substances were grouped according to their most likely mechanism of reaction with skin proteins. Where no mechanism could be identified, structural alerts were created for groups of chemicals with similar functional groups. However, this system does not account for stratum corneum penetration, unlike the classification and rank models described above. Therefore, structural alerts may be identified in a particular query, when in practice the test substance's skin permeability is too low to cause sensitization.

A futher FDA rule set pertains to structural alerts for toxophores considered of interest to the U.S. Food and Drug Administration.

The DEREK model has been tested on a dataset of 37 chemicals, of which 25 were skin sensitizers and 12 non-sensitizers. Of the sensitizers, DEREK correctly identified 24 containing structural alerts for skin sensitization. Of the 12 non-sensitizers, 2 were incorrectly identified as skin sensitizers, perhaps a result of the inability of the system to consider percutaneous penetration.[68]

TABLE 1.4
QSAR Decsriptors

Determinant	Descriptor	Definition	Variable
Protein Reactivity			
	IX	Reactive aliphatic/aromatic halides	Dichotomous
	IUNIQ	Strong nucleophiles/reactive electrophiles	Dichotomous
	ICOOR	Simple aliphatic esters	Dichotomous
	ICONJ	Conjugated olefins: activated by SH/NH_2 Michael addition	Dichotomous
	IArOH	Easily metabolized phenols of quinone	Dichotomous
	IArNH	Easily metabolized anilines precusors	Dichotomous
	IOH	Oxidizable primary alcohols	Dichotomous
Transport/Binding			
	MR	Polarizable molecular volume	Continuous
	PL	Lipophillic contribution of log *P*	Continuous
	HBA	Hydrogen bond acceptor groups (count of electron pairs on N and O)	Ordinal
	HBD	Hydrogen bond donor groups (count of N-H and O-H)	Ordinal

1.5 CONCLUSION

The predictive power of QSAR depends directly on the quality of the database and our understanding of the mechanisms of skin sensitization. As knowledge in both spheres expands, so shall the power of predictive modeling. Clinical reports of allergy must be reliable to ensure that compounds entered into the database are true allergens. Diagnostic patch tests with adequate controls are required to make the diagnosis of allergic contact dermatitis to a particular allergen.

The above studies demonstrate that QSAR is rapidly developing into a useful technique in the sphere of contact allergy. Accurate predictive mathematical modeling may allow us to forego the lengthy, cumbersome, labor-intensive experimental techniques for skin senstization. Prospective studies of novel compounds are necessary to judge the usefulness of this technique in the real-life setting.

REFERENCES

1. **Draize, J., Woodward, G., and Calvery, O.,** Methods for the study of irritation and toxicity of substances applied topically to the skin and mucous membrane, *J. Pharmacol. Exp. Ther.*, 82, 377, 1944.
2. **Patrick, E. and Maibach, H.,** Comparison of the time course, dose response and mediators of chemically induced skin irritation in three species, in *Current Topics in Contact Dermatitis*, P. J. F. et al, Eds, Springer-Verlag, New York, 1989, 399–402.
3. **Office of the Federal Registrar, N. A. O. R.**, *Code of Federal Regulations*, Title 16, parts 1500.40-1500.42, 1985.
4. **Patil, S., Patrick, E., and Maibach, H.,** Animal, human and *in vitro* test methods for predicting skin irritation, in *Dermatotoxicology Methods: The Laboratory Worker's Vade Mecum*, Marzulli, F. and Maibach, H., Eds., Taylor & Francis, Washington, D.C., 1998, 89.

5. **Justice, J., Travers, J., and Vinson, L.,** The correlation between animal tests and human tests in assessing product mildness, *Proc. Sci. Sect. Toilet Goods Assoc.*, 35, 12, 1961.
6. **Frosch, P., Schulze-Dirks, A., Hoffmann, M., Axthelm, I., and Kurte, A.,** Efficacy of skin barrier creams (I). The repetitive irritation test (RIT) in the guinea pig, *Contact Dermatitis*, 28, 94, 1993.
7. **Frosch, P., Schulze-Dirks, A., Hoffmann, M., and Axthelm, I.,** Efficacy of skin barrier creams (II). Ineffectiveness of a popular "skin protector" against various irritants in the repetitive irritation test in the guinea pig, *Contact Dermatitis*, 29, 74, 1993.
8. **Phillips, L., Steinberg, M., Maibach, H., and Akers, W.,** A comparison of rabbit and human skin responses to certain irritants, *Toxicol. Appl. Pharmacol.*, 21, 369 – 382, 1972.
9. **Wahlberg, J.,** Measurement of skin fold thickness in the guinea pig. Assessment of edema-inducing capacity of cutting fluids acids, alkalis, formalin and dimethyl sulfoxide, *Contact Dermatitis*, 28, 141, 1993.
10. **Marzulli, F. and Maibach, H.,** The rabbit as a model for evaluating skin irritants: A comparison of results obtained on animals and man using repeated skin exposure, *Food Cosmet. Toxicol.*, 13, 533, 1975.
11. **Anderson, C., Sundberg, K., and Groth, O.,** Animal model for assessment of skin irritancy, *Contact Dermatitis*, 15, 143, 1986.
12. **Opdyke, D. and Burnett, C.,** Practical problems in the evaluation of the safety of cosmetics, *Proc. Sci. Sect., Toilet Goods Assoc.*, 44, 3, 1965.
13. **Calandra, J.,** Comments on the guinea pig immersion test, *CTFA Cosmet. J.*, 3, 47, 1971.
14. **Opdyke, D.,** The guinea pig immersion test – A 20 year appraisal, *CTFA Cosmet. J.*, 3, 46, 1971.
15. **MacMillan, F., Ram, R., and Elvers, W.,** A comparison of the skin irritation produced by cosmetic ingredients and formulations in the rabbit, guinea pig, beagle dog to that observed in the human, in *Animal Models in Dermatology*, Maibach, H., Ed., Churchill Livingstone, Edinburgh, 1975, 12.
16. **Gupta, B., Mathur, A., Srivastava, A., Singh, S., Singh, A., and Chandra, S.,** Dermal exposure to detergents, *Vet. Hum. Toxicol.*, 34, 405, 1992.
17. **Uttley, M. and Van Abbe, N.,** Primary irritation of the skin: mouse ear test and human patch test procedures, *J. Soc. Cosmet. Chem.*, 24, 217 – 227, 1973.
18. **Patrick, E. and Maibach, H.,** A novel predictive assay in mice, *Toxicologist*, 7, 84, 1987.
19. **Inoue, H., Asaka, T., Nagata, N., and Koshihara, Y.,** Mechanism of mustard oil-induced skin inflammation in mice, *Eur. J. Pharmacol.*, 333, 231, 1997.
20. **Humphrey, D.,** Measurement of cutaneous microvascular exudates using Evans blue, *Biotech. Histochem.*, 68, 342, 1993.
21. **Trush, M., Egner, P., and Kensler, T.,** Myeloperoxidase as a biomarker of skin irritation and inflammation, *Food Chem. Toxicol.*, 32, 143, 1994.
22. **National Academy of Sciences, C. F. T. R. O. N. P.,** Principles and Procedures for Evaluating the Toxicity of Household Substances, Washington, D.C., 1977.
23. **Kligman, A. M. and Wooding, W.M.,** A method for the measurement and evaluation of irritants on human skin, *J. Invest. Dermatol.*, 49, 78 – 94, 1967.
24. **Robinson, M. K., Perkins, M. A., and Basketter, D. A.,** Application of a 4-h human patch test method for comparative and investigative assessment of skin irritation, *Contact Dermatitis*, 38, 194, 1998.

25. **McFadden, J. P., Wakelin, S. H., and Basketter, D. A.,** Acute irritation thresholds in subjects with type I–type VI skin, *Contact Dermatitis*, 38, 147, 1998.
26. **Lanman, B., Elvers, W., and Howard, C.,** The role of human patch testing in a product development program, in *Proc. Joint Conf. Cosmet. Sci.*, Toilet Goods Association, Washington, D.C., 1968, 135 – 145.
27. **Battista, C. and Rieger, M.,** Some problems of predictive testing, *J. Soc. Cosmet. Chem.*, 22, 349 – 359, 1971.
28. **Wigger-Alberti, W., Hinnen, U., and Elsner, P.,** Predictive testing of metalworking fluids: A comparison of 2 cumulative human irritation models and correlation with epidemiological data, *Contact Dermatitis*, 36, 14, 1997.
29. **Frosch, P. and Kligman, A.,** The chamber scarification test for testing the irritancy of topically applied substances, in *Cutaneous Toxicity*, Drill, L. and Lazar, P., Eds., Academic Press, New York, 1977, 150.
30. **Frosch, P. and Kligman, A.,** The chamber scarification test for irritancy, *Contact Dermatitis*, 2, 314 – 324, 1976.
31. **Kooyman, D. and Snyderm, F.,** The test for mildness of soaps, *Arch. Dermatol. Syphilol.*, 46, 846, 1942.
32. **Clarys, P., Manou, I., and Barel, A.,** Influence of temperature on irritation in the hand/forearm immersion test, *Contact Dermatitis*, 36, 240, 1997.
33. **Allenby, C. F., Basketter, D. A., Dickens, A., Barnes, E. G., and Brough, H. C.,** An arm immersion model of compromised skin (I). Influence on irritation reactions, *Contact Dermatitis*, 28, 84, 1993.
34. **Allenby, C. F. and Basketter, D. A.,** An arm immersion model of compromised skin: (II). Influence on minimal eliciting patch test concentrations of nickel, *Contact Dermatitis*, 28, 129, 1993.
35. **Zhai, H., Willard, P., and Maibach, H.,** Evaluating skin-protective materials against contact irritants and allergens. An *in vivo* screening human model, *Contact Dermatitis*, 38, 155, 1998.
36. **Zhai, H. and Maibach, H.,** Moisturizers in preventing irritant contact dermatitis: an overview, *Contact Dermatitis*, 38, 241, 1998.
37. **Andersen, P. and Maibach, H.,** Skin irritation in man: a comparative bioengineering study using improved reflectance spectroscopy, *Contact Dermatitis*, 33, 315, 1995.
38. **Klecak, G.,** Allergic contact dermatitis in animals, in *Dermatotoxicology Methods: The Laboratory Worker's Vade Mecum*, Marzulli, F. and Maibach, H., Eds., Taylor & Francis, Washington, D.C., 1998, chap. 11.
39. **Magnusson, B. and Kligman, A.,** The identification of contact allergens by animal assay, the guinea pig maximization test, *J. Invest. Dermatol.*, 52, 268, 1969.
40. **Magnusson, B. and Kligman, A.,** *Allergic Contact Dermatitis in the Guinea Pig*, Thomas, Springfield, 1970.
41. **Kligman, A. and Basketter, D.,** A critical commentary and updating of the guinea pig maximization test, *Contact Dermatitis*, 32, 129, 1995.
42. **Maguire, H. J.,** The bioassay of contact allergens in the guinea pig, *J. Soc. Cosmet. Chem.*, 24, 121, 1973.
43. **Maguire, H. J.,** Estimation of the allergenicity of prospective human contact sensitizers in the guinea pig, in *Animal Models in Dermatology*, Maibach, H., Ed., Churchill Livingstone, New York, 1975, 67.
44. **Maguire, H. J. and Cipriano, D.,** Allergic contact dermatitis in laboratory animals, in *Cutaneous Toxicicity*, Drill, L., Ed., Raven Press, New York, 1984, 55.
45. **Maurer, T.,** The optimization test, *Curr. Probl. Dermatol.*, 14, 114, 1985.

46. **Klecak, G.,** The Freund's complete adjuvant test and the open epicutaneous test. A complementary test procedure for realistic assessment of allergenic potential, *Curr. Probl. Dermatol.*, 14, 152, 1985.
47. **Johnson, A. and Goodwin, B.,** The Draize test and modifications in contact allergy. Predictive tests in guinea pigs, *Curr. Probl. Dermatol.*, 14, 31, 1985.
48. **Buehler, E. V.,** A rationale for the selection of occlusion and elicit delayed contact hypersensitivity in the guinea pig. A prospective test, *Curr. Probl. Dermatol.*, 14, 39, 1985.
49. **Kimber, I.,** The local lymph node assay, in *Dermatotoxicology Methods: The Laboratory Worker's Vade Mecum*, Marzulli, F. and Maibach, H., Eds., Taylor & Francis, Washington, D.C., 1998, chap. 12.
50. **National Institutes of Health**, The murine local lymph node assay: a test method for assessing the allergic contact dermatitis potential of chemicals/compounds, National Institutes of Health, 99-4494, 1999.
51. **Kligman, E. and Epstein, W.,** Updating the maximization test for identifying contact allergens, *Contact Dermatitis*, 1, 231, 1975.
52. **Marzulli, F. and Maibach, H.,** Antimicrobials: Experimental contact sensitization in man, *J. Soc. Cosmet. Chem.*, 20, 67, 1973.
53. **Marzulli, F. and Maibach, H.,** The use of graded concentrations in studying skin sensitizers: Experimental contact sensitization in man., *Food Cosmet. Toxicol.*, 12, 219, 1974.
54. **Marzulli, F. and Maibach, H.,** Test methods for allergic contact dermatitis in humans, in *Dermatotoxicology Methods: The Laboratory Worker's Vade Mecum*, Marzulli, F. and Maibach, H., Eds., Taylor & Francis, Washington, D.C., 1998, chap. 13.
55. **Shelanski, H.,** Experience with and considerations of the human patch test method, *J. Soc. Cosmet. Chem.*, 2, 324, 1951.
56. **Kligman, A.,** The SLS provocative patch test in allergic contact sensitization, *J. Invest. Dermatol.*, 46, 571, 1966.
57. **Spier, H. and Sixt, I.,** Untersuchungen uber die Abhangigkeit des Ausfalles der exzem Lappshenpraben von der Hornschichtdicke, *Hautarzt*, 6, 152, 1955.
58. **Epstein, W., Kligman, A., and Senecal, I.,** Role of regional lymph nodes in contact sensitization, *Arch. Dermatol.*, 88, 789, 1963.
59. **Freund, J.,** Effect of paraffin oil and mycobacteria on antibody formation and sensitization; review, *Am. J. Clin. Pathol.*, 6, 227, 1951.
60. **Hostýnek, J., Lauerma, A., Magee, P., Bloom, E., and Maibach, H.,** A local lymph-node assay validation study of a structure-activity relationship model for contact allergens [published erratum appears in *Arch. Dermatol. Res.*, 1995; 287(8):767], *Arch. Dermatol. Res.*, 287, 567, 1995.
61. **Hostýnek, J., Magee, P., and Maibach, H.,** QSAR predictive of contact allergy: scope and limitations, in *Prevention of Contact Dermatitis*, Vol. 25, P. Elsner, J. Lachapelle, and Wahlberg, J., Eds., Karger, Basel, 1996, 18-27.
62. **Hostýnek, J.,** Structure-Activity Relationships in Contact Sensitization, in *Dermatotoxicology Methods: The Laboratory Worker's Vade Mecum*, Marzulli, F. and Maibach, H., Eds., Taylor & Francis, Washington, D.C., 1998.
63. **Magee, P. S., Hostýnek, J.J., and Maibach, H.I.,** A classification model for contact dermatitis, *Quant. Struct.-Act. Relat.*, 13, 22, 1994.
64. **Magee, P., Host'nek, J., and Maibach, H.,** Modeling allergic contact dermatitis, in *In Vitro Skin Toxicology: Irritation, Phototoxicity, Sensitization*, Rougier, A., Goldberg, A.M., and Maibach, H.I, Eds., Mary Ann Liebert, New York, 1994, 281.

65. **Magee, P., Hostýnek, J., and Maibach, H.,** Toward a predictive model for allergic contact dermatitis, in *Animal Test Alternatives: Refinement-Reduction-Replacement*, Salem, H., Ed., Marcel Dekker, New York, 1994, 159.
66. **Enslein, K., Gombar, V., Blake, B., Maibach, H., Hostýnek, J., Sigman, C., and Bagheri, D.,** A quantitative structure-toxicity relationships model for the dermal sensitization guinea pig maximization assay, *Food Chem. Toxicol.*, 35, 1091, 1997.
67. **Sanderson, D. and Earnshaw, C.,** Computer prediction of possible toxic action from chemical structure; the DEREK system, *Hum. Exp. Toxicol.*, 10, 261, 1991.
68. **Ridings, J. E., Barratt, M. D., Cary, R., Earnshaw, C. G., Eggington, C. E., Ellis, M. K., Judson, P. N., Langowski, J. J., Marchant, C. A., Payne, M. P., Watson, W. P., and Yih, T. D.,** Computer prediction of possible toxic action from chemical structure: an update on the DEREK system, *Toxicology*, 106, 267, 1996.

2 Basic Patterns in Clinical Dermatology

Homero A. Penagos

CONTENTS

2.1 INTRODUCTION

The skin conditions produced by pesticides do not differ from those produced by other chemicals. This chapter will describe basic patterns of several skin conditions. In addition to contact dermatitis, these include chloracne, porphyria cutanea, immunologic syndromes including lupus-like syndromes, toxic epidermal necrolysis (TEN), and skin cancer.[1]

1-56670-293-3/00/$0.00+$.50

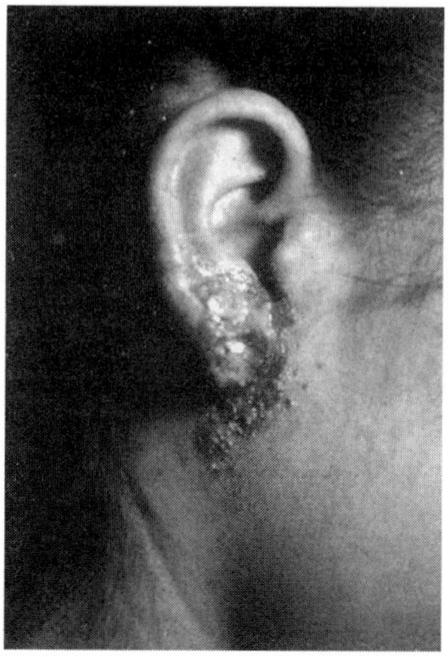

FIGURE 2.1 Contact burn due to misapplication of paraquat as a medication.

2.2 BASIC CLINICAL ISSUES

2.3 CONTACT DERMATITIS (CD)

Contact dermatitis is a frequently encountered clinical problem, found in about 2% of the U.S. population in prevalence surveys.[2] The hallmark of this condition is the agreement between cutaneous exposure to the agent in question and the distribution of the resulting lesions.[3] These typically include erythema and edema, often followed by vesicles and bullae in the area of contact of the suspected agent. Late lesions include weeping, crusting, secondary infection, post-inflammatory hyperpigmentation, hypopigmentation, and other pigmentary disorders.

2.3.1 Irritant CD (ICD)

Irritant contact dermatitis may be defined as a non-immunological inflammatory reaction of the skin to an external agent. Either an active ingredient or any other substance in the formulation may be the causative factor of the ICD.

The severity of ICD is proportional to the concentration of the substance and the condition of the skin affected. Mild symptoms may be a stinging or itching of the skin which, if untreated, will develop into skin blistering or peeling.

Some materials may produce irritation primarily in a follicular pattern. Others may produce frank acneform reactions, miliaria, or granulomatous reactions. In individuals with endogenous skin conditions such as psoriasis or atopic eczema, exposure to irritant or corrosive agents may produce eruptions of typical lesions of the endogenous condition (e.g., psoriatic plaques), in a contact pattern (Koebner phenomenon).

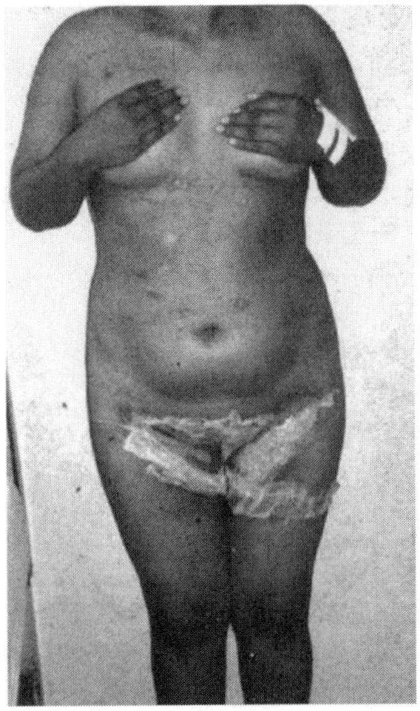

FIGURE 2.2 Burns on a 54-year-old housewife secondary to misapplication of paraquat as a topical medication.

2.3.1.1 Subtypes of ICD

Acute ICD from a single contact with a strong chemical substance can cause a toxic reaction similar to a burn. Erythema, blistering, and ulceration occur at the site almost immediately after contact. Chemicals that cause these corrosive reactions can usually be identified with predictive testing in animals (see chapter on dermatotoxicology).

Corrosive reactions may occur when the irritant agent is trapped by clothing or footwear. Following contact with uncovered areas, the agent may penetrate less deeply into the skin and cause less severe irritation.

A broader group of chemicals damage the skin only after multiple contacts, repeated and prolonged, which result in delayed or cumulative ICD. Common clinical findings include erythema, dryness and thickening, hyperkeratosis with itching, and painful fissuring. Among the important contributing factors are friction, occlusion, minor lacerations, and excessive environment heat (such as tropical countries) or cold. Because there are similar clinical findings, it is often confused with allergic CD.

2.3.1.1.1 Case examples

Paraquat burns in nonoccupational settings:

Case Reports 1 and 2 — A 19-year-old girl (Figure 2.1) and a 54-year-old woman (Figure 2.2), daughter and wife, respectively, of a Panamanian banana plantation worker, complained that they had "fungous skin disease." The worker brought home a container full of paraquat because it was a "good medicine."The following day both women developed an irritant acute dermatitis or a chemical burn as demonstrated by the figures. No evidence of systemic paraquat poisoning, as reported in the literature,[4,5] developed in either case.

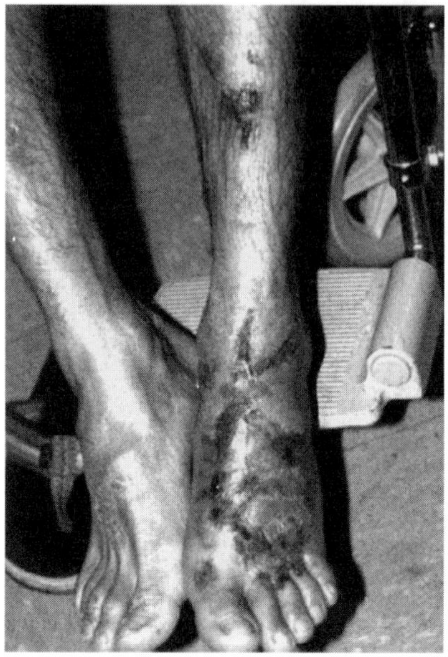

FIGURE 2.3 Contact burn secondary to paraquat in an applicator employed on a banana plantation in Panama.

Case Report 3 — A Panamanian banana plantation worker reported at the Hospital Regional in Puerto Armuelles with severe chemical burn, which involved the herbicide paraquat. The material leaked into an employee's boots when he applied the herbicide to the field. The legs and feet became erythematous, edematous, burning, and ulcerated several days before he sought medical care (Figure 2.3). Delays in seeking medical treatment are common in Panama and may be as common in other developing countries.

Case Report 4 — A patient sought medical care after using a leaking backpack sprayer to apply paraquat on a banana plantation. Exposed areas included the genitalia as well as the back. The patient delayed seeking treatment for a week after the incident and was found on initial examination to have ulceration of the scrotum (Figure 2.4) and a secondary infection that required hospital treatment.

2.3.2 Allergic CD

The clinical history usually reveals repeated contact with the suspected allergen prior to the onset of sensitization. However, for potent sensitizers such as urushiol, the antigen from *Rhus*, or dinitrochlorobenzene (DNCB), only a single episode of contact with the suspected compound may produce sensitization. Once the allergy is fully developed, skin lesions begin 1–2 days after contact. This characteristic delay between contact and onset reaction indicates to the clinician that the eruption is due to a type IV immune reaction.

The rash usually begins with erythema at the site of contact and may quickly evolve into a bullous eruption. After a few days, new lesions may appear at distant,

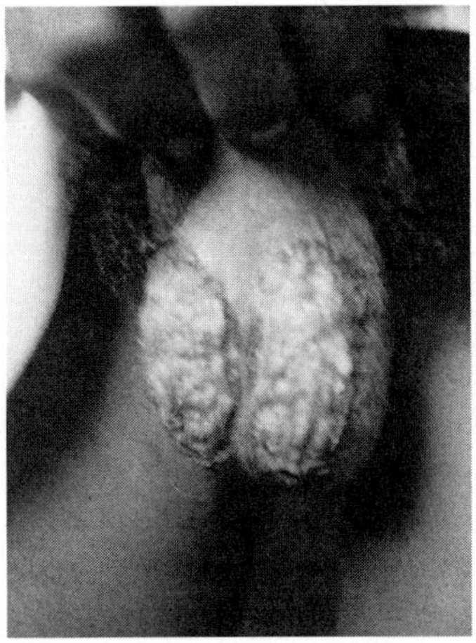

FIGURE 2.4 Contact burn of the scrotum due to a leaking backpack sprayer.

seemingly unrelated sites, usually because of transfer of the pesticide from the hands. With repeated contact, a subacute or chronic dermatitis may develop, characterized by thickened, fissured skin that occasionally erupts into more acute dermatitis because of the re-exposure to the pesticide involved.

Allergic CD can be confirmed by patch testing, a procedure that takes five days to complete. Patients wear non-irritating concentrations of test antigens on their upper backs for 48 h. After removal, an initial reading is done, and a final reading is done at 96 h. Reactions are scored as 1+ (weak reaction, macular erythema), 2+ (strong reaction, edematous, or vesicular), or 3+ (extreme reaction, spreading, bullous, or ulcerative).[6,7]

2.3.2.1 Case Examples

Case Report 5 — A 35-year-old worker, with 8 years of experience in a banana packing plant, worked as a banana bunch "seleccionador" at a "pila" (packing) operation. After a splash exposure to water containing several fungicides and antimicrobial compounds (imazalil, thiabendazole, aluminum hydroxide, formaldehyde, and chlorine) used to keep the bananas in good condition during transport, the worker developed dermatitis on his face, upper neck, and hand. The following day he had itchy, red, swollen skin lesions without sharp borders and peeling on his chin and the dorsum of his hands (Figure 2.5). After the dermatitis became quiescent, the patient was patch tested with the Spanish Research Group standard (J. Marti Tor) and with a pesticide series developed for use in Panama. The Spanish Research Group standard antigen series was negative; Mertect® (thiabendazole 1% Vaseline) and buffer* showed positive reaction (++ and +, respectively).

Conclusion: Occupational pesticide allergic contact dermatitis to thiabendazole.

Case Report 6 — A 28-year-old seamstress was seen at a hospital bed with a disseminated maculopapular erythematous rash and facial edema. She had previ-

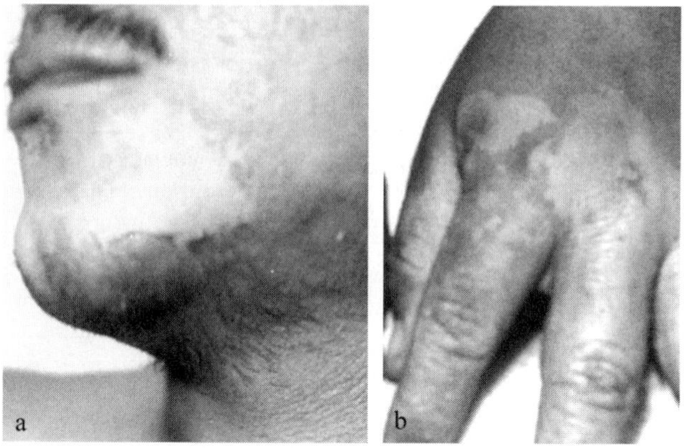

FIGURE 2.5 Peeling skin on face and hands.

ously visited the emergency room three times with her complaints. Ten days earlier the National Vector Control Program had sprayed her house with Sumithion® (fenitrothion). That evening she developed itching and discomfort at sites of contact with spray residue, including the lower extremities (Figure 2.6). The problem became worse the next day, her dermatitis worsened, and she sought medical care. Two months after her recovery, she underwent testing with the European standard series (Hermal) and with a pesticide series, including fenitrothion 1% Vaseline. The only positive (++) reaction was to fenitrothion 1% Vaseline.
Conclusion: Allergic contact dermatitis to fenitrothion.

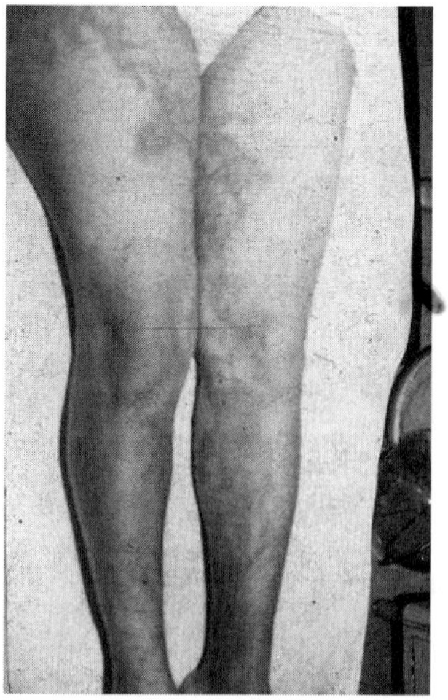

FIGURE 2.6 Dermal reaction following exposure to fenitrothion. (Photo, H. Penagos).

2.4 PHOTOTOXIC AND PHOTOALLERGIC DERMATITIS, ACTINIC DERMATITIS, AND SKIN PIGMENTATION DISORDERS

Phototoxic dermatitis occurs when a proto-irritant material is deposited on the skin and a complete irritant develops after exposure to light of the correct wavelength. The absorption spectrum causing phototoxicity is usually in the range of 280-430 nm, although some dyes (eosin) sensitize with visible light. The reaction often occurs in a well-demarcated area with a cutoff at the clothing lines, such as the "V" of the neck and arms beyond the sleevelines. Sparing of skin under the chin and upper eyelids strongly suggests a photoallergic or phototoxic mechanism. Phototoxic reactions appear as an exaggerated sunburn reaction, with erythema, bullae in severe cases, and hypo- or hyperpigmentation of the skin. Classic occupational phototoxic dermatitis arises from contact with psoralen-containing plants, such as celery, lime, and parsnips. The pattern of lesions with occupational contact phototoxins is far less homogenous than the pattern seen in phototoxic drug eruptions. This is illustrated in photographs taken during NIOSH investigations[8,9] of photodermatitis in grocery workers during the 1980s (Figure 7.4).

Photoallergic reactions are immunologically mediated and, consequently, less common than phototoxic dermatitis. Clinically, these reactions usually appear as an abrupt onset dermatitis, but may later become lichenoid at involved sites and spread to adjacent areas.

Chronic actinic dermatitis is a condition that persists long after contact with the exciting chemical (e.g., antimicrobial compounds such as hexachlorophene or fentichlor), and in some patients may persist for many years. Typical patients are men over the age of 40, with a persistent, eczematous rash, predominantly involving sun-exposed skin. In severe cases, the rash may spread to covered areas and even progress to erythroderma.[10] Some may present with depigmented, rather than pigmented lesions.[11]

> Case Report 7 — A 52-year-old man was employed as a manual laborer by a local municipality for the last 5 years. His work involved landscape maintenance and this sometimes included cleaning ditches contaminated with pesticide runoff from adjacent banana plantations.
>
> Ten years previously he had worked on a banana plantation as a field worker and was exposed to several pesticides (fungicides, insecticides, and herbicides). When examined for the first time, he had a severe rash with pruritus and erythema on exposed areas of skin, including the face, neck, and "V" of the neck. He had a less pronounced eruption on the forearms, but non-exposed skin under the jaw and behind the ears was unaffected.
>
> He was patch tested with the Spanish Research Group standard (J. Marti Tor) and with a local pesticide series. The standard antigen series was negative, but he had a (++) reaction to the fungicide captan. Although it was not possible to conduct photopatch tests, clinically he was considered to be a case of persistent light reactivity.

Less persistent hyperpigmentation following contact dermatitis has been associated with a number of compounds including cosmetics, azo textile dyes, and rubber products.[12] Pesticide-induced dyschromic changes are discussed extensively in Section 8.3.6 (review of the effects of chlorothalonil) and in the review of skin conditions occurring in Latin America (Section 9.8).

Contact leukoderma has been classically associated with phenol and catechol compounds,[13] but it has also been reported following contact dermatitis due to dental acrylics,[14] thiram,[15] and hair dyes.[16] Pesticides associated with loss of skin pigment are discussed in Sections 8.5.8, 9.2.5, 9.4.2, and 9.6. The individual compounds involved include carbyne,[17] paraquat,[18] dinoterbe,[19] thiram,[15] and mancozeb.[20]

2.4.1 Urticaria and Contact Urticaria

Maibach et al.[21] defined contact urticaria syndrome as local and systemic immediate reactions precipitated by contact urticarial agents. The insect repellent diethyltoluamide (see chapter on literature review of herbicides) was among the agents identified in the initial report. Local wheal and flare is the principal sign of contact urticaria. Generalized urticaria and systemic manifestations such as asthma and/or anaphylactoid can be seen with strong urticant allergens.

A full medical (included occupational and hobby issues) history, a radioallergosorbent test (RAST), if there was a risk of anaphylaxis, and some immediate skin tests (open test, scratch test, and prick test) are necessary to diagnose the contact urticaria. Please note if you decide to perform scratch or prick test, resuscitation facilities should be immediately available.

> Case Report 8 — A 39-year-old woman, who worked in an agricultural store for 2 years prior to evaluation, presented a clinical history of urticarial plaques especially on neck and face beginning four months earlier. She developed pruritus, itching, and wheals at work. The patient was patch tested with the European standard series (Hermal) and with a pesticide series developed according to the recommendations of the International Contact Dermatitis Research Group (ICDRG). The European standard series and pesticides series were negative. A prick test with chlorothalonil (0.001% acetone) and a control test with saline on the forearm were made. After 30 min an urticarial lesion at the site of the chlorothalonil test was observed.

2.4.2 Erythema Multiforme and TEN

Erythema multiforme (EM) and TEN represent systemic allergies, although most commonly due to drug reactions. They can also occur following contact reaction and perhaps are facilitated by systemic absorption following skin contact. Clinical characteristics of EM include target-like, erythematovesicular or urticarial lesions and are usually associated with a positive patch test reaction in the patient. Classical EM tends to have an acral distribution, which appear on crops and are almost all target-like. EM-like syndromes tend to have central (trunk) localization and have occasional target-like lesions.[22] Provocative causes include pharmaceuticals and plants (see Section 9.2). Case reports discribing EM provoked by pesticides are discussed in Section 8.2.1. Compounds involved include methyl parathion[23] and formaldehydes.[24]

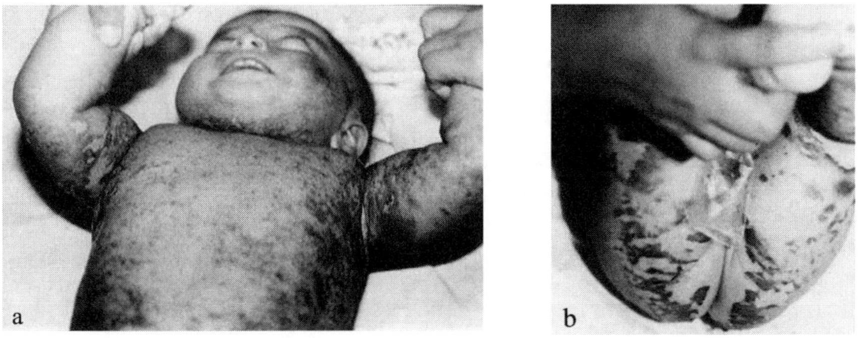

FIGURE 2.7 Toxic epidermal necrolysis — a systemic reaction to the use of topical lindane for control of scabies.

Case Report 9 — A 28-year-old male working at a banana packing house as a "demandador" developed a papular eruption principally on his trunk, with erythema, some scarce target iris lesions, and highly pruritic. He was diagnosed as erythema multiforme possibly induced by an external chemical and was treated with systemic steroids and hydroxyzine. One month after recovery he was patch tested with European standard series (Hermal) and a pesticide patch test series developed by us. He tested positive at 96 h to imazalil 1% Vaseline, and thiabendazole 1% Vaseline.

TEN is a severe, idiosyncratic, blistering, and exfoliative disease of skin and mucous membranes.[25] Pesticides implicated include thiocarbamate fungicides[26,27] and fumigants.[28]

Case Report 10 — A 9-month-old native Amerindian boy from Panama was hospitalized with generalized dermatosis, 2 weeks after being treated for scabies with topical lindane (Figure 2.7). The boy had skin excoriation because of his severe pruritus and his mother applied the lindane twice daily for one week. The child developed general malaise, fever, severe pruritus, and skin erythema, and the boy's mother brought him to the emergency room. The next day he was hospitalized because of disseminated bullae and fever. He remained hospitalized for three weeks, but was well on discharge. Biopsy showed epidermal necrosis compatible with TEN.

2.5 HAIR AND NAIL DISEASE

2.5.1 Alopecia

Alopecia appears to be an uncommon manifestation of topical exposure to environmental chemicals, but a cluster of alopecia has previously been reported among petroleum refinery workers. Although the circumstances suggested a chemical exposure as the cause, no particular chemical was identified as the cause on investigation.[29]

Alopecia is a well-recognized side effect of chemotherapeutic agents for cancer that work by inhibiting cell division. It is also a characteristic sign of poisoning with

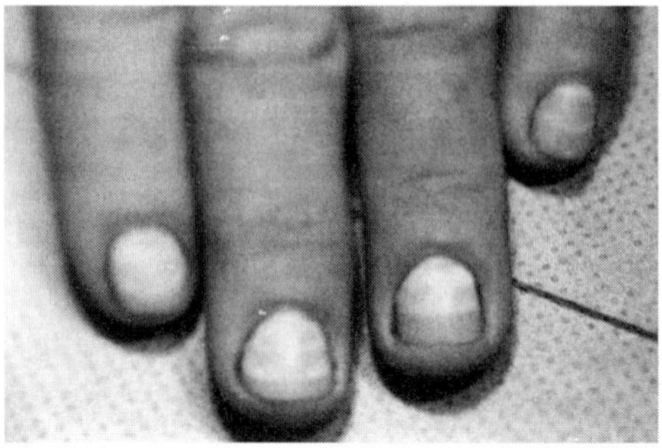

FIGURE 2.8 Transverse leukonychia associated with handling paraquat.

thallium.[30-32] Alopecia associated with boric acid[33] typically occurs after cases of intentional or accidental ingestion. Systemic use of warfarin as an anticoagulant has also been described as a cause of alopecia in a case report.[34] However, there is no association between contact exposure to warfarin and the occurrence of alopecia. Alopecia associated with exposure to DDT has been reported by Kwiatkowska et al.,[35] but not confirmed in any cohort data, case control study, or other case reports.

2.5.2 Nail disease

As summarized by Baran,[36] nails are subject to a variety of occupational injuries and diseases. Among chemical agents, particular problems are noted with corrosive agents, such as hydrofluoric acid, that readily penetrate the skin. Pesticides may be known for discoloring (chloronitrophenols[37]) or pitting of the nails or causing a form of tranverse leukonychia.[38-42]

> Case Report 11 — A 38-year-old man, herbicide applicator, consulted us because 2 weeks before he had a "fungus disease" on his hands, but his principal complaint now was that "he had fungous on his nails." He prepared dilutions of paraquat for two months without gloves or other protective measures. Physical examination showed an otherwise healthy man with resolving hand dermatitis, especially in the periungual zone, and very notable transverse leukonychia (Figure 2.8). He lost all ten of his fingernails, but eventually recovered.

2.6 CHLORACNE

In 1899, Herxheimer[43] described severe acne in four men working in a plant producing hypochlorite from potash, and cases associated with chlornaphthalenes and chlordiphenyls (PCBs) were also described by the U.S. Public Health Service during the 1940s.[44] The cases differed from oil folliculitis in the copious production of straw-colored cysts and the comparative histologic absence of inflammatory infiltration.

Areas most frequently affected included the malar crescent and the back of the ear. Unlike acne vulgaris, the nose is frequently spared and may appear as "an island in a sea of lesions." The genitalia, both penis and scrotum (principally), are also frequent sites of occurrence. Lesions also may appear on the shoulders, chest, back, and even the buttocks and abdomen, depending on the level of exposure and the toxic reaction. The axillae have been commonly involved in those patients who have ingested or inhaled the chloracnegens and this is a principal route of exposure. Less frequently affected areas include the hands, forearms, feet, legs, and thighs. Crow[45] reported that at Seveso, Italy, the axillary involvement was seen in those few patients who were enveloped in the toxic cloud.

The condition is generally systemic, rather than topical, as demonstrated by its occurrence following ingestion of oils contaminated by PCBs, but the occurrence of lesions in a contact pattern has occasionally been reported. The disease will regress over a 4-6 month period or in more severe cases and/or with re-exposure to chemical it may persist in an active form for 15-30 years or even longer.[45] Finally, scarring and fine pitting may occur as the end stage of moderate to severe chloracne, akin to atrophodermia vermiculata, and to the scars caused by severe acne vulgaris. Chloracne associated with the production of herbicides is discussed in Sections 8.5.2 and 8.5.3.

2.6.1 Related Systemic Illnesses

As previously noted, chloracne is a systemic illness and may be accompanied by liver injury and occasionally by peripheral neuropathy with pain and numbness. Neuropathy has been confirmed by the delayed conduction velocity test.[46] This neuropathic effect appears to be confined to those with unusually intense exposure[47] and the prevalence of peripheral neuropathy is variably increased in cohorts of workers exposed to dioxins.[48,49]

2.6.2 Histopathology

Hambrick[50] described the primary lesion of chloracne in animal studies and in human volunteers using as chloracnegens 3% Halowax 1014 and 3% hexachloronaphthalene as keratotic plugging of the hair follicle with hyperkeratosis in the follicular ostium. He also observed thickening of the follicular epithelium followed by the posterior disappearance of sebaceous glands and their replacement by keratinous cysts, accompanied by minimal inflammatory reaction. Cunliffe emphasized the similarity between chloracne and acne vulgaris in ultrastructural histologic findings, but found that industrial patients had significantly fewer C. acnes than the patients with ordinary acne. They also found increases in the surface lipid percentage composition of squalene, wax ester, and cholesterol.

> Case Report 12 — A 28-year-old black man working as a herbicide applicator handling a mixture of the herbicides 2,4,5-T, 2,4-D, and paraquat had a spill on his back. He did not have any immediate problem, but sought medical attention 3 weeks later because of a case of severe "acne," despite having no prior history of acne vulgaris (Figure 2.9). He had an apparent contact reaction on the back and also a systemic reaction, evidenced by comedones in distant sites such as the thorax and the face (with central sparing or absence of comedones in the area around the nose).

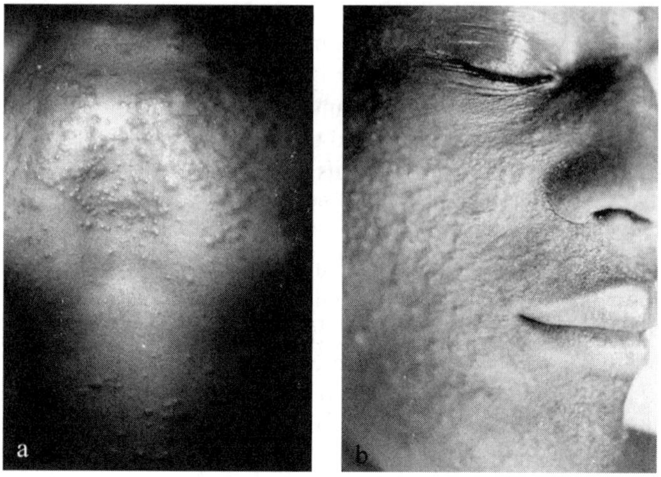

FIGURE 2.9 Chloracne in a Panamanian herbicide sprayer

2.7 OTHER IMMUNOLOGIC EFFECTS

Pesticides have been hypothesized to cause alterations of immune function. Evidence in favor of this hypothesis includes the *in vitro* effect of malathion in suppressing cytotoxic T-lymphocytes, the association of DDT with experimental thymus atrophy,[51] and reduction of lymphocyte subsets in farmers exposed to 2,4-D.[52] Nevertheless, it is difficult to find any consistent evidence for immune effects of clinical significance apart from the well-recognized mechanisms associated with the occurrence of immediate and delayed allergy. Many traditionally recognized immunologic disorders have well-documented associations with pharmacologic agents (e.g., between drugs such as the anesthetic procainamide and the anti-hypertensive hydralazine and a condition resembling systemic lupus).[53] There is thus ample precedence for the role of exogenous compounds in causing these conditions. It is also appropriate to recognize that pharmacologic doses used in medicine (as high as 1-2 g per day) often far exceed the absorbed doses associated with environmental or occupational exposures (1-2 mg per day).

2.7.1 Lupus

Chemically induced forms of lupus have principally been associated with the use of a broad array of medications.[63] There have also been case reports attributing the onset of systemic lupus erythematosus (SLE) to a variety of environmental agents, but no firm occupational association has yet been made.[54] Evidence for association with pesticides is limited to case reports, including permethrin[55] and thallium sulfate.[32] Thrombocytopenic purpura, associated with the use of many different types of drugs, has also been associated in a case report exposure to the insecticide permethrin.[56]

2.7.2 Scleroderma-like conditions

Scleroderma has been firmly associated with occupational exposure to both silica and vinyl chloride. Case reports have linked it to aromatic hydrocarbons, and a large outbreak was reported in Japan among workers exposed to an amine derivative bis(4-amino-3-methylcylohexyl)methane (CAS# 6864-37-5), used as an epoxy resin cross-linking agent.[57] Scleroderma has been associated with use of pesticides in scattered case reports.[58 59]

As with other autoimmune disorders, pemphigus vulgaris (PV) -likediseases are frequently reported following reactions to drugs. Lambert et al[60] reported two patients who developed PV-like diseases after chronic exposure to pentachlorophenol (PCP). Interestingly, the clinical course and the titer of pemphigus antibodies tightly correlate with the PCP levels in serum. Tsaunkov[61] reported a case of pemphigus, possibly pemphigus vulgaris, induced by phosphamide (dimethoate).

2.7.3 Skin Cancer

Pesticides with carcinogenic effects in animal studies include amitrazole, nitrofen, sulfallate, aramite, chlordane, organochlorines (chlordecone, DDT, dichlorvos, mirex, hexachlorobenzene), captafol, and formaldehyde.[62] Human studies have focused on occupational exposures of agricultural workers[63-66] and exposures of children to household insecticides.[67,68] Specific pesticides including creosote,[69] arsenic,[70] and paraquat[71] and a related manufacturing precursor are associated with the occurrence of non-melanoma skin cancers.[72]

Natural sources of arsenic such as ground water (especially near geothermal activity) are the main routes of exposure to the general population. Arsenic toxicity with human skin importance caused by well water contaminated by natural sources, such as arsenic-containing bed rocks, have been reported in areas of the U.S. (Southwest, Northwest, Alaska), Germany, Argentina, Chile, Taiwan, and the U.K. Studies in Taiwan have shown a dose-response relationship between arsenic concentration in well water and in the occurrence of skin cancer.[73] Another less important source of arsenic is by inhalation of small amounts of arsenic as a result of pesticide residue on tobacco leaves and by drinking wine made from grapes sprayed with arsenic-containing pesticides.[74] Exposures to applicators and other agricultural workers were a major occupational health issue in the 1970s when arsenic had major uses as an herbicide, insecticide, and rodenticide.[75]

Clinically, arsenic-induced skin cancer is characterized by multiple lesions, mostly in portions of skin with minimal sun exposure (trunk, palms, and soles). More than one type of skin cancer may occur in a patient, including Bowen's disease, squamous cell carcinoma, basal cell carcinoma, and mixed types. Jackson et al.[76] reported seven patients who developed palmoplantar keratosis many years after ingestion of trivalent inorganic arsenic. Six patients had basal cell carcinoma (superficial multicentric type in five) and one carcinoma *in situ* or squamous cell carcinoma of the skin, and two had systemic carcinoma, one bilateral breast adenocarcinoma, and one, carcinoma of the colon. Leyh et al.[77] reported a case of multiple keratosis and chronic lymphatic leukemia after arsenic poisoning 30 years ago during a one-year exposure to acetoarsenate in a pesticide factory.

2.8 CONCLUSIONS

Pesticides include diverse classes of chemical compounds that cause skin reactions in humans by means of both contact action and systemic absorption. Potential targets of damage include skin and mucous membranes, hair, and nails.

A single pesticide may cause several forms of skin disease (e.g., chlorothalonil is associated with allergic contact dermatitis, pigmented allergic contact dermatitis, photoallergy and photoirritation, toxic epidermal necrolysis, and death). On the other hand, the same clinical picture may be produced by different pesticides, by non-pesticidal compounds, and sometimes, by endogenous skin diseases. As discussed in Chapter 3, proper diagnosis may be one of the most complex issues in the clinical management of skin reactions to pesticides.

REFERENCES

1. **Li, Q., Wang, Z. Y., Inagaki, H., Li, Y. J., and Minami, M.,** Evaluation of contact sensitivity to formaldehyde and tetramethylthiuram monosulfide using a modified lymphocyte transformation test, *Toxicology*, 104, 17, 1995.
2. **National Health Assessment Nutrition and Examination Survey**, Skin conditions and related need for medical care among persons 1-74 years. United States, 1971-1974, Hyattsville, MD, 1978, U.S. DHEW.
3. **Cronin, E.,** *Contact Dermatitis*, Churchill Livingstone, Edinburgh, 1980.
4. **Smith, J. G.,** Paraquat poisoning by skin absorption: A review, *Hum. Toxicol.*, 7, 15, 1988.
5. **Garnier, R., Chataigner, D., Efthymiou, M. L., Moraillon, I., and Bramary, F.,** Paraquat poisoning by skin absorption report of two cases, *Vet. Hum. Toxicol.*, 36, 313, 1994.
6. **RM, A.,** W.B. Saunders, Philadelphia, 1990, 223.
7. **Adams, R. and Fischer, T.,** in *Occupational Skin Disease, 2nd Edition*, Adams, R., Ed., W.B. Saunders, Philadelphia, 1990, 223.
8. **NIOSH,** Phytophotodermatitis among grocery workers — Ohio, *MMWR,* 34, 11, 1985.
9. **Seligman, P. J., Mathias, C. G., O'Malley, M. A., Beier, R. C., Fehrs, L. J., Serrill, W. S., and Halperin, W. E.,** Phytophotodermatitis from celery among grocery store workers, *Arch. Dermatol.*, 123, 1478, 1987.
10. **Bleehen, S. S.,** Light, chemicals and the skin: contact photodermatitis, *Br. J. Dermatol.*, 105 Suppl. 21, 23, 1981.
11. **von den Driesch, P., Fartasch, M., and Hornstein, O. P.,** Chronic actinic dermatitis with vitiligo-like depigmentation, *Clin. Exp. Dermatol.*, 17, 38, 1992.
12. **Trattner, A., Hodak, E., and David, M.,** Screening patch tests for pigmented contact dermatitis in Israel, *Contact Dermatitis*, 40, 155, 1999.
13. **Gawkrodger, D., Cork, M., and Bleehen, S.,** Occupational vitiligo and contact sensitivity to para-tertiary butyl catechol, *Contact Dermatitis*, 25, 200, 1991.
14. **Kanerva, L. and Estlander, T.,** Contact leukoderma caused by patch testing with dental acrylics, *Am. J. Contact Dermatitis*, 9, 196, 1998.
15. **Riordan, A. T. and Nahass, G. T.,** Occupational vitiligo following allergic contact dermatitis, *Contact Dermatitis*, 34, 371, 1996.
16. **Taylor, J., Maibach, H., Fisher, A., and Bergfeld, W.,** Contact leukoderma associated with the use of hair colors, *Cutis*, 52, 273, 1993.

17. **Brancaccio, R. R. and Chamales, M. H.,** Contact dermatitis and depigmentation produced by the herbicide Carbyne, *Contact Dermatitis*, 3, 108, 1977.
18. **George, A. O.,** Contact leucoderma from paraquat dichloride?, *Contact Dermatitis*, 20, 225, 1989.
19. **Sabouraud, S., Testud, F., Rogerie, M. J., Descotes, J., and Evreux, J. C.,** Occupational depigmentation from dinoterbe, *Contact Dermatitis*, 36, 227, 1997.
20. **Peluso, A. M., Tardio, M., Adamo, F., and Venturo, N.,** Multiple sensitization due to bis-dithiocarbamate and thiophthalimide pesticides, *Contact Dermatitis*, 25, 327, 1991.
21. **Maibach, H. I. and Johnson, H. L.,** Contact urticaria syndrome. Contact urticaria to diethyltoluamide (immediate-type hypersensitivity), *Arch. Dermatol.*, 111, 726, 1975.
22. **Radimer, G. F., Davis, J. H., and Ackerman, A. B.,** Fumigant-induced toxic epidermal necrolysis, *Arch. Dermatol.*, 110, 103, 1974.
23. **Bhargava, R. K., Singh, V., and Soni, V.,** Erythema multiforme resulting from insecticide spray [letter], *Arch. Dermatol.*, 113, 686, 1977.
24. **Goh, C.,** *Textbook of Contact Dermatitis*, Rycroft, R., Menne, T., and Frosch, P., Eds., Springer-Verlag, Berlin, 1995, chap. 9.
25. **Avakian, R., Flowers, F. P., Araujo, O. E., and Ramos-Caro, F. A.,** Toxic epidermal necrolysis: A review, *J. Am. Acad. Dermatol.*, 25, 69, 1991.
26. **Copeman, P. W.,** Toxic epidermal necrolysis caused by skin hypersensitivity to monosulfiram, *Br. Med. J.*, 1, 623, 1968.
27. **Lord, J., Moats, R., and Jones, J.,** Too much golf, *J. Forensic Sci. Soc.*, 24, 359, 1984.
28. **Heng, M. C.,** Drug-induced toxic epidermal necrolysis, *Br. J. Dermatol.*, 113, 597, 1985.
29. **Williams, N. and Riegert, A.,** Epidemic alopecia areata. An outbreak in an industrial setting., *J. Occup. Med.*, 13, 535, 1971.
30. **García Pérez, A.,** *Tratado de dermatosis profesionales*, García, A., Conde-Salazar, L., and Giménez Camarasa, J. M., Eds., EUDEMA, Madrid, 1986, chap. 21.
31. **Estrella, R., Azulay, R., Peixoto, P., and Marinho, D.,** Alopecia caused by thallium sulfate, *An. Bras. Dermatol.*, 45, 333, 1970.
32. **Montoya-Cabrera, M., Saucedo-Garcia, J., Escalante-Galindo, P., and Lopez-Morales, E.,** Thallium poisoning which stimulated systemic lupus erythematosus in a child, *Gac. Med. Mex.*, 127, 133, 1991.
33. **Stein, K., Odom, R., Justice, G., and Martin, G.,** Toxic alopecia from ingestion of boric acid, *Arch. Dermatol.*, 108, 95, 1973.
34. **Umlas, J. and Harken, D. E.,** Warfarin-induced alopecia, *Cutis*, 42, 63, 1988.
35. **Kwiatkowska, E. J. and Plonka, T. J.,** A case of alopecia due to DDT, *Przegl. Dermatol.*, 58, 187, 1971.
36. **Baran, R.,** *Occupational Skin Disease*, Adams, R., Ed., W.B. Saunders, Philadelphia, 1990, 160.
37. **Wahlberg, J.,** Yellow staining of hair and nails and contact sensitivity to dinobuton, *Contact Dermatitis Newsletter*, 16, 481, 1974.
38. **Samman, P. and Johnston, E.,** Nail damage associated with handling of paraquat and diquat, *Br. Med. J.*, 1, 818, 1969.
39. **Dobbelaere, F. and Bouffioux, J.,** Leukonychia in bands due to paraquat, *Arch. Belg. Dermatol.*, 30, 283, 1974.
40. **Botella, R., Sastre, A., and Castells, A.,** Contact dermatitis to paraquat, *Contact Dermatitis*, 13, 123, 1985.

41. **Hearn, C. E.,** A review of agricultural pesticide incidents in man in England and Wales, 1952-71, *Br. J. Ind. Med.*, 30, 253, 1973.
42. **Baran, R.,** Letter: Nail damage caused by weed killers and insecticides, *Arch. Dermatol.*, 110, 467, 1974.
43. **Herxheimer, K.,** Uber chlorakne, *Munch. Med. Wchst.*, 46, 278, 1899.
44. **Schwartz, L. and Peck, S.,** Occupational acne, *NY State J. Med.*, 43, 1711, 1943.
45. **Crow, K. D.,** Chloracne and its potential clinical implications, *Clin. Exp. Dermatol.*, 6, 243, 1981.
46. **Tindall, J. P.,** Chloracne and chloracnegens, *J. Am. Acad. Dermatol.*, 13, 539, 1985.
47. **Moses, M., Lilis, R., Crow, K. D., Thornton, J., Fischbein, A., Anderson, H. A., and Selikoff, I. J.,** Health status of workers with past exposure to 2,3,7,8-tetrachlorodibenzo-p-dioxin in the manufacture of 2,4,5-trichlorophenoxyacetic acid: Comparison of findings with and without chloracne, *Am. J. Ind. Med.*, 5, 161, 1984.
48. **Thomke, F., Jung, D., Besser, R., Roder, R., Konietzko, J., and Hopf, H. C.,** Increased risk of sensory neuropathy in workers with chloracne after exposure to 2,3,7,8-polychlorinated dioxins and furans, *Acta Neurol. Scand.*, 100, 1, 1999.
49. **Sweeney, M. H., Fingerhut, M. A., Arezzo, J. C., Hornung, R. W., and Connally, L. B.,** Peripheral neuropathy after occupational exposure to 2,3,7,8-tetrachlorodibenzo-p-dioxin (TCDD), *Am. J. Ind. Med.*, 23, 845, 1993.
50. **Hambrick, G. and Blank, H.,** A microanatomical study of the response of the pilosebaceous apparatus of the rabbit's ear canal, *J. Invest. Dermatol.*, 26, 185, 1956.
51. **Street, J. C. and Sharma, R. P.,** Alteration of induced cellular and humoral immune responses by pesticides and chemicals of environmental concern: Quantitative studies of immunosuppression by DDT, aroclor 1254, carbaryl, carbofuran, and methylparathion, *Toxicol. Appl. Pharmacol.*, 32, 587, 1975.
52. **Faustini, A., Settimi, L., Pacifici, R., Fano, V., Zuccaro, P., and Forastiere, F.,** Immunological changes among farmers exposed to phenoxy herbicides: preliminary observations, *Occup. Environ. Med.*, 53, 583, 1996.
53. **Hahn, B.,** *Harrison's Principles of Internal Medicine*, Wilson, J., Braunwald, E., Isselbacher, J., Petersdorf, R., Martin, J., Fauci, A., and Root, R., Eds., McGraw-Hill, New York, 1991, 1432.
54. **Adams, R.,** W.B. Saunders, Philadelphia, 1990, 180.
55. **Beer, K. R., Lorincz, A. L., Medenica, M. M., Albertini, J., Baron, J., Drinkard, L., and Swartz, T.,** Insecticide-induced lupus erythematosus, *Int. J. Dermatol.*, 33, 860, 1994.
56. **Hay, A. and Singer, C. R.,** Wood preservatives, solvents, and thrombocytopenic purpura [letter], *Lancet*, 338, 766, 1991.
57. **Adams, R.,** W.B. Saunders, Philadelphia, 1990, 178.
58. **Dunnill, M. and Black, M.,** Sclerodermatous syndrome after occupational exposure to herbicides — response to systemic steroids, *Clin. Exp. Dermatol.*, 19, 1994.
59. **Poskitt, L. B., Duffill, M. B., and Rademaker, M.,** Chloracne, palmoplantar keratoderma and localized scleroderma in a weed sprayer, *Clin. Exp. Dermatol.*, 19, 264, 1994.
60. **Lambert, J., Schepens, P., Janssens, J., and Dockx, P.,** Skin lesions as a sign of subacute pentachlorophenol intoxication, *Acta. Derm. Venereol.*, 66, 170, 1986.
61. **Tsankov, N., Dimitrowa, J., Obreschkowa, E., and Lasarowa, A.,** Induced pemphigus caused by the pesticide phosphamide, *Z. Hautkr.*, 62, 196, 1987.
62. **Rosenberg, J. and O'Malley, M.,** *Occupational Medicine*, LaDou, J., Ed., Lange-Appleton, 1997, chap. 32.
63. **Zahm, S., Ward, M., and Blair, A.,** Pesticides and cancer, *Occup. Med., State Art Rev.*, 12, 269, 1997.

64. **Blair, A. and Zahm, S.,** Agricultural exposure and cancer, *Environ. Health Perspect.*, 103 (Suppl. 8), 205, 1995.
65. **Blair, A. and Zahm, S.,** Cancer among farmers, *Occup. Med. State. Art. Rev.*, 6, 335, 1991.
66. **Pearce, N. and Reif, J.,** Epidemiologic studies of cancer in agricultural workers, *Am. J. Ind. Med.*, 18, 133, 1990.
67. **Reeves, J.,** Household insecticide-associated blood dyscrasias in children, *Am. J. Hematol.*, 4, 438, 1982.
68. **Reeves, J. D., Driggers, D. A., and Kiley, V. A.,** Household insecticide associated aplastic anaemia and acute leukaemia in children [letter], *Lancet*, 2, 300, 1981.
69. **Karlehagen, S., Andersen, A., and Ohlson, C.,** Cancer incidence among creosote-exposed workers, *Scand. J. Work Environ. Health*, 18, 26, 1992.
70. **Hogan, D.,** *Occupational Skin Disease*, Adams, R., Ed., W.B. Saunders, Philadelphia, 1990, chap. 31.
71. **Jee, S., Kuo, H., Su, W., Chang, C., Sun, C., and Wang, J.,** Photodamage and skin cancer among paraquat workers, *Int. J. Dermatol.*, 334, 466, 1995.
72. **Bowra, G., Duffield, D., Osborn, A., and Purchase, I. F. H.,** Premalignant and neoplastic skin lesions associated with occupational exposure to "tarry" byproduct during manufacture of 4, 4'-bipyridyl, *Br. J. Ind. Med.*, 39, 76, 1982.
73. **Kosnett, M.,** *Environmental Medicine: Integrating a Missing Element into Medical Education*, Pope, A. M., Rall, D. P., Eds., National Academy Press, Washington, D. C., 1995.
74. **Sanon, R. and Stragen, D.,** Arsenic-induced skin toxicity, *Hum. Toxicol.,* 8, 99, 1989.
75. **Peoples, S. A., Maddy, K. T., Peifer, W. R., and Edmiston, S.,** Occupational exposures to pesticides containing organoarsenicals in California, *Vet. Hum. Toxicol.*, 21, 417, 1979.
76. **Jackson, R. and Grainge, J.,** Arsenic and cancer, *Can. Med. Assoc. J.*, 113, 396, 1975.
77. **Leyh, F. and Rothlaender, J.,** Multiple bowenoid arsenic keratoses, *Derm. Beruf. Umwelt.*, 33, 99, 1985.

3 Diagnostic Methods

A.L. Chew and H.I. Maibach

CONTENTS

3.1 INTRODUCTION

Farmers and other agricultural workers are exposed to a spectrum of occupational hazards (chemical, biologic, and physical), which may result in various occupational diseases. Skin diseases comprise a significant proportion of agricultural occupational disease.[1] Contact with pesticides and other agricultural chemicals results in dermatologic disorders, the most important ones being contact dermatitis, photoallergic contact dermatitis, and contact urticaria. Diagnostic *in vivo* skin tests are used to identify the exogenous chemical agent that causes these skin disorders, and hence are critical in their scientific documentation. This chapter briefly describes the *in vivo* diagnostic test methods commonly employed.

3.2 CONTACT DERMATITIS

Contact dermatitis may be divided into irritant contact dermatitis and allergic contact dermatitis.

3.2.1 Irritant Contact Dermatitis

This is the more common; it is initiated by non-immunological toxic mechanisms, therefore it can occur with first contact with a causative agent. Diagnosis is still, in

1-56670-293-3/00/$0.00+$.50

TABLE 3.1.
Patch Test

Materials:	Allergens in vehicle Vehicle (e.g., petrolatum, ethanol, water) Aluminum chambers (Finn Chambers) Adhesive tape (e.g., Scanpore) Filter paper OR Ready-made patch test series (e.g., TRUE test)
Method:	The allergen is applied in a suitable concentration and suitable vehicle onto a test unit placed on adhesive tape which is then applied onto the skin. The patches are removed after 48 h.
Reading time:	48 and 96 h
Interpretation:	Erythema, edema, or bullae is positive Distinguishing between allergic and irritant are important Allergic: reaction extends beyond boundaries of patch site Irritant: reaction peaks at 48 h then fades rapidly
Precautions:	Intense skin reactions possible Active sensitization possible
Controls:	Required

Source: From Amin, S., Laurema, A., and Maibach, H., in *Dermatotoxicology Methods: The Laboratory Worker's Vade Mecum,* Vol. 15, Marzulli, F. and Maibach, H., Eds., Taylor & Francis, Washington, D.C., 1997, 177.

most cases, a process of exclusion, as there is no diagnostic test available. Thus a detailed history, including exposure conditions, is crucial.[2] Irritants encountered in agriculture include pesticides, fertilizers, secretions from animals, and animal feed.[3]

3.2.2 Allergic Contact Dermatitis (ACD)

This requires topical sensitization to a low molecular weight hapten; therefore, at least one prior exposure to the causative agent is required, and, in subsequent topical contact, a skin reaction develops characterized by erythema, edema, and vesiculation.[4] ACD is mediated by delayed hypersensitivity (type IV) mechanisms. Examples of allergenic compounds include rubber, feed additives, and medicaments, as well as numerous pesticides, for example, dazomet, captan, and Carbyne.[3] ACD may be diagnosed by patch testing (Table 3.1), intradermal tests (Table 3.2), or open tests (Table 3.3). Of these methods, patch testing is the most established diagnostic test for ACD.[5]

3.3 PATCH TESTING

Numerous commercial patch testing methods exist. Two widely used and reliable methods are the TRUE test and Finn Chamber methods.

TABLE 3.2
Intradermal Test

Materials:	Allergens in vehicle Vehicle (isotonic solution, e.g., normal saline) Tuberculin (1-cc) syringes and needles
Method:	0.05-0.1 ml of allergen solution and vehicle solution is applied intradermally to the skin of the volar aspects of the forearms, using a tuberculin needle and syringe
Reading time:	30 min, 24 and 48 h
Interpretation:	Erythema and edema at 30 min suggest immediate type I allergy, in the absence of a reaction in the control; Arthus reaction with polymorphonuclear leukocyte infiltration appearing in 2-4 h suggests cytotoxic type III reaction
Precautions:	Risk of anaphylaxis is higher than in prick or scratch tests because of larger amount of allergen introduced; therefore, a physician should always be available. The risk is greater in asthmatic patients. The patient should remain at the premises for 30 min after the test.
Controls:	Required

Source: From Amin, S., Laurema, A., and Maibach, H., in *Dermatotoxicology Methods: The Laboratory Worker's Vade Mecum*, Vol. 15, Marzulli, F. and Maibach, H., Eds., Taylor & Francis, Washington, D.C., 1997, 177.

TRUE Test is an example of a ready-to-use patch test system. All necessary material is prepared in advance. The investigator only removes the covering material and applies the test strips. Finn-Chamber provides flexibility for the dermatologist to test substances not in routine patch test use.

TABLE 3.3
Open Test (Provocative Use Test or Repeated Open Application Test)

Materials:	Allergen in vehicle (e.g., petrolatum, ethanol, water) Vehicle Applicator (to spread the preparations)
Methods:	Patient applies allergen and vehicle on antecubital fossa (outpatient) or shoulder regions of upper back, twice a day for 14 days or until a positive reaction appears
Reading time:	Patient reports whether positive reaction appears At 7 and 14 days, the patient returns for reading of test site
Interpretation:	Erythema and edema are positive
Precautions:	Active sensitization possible, but not yet documented
Controls:	May be required

Source: From Amin, S., Laurema, A., and Maibach, H., in *Dermatotoxicology Methods: The Laboratory Worker's Vade Mecum,* Vol. 15, Marzulli, F. and Maibach, H., Eds., Taylor & Francis, Washington, D.C., 1997. 177.

3.3.1 Patch Testing Limitations

- False-negative results caused by insufficient concentration or penetration of the allergen, or if patch testing is performed in the lower back or forearms
- False-positive results resulting from irritation from the test compound, as well as irritation or sensitization from other substances, e.g., the vehicle or adhesive tape
- May cause exacerbation of eczema in other skin sites (excited skin syndrome)
- Active sensitization to patch compounds may occur[6]

3.4 INTRADERMAL TEST

Although intradermal testing may have been used earlier in parallel with patch testing, it has gradually been replaced by patch testing alone.[7] Nowadays, intradermal testing is limited to systemically introduceable antigens such as local anesthetics and collagen. Recently, it has also been thought to be more sensitive than patch testing in diagnosing hydrocortisone contact allergy.[8]

Intradermal testing is underutilized mainly because it is an invasive procedure and application of the allergen is therefore painful and more complicated than patch testing. It also carries the risk of anaphylactic reactions, general skin reactions, and active sensitization. The advantage, however, is that delivery of hapten to the skin is ascertained. Therefore, haptens that are minimal penetrants may be easier to test.

3.5 REPEATED OPEN APPLICATION TEST (ROAT) OR PROVOCATIVE USE TEST (PUT)

This method is indicated when a patch test result is either questionable or negative, while the history suggests contact allergy, or if a patch test result with a little known or new possible contact allergen is positive.[9]

3.6 PHOTOALLERGIC CONTACT DERMATITIS

Photoallergic contact reactions are clinically identical to any other type of allergic contact dermatitis. The spectrum of possible responses ranges from erythema, to an eczematous picture, to a severe vesiculobullous eruption.[5] Involvement of the sun-exposed areas of skin are most common. However, the eruption may extend to unexposed parts of the body, becoming generalized in severe cases.

The diagnosis of photoallergic contact dermatitis is suspected by the clinical picture, including the character and distribution of the eruption and the histology. Confirmation and identification of the offending chemical involves photopatch testing (Table 3.4).

TABLE 3.4
Photopatch Test

Materials:	Ultraviolet radiation source Patch test materials (see Table 3.1)
Method:	Two sets of patch test are applied for 48 h; after removal, one set is irradiated with UVA at a dose below minimal erythematous dose (MED); (5-10 J/cm^2 or 50% of MED) and the other set is protected from the UV dose
Reading time:	48 and 96 h
Interpretation:	Reaction only at the irradiated site suggests photoallergy, while reaction at both sites suggests contact allergy; reaction at both sites with a stronger reaction at the irradiated sites suggests both contact allergy and photoallergy
Controls:	Required

Source: From Amin, S., Laurema, A., and Maibach, H., in *Dermatotoxicology Methods: The Laboratory Worker's Vade Mecum*, Vol. 15, Marzulli, F. and Maibach, H., Eds., Taylor & Francis, Washington, D.C., 1997, 177.

3.7 CONTACT URTICARIA

The contact urticaria syndrome (immediate contact reactions), first described in 1975,[10] comprises a heterogeneous group of inflammatory reactions that appear, usually within minutes, after contact with the eliciting substance. Symptoms include pruritus, tingling, erythema, and a wheal and flare reaction. Extracutaneous reactions such as wheezing and anaphylactic shock may occur in severe cases The mechanisms underlying contact urticaria syndrome are divided into immunologic contact urticaria (ICU) (type I, IgE mediated) and non-immunologic contact urticaria (NICU).[11]

TABLE 3.5
Open Test

Materials:	Allergen in vehicle (petrolatum, ethanol, water) Vehicle Applicators (to spread the preparations)
Methods:	Allergen and vehicle are applied to the skin
Reading time:	Up to 1 h
Interpretation:	Urticarial reaction is positive
Precautions:	See Prick Test
Controls:	Required to aid in discriminating ICU from NICU In NICU, the reaction will be noted in most controls

Source: Amin, S., Lanti, A., and Maibach, H. *Contact Urticaria Syndrome*, CRC Press, Boca Raton, FL, 1997

TABLE 3.6
Prick Test

Materials:	Allergens in vehicles Vehicles Histamine in 0.9% NaCl (positive control) Prick lancets
Method:	One drop of each test allergen in their vehicles, and histamine control is applied to the volar aspect of the forearms; the test site is pierced with a lancet to introduce the allergen into the skin
Reading time:	15-30 min
Interpretation:	An edematous reaction (wheal) of at least 3 mm in diameter and at least half the size of the histamine control is considered positive, in the absence of such a reaction in the vehicle control
Precautions:	General anaphylaxis unlikely, because of the small amount of allergen introduced, but a physician should always be available; the patient should not leave the premises for 30 min after the test
Controls:	Required

Source: Amin, S., Lanti, A., and Maibach, H., *Contact Urticaria Syndrome*, CRC Press, Boca Raton, FL, 1997.

3.7.1 ICU

This is usually tested by *in vivo* methods: prick test (Table 3.6) and open application test (single application) (Table 3.5) are more popular, and scratch test and chamber scratch test are seldom used now. *In vitro* methods are also used, most commonly RAST, which detects antigen-specific IgE molecules from the patient's serum.[12]

3.7.2 NICU

The diagnostic test for NICU is the open application test (single application) (Table 3.5).

3.8 CONCLUSIONS

Diagnostic tests employed in evaluation of skin reactions to exogenous substances depend upon the nature of reaction being evaluated. Delayed and immediate reactions require different methods of evaluation. No simple test is available for evaluation of irritant skin reactions.

REFERENCES

1. **Hogan, D.,** *Occupational Skin Disease*, Adams, R., Ed., W. B. Saunders, Philadelphia, 1990, chap. 31.
2. **Van der Valk, P. and Maibach, H.,** *The Irritant Contact Dermatitis Syndrome*, CRC Press, Boca Raton, FL, 1996.

3. **Veien, N. K.,** Occupational dermatoses in farmers, *Occupational and Industrial Dermatology, Second Edition,* Maibach, H. I., Ed., 1987.
4. **Ale, S. and Maibach, H.,** Clinical relevance in allergic contact dermatitis: an algorithmic approach, *Dermatosen*, 43, 119, 1995.
5. **Rycroft, R., Menne, T., and Frosch, P.,** *Textbook of Contact Dermatitis*, 2nd ed., Springer-Verlag, Berlin, 1995.
6. **Fischer, T. and Maibach, H.,** *Exogenous Dermatoses: Environmental Dermatitis*, Menne, T. and Maibach, H., Eds., CRC Press, Boca Raton, FL, 1991.
7. **Herbst, R., Lauerma, A., and Maibach, H.,** Intradermal testing in the diagnosis of allergic contact dermatitis. A reappraisal, *Contact Dermatitis*, 29, 1, 1993.
8. **Wilkinson, S., Cartwright, P., and English, J.,** Hydrocortisone: an important cutaneous allergen, *Lancet*, 337, 761, 1991.
9. **Hannuksela, M. and Salo, H.,** The Repeated Open Application Test (ROAT), *Contact Dermatitis*, 14, 221, 1986.
10. **Maibach, H. I. and Johnson, H. L.,** Contact urticaria syndrome. Contact urticaria to diethyltoluamide (immediate-type hypersensitivity), *Arch. Dermatol.*, 111, 726, 1975.
11. **Liden, C.,** Facial dermatitis caused by chlorothalonil in a paint [see comments], *Contact Dermatitis*, 22, 206, 1990.
12. **Lahti, A. and Maibach, H.,** *Dermatology*, Moschella, S. and Hurley, H., Eds., W.B. Saunders, Philadelphia, 1992.
13. **Amin, S., Lauerma, A., and Maibach, H.,** *Dermatotoxicology Methods: The Laboratory Worker's Vade Mecum*, Vol. 15, Marzulli, F. and Maibach, H., Eds., Taylor & Francis, Washington, D.C., 1997, 177.
14. **Amin, S., Lahti, A., and Maibach, H.,** *Contact Urticaria Syndrome*, CRC Press, Boca Raton, FL, 1997.

4 Treatment and Case Management

Michael O'Malley

CONTENTS

4.1 INTRODUCTION

This chapter covers some basic issues in treatment of contact dermatitis and other skin conditions related to exposure to exogenous chemicals. This excludes issues of diagnostic patch testing covered in Chapter 3.

4.2 TREATMENT WITH TOPICAL STEROIDS

Treatment of acute contact dermatitis is straightforward, employing use of topical steroids (e.g., fluocinonide, Lidex®, triamcinolone, Kenalog®, etc.) two or three times daily. For chronic dermatitis there are potential complications: allergies to topical steroids develop in a small percentage of patients[1] and chronic use of topical steroids may also be associated with local skin irritation (burning, itching, irritation, dryness), hypertrichosis, folliculitis and acneiform eruptions, hypopigmentation, perioral dermatitis, allergic contact dermatitis, maceration of the skin, secondary infection, skin atrophy, striae, and miliaria.[2] These complications can be minimized by avoiding use of fluorinated steroids on the sensitive skin of the face and genitalia. Effective non-fluorinated steroids include 2.5% hydrocortisone and 0.05% desonide (Tridesilon®).

1-56670-293-3/00/$0.00+$.50

4.3 SYSTEMIC STEROIDS

Systemic steroids may be necessary in acute generalized contact dermatitis. Regimens vary from simple one dose injections, (e.g., 1 cc of IM Kenalog-40 (40-40 mg of triamcinolone) to treatment with oral prednisone in decreasing doses over a 7-14 day period, (e.g., 60 mg daily ∞ 3 days, 40 mg daily ∞ 3 days, 20 mg daily ∞ 3 days.

Gbucase intolerance and gastric ulcer complications of systemic steroids resemble those reported for topical treatment, occurring principally following chronic use. Additional complications include adrenal suppression. The occasional occurrence of allergy to the synthetic systemic prednisolone has also been reported.[3]

4.3.1 Use of Steroids in Irritant Dermatitis

The benefit of using topical or systemic steroids in irritant dermatitis has been demonstrated for subacute or chronic irritation for model compounds such as sodium lauryl sulfate.[4,5] Whether steroid treatment is useful for more acute forms of irritation is still a matter of conjecture and it does not seem inherently plausible that steroids substantially modify the effects of acutely corrosive materials. Nevertheless, most skin injuries cause the release of cytokines and some secondary inflammation that may respond to the use of topical or systemic steroids.[6]

4.3.2 Alternatives to Systemic Steroids

For chronic dermatitis, azathioprine is an alternative to steroids.[7] In a study of 15 patients from India with chronic parthenium dermatitis, all but one showed a response and most (66.6%) showed nearly complete clearance. Side effects in the limited study were reported to be minimal. European studies have also demonstrated immunomodulation effects with extracts of ginkgo biloba in treatment of purely experimental allergic contact dermatitis.[8]

Animal studies have demonstrated anti-inflammatory effects for other classes of compounds such as calcium channel blockers.[9] Animal studies have also shown immunosuppressive properties for other classes of compounds such as urocanic acid isomers.[10] Cutaneous exposure to ultraviolet radiation suppresses the induction of T-cell-mediated responses such as contact and delayed type hypersensitivity (DTH) by altering the function of immune cells in the skin and causing the release of immunoregulatory cytokines. Extracts of crude *Aloe barbadensis* gel prevent this photosuppression.

4.4 SUPPORTIVE TREATMENT

Supportive treatment with moisturizing creams (Eucerin®, Lubriderm®, etc.) is very important. This may be particularly so in cases of irritant contact dermatitis.[11, 12]

4.5 MANAGEMENT OF ACUTE CORROSIVE DERMATITIS

It may be necessary to treat cases with extensive areas of skin involved like burn patients, and a portion of these cases may require hospitalization for intensive nursing care. Use of topical and systemic steroids in these cases is often done, but may not be of benefit.

4.6 RECOMMENDATIONS FOR PROTECTIVE CLOTHING

On average, protective clothing reduces pesticide contact with the skin by 90-95%.[13, 14] This may be sufficient protection against irritant dermatitis in many situations, but may not provide adequate protection against allergic contact dermatitis for sensitized individuals.

4.7 MANAGEMENT OF WORK RESTRICTIONS AND PREVENTION OF RECURRENT EXPOSURE

In addition to managing purely medical issues, physicians often have an obligation to manage workplace restrictions — to what degree workers can handle re-exposure to a given contact irritant. To adequately protect the worker requires detailed knowledge of the job duties and work hazards. If alternative work is not available, it is sometimes necessary to prohibit work altogether. This can be costly to both the worker who recovers only a portion of the usual income and also to the employer who must employ replacement labor, and may also pay medical expenses and compensation costs. A thorough understanding of the work environment (see Chapter 5) is necessary to properly advise workers and employers regarding chronic occupational skin conditions.

Evaluation of the role of an allergy in provoking dermatitis is often critical in deciding what level of restriction to put in place. The role of provocation tests in performing this evaluation is covered in detail by Dr. Chew and Dr. Maibach in Chapter 3 and recommended test concentrations for many commonly used pesticides are given in Appendix II. Once an allergy is documented by provocation tests, the physician, patient, and employer must jointly decide what level of restrictions are warranted. An ideal approach is to exclude the allergen, rather than the patient, from the work environment.

4.8 COMMUNICATION ISSUES

Communication between medical providers and agricultural workers is often complicated by the variable nature of the work tasks and the intermittent nature of chemical exposures. Field workers or other hired laborers may not know the nature of chemicals applied to the fields in which they work. Judicious communication with the employer is therefore necessary. Restrictions written for pesticide handlers can be simpler to write, but should also be communicated to the employer directly if possible. Compliance with the restrictions should also be investigated during follow-up visits.

Compensation issues and other "secondary gain" issues often complicate evaluation of occupational skin disease and other cases of contact dermatitis. These also complicate communication with the employer. Most employers find acute injuries and illnesses simple to accept, but may be reluctant to accept complaints of cumulative irritation, allergy, or other complaints of "sensitivity."

4.9 CONCLUSIONS

In addition to use of appropriate medication, treatment of skin conditions related to pesticides includes limiting recurrent exposures. This requires differentiation between irritation and allergy (see Chapter 3) and knowledge of the circumstances of exposure (Chapter 5).

REFERENCES

1. **Vestergaard, L. and Andersen, K.,** Contact allergy to local steroids. Contact allergy to corticosteroids among consecutively tested patients with eczema, *Ugeskr. Laeg.*, 159, 5662, 1997.
2. **Medical Economics,** *1999 PDR Electronic Library Version 5.00A.*, Medical Economics, Montvale, NJ, 1999.
3. **Bircher, A., Levy, F., Langauer, S., and Lepoittevin, J.,** Contact allergy to topical corticosteroids and systemic contact dermatitis from prednisolone with tolerance of triamcinolone, *Acta Dermat. Venereol.*, 75, 490.
4. **Le, T., De Mon, P., Schalkwijk, J., and van der Valk, P.,** Effect of a topical corticosteroid, a retinoid and a vitamin D3 derivative on sodium dodecyl sulphate induced skin irritation, *Contact Dermatitis*, 37, 19, 1997.
5. **Ramsing, D. and Agner, T.,** Efficacy of topical corticosteroids on irritant skin reactions. Efficacy of topical corticosteroids on irritant skin reactions, *Contact Dermatitis*, 32, 293, 1995.
6. **Berardesca, E. and Distante, F.,** The modulation of skin irritation, *Contact Dermatitis,*, 31, 281, 1994.
7. **Sharma, V.K., Chakrabarti, A., and Mahajan, V.,** Azathioprine in the treatment of parthenium dermatitis, *Int. J. Dermatol.*, 37, 299, 1998.
8. **Castelli, D., Colin, L., Camel, E., and Ries, G.,** Pretreatment of skin with a Ginkgo biloba extract/sodium carboxymethyl- beta-1,3-glucan formulation appears to inhibit the elicitation of allergic contact dermatitis in man [see comments], *Contact Dermatitis*, 38, 123, 1998.
9. **Katoh, N., Hirano, S., Kishimoto, S.,Yasuno, H., Katoh, N., Hirano, S., Kishimoto, S., asuno, H.** Calcium channel blockers suppress the contact hypersensitivity reaction (CHR) by inhibiting antigen transport and presentation by epidermal Langerhans cells in mice, *Clin. Exp. Immunol.*, 108, 302, 1997.
10. **el-Ghorr, A. and Norval, M.,** The effect of chronic treatment of mice with urocanic acid isomers, *Photochem. Photobiol.,*, 65, 866, 1997.

11. **Loden, M.,** Barrier recovery and influence of irritant stimuli in skin treated with a moisturizing cream, *Contact Dermatitis*, 36, 256, 1997.
12. **Heenan, A.,** Emollient applications for chronic skin problems, *Prof. Nurse*, 11, 747, 1996.
13. **Stewart, P., Fears, T., Kross, B., Ogilvie, L., and Blair, A.,** Exposure of farmers to phosmet, a swine insecticide, *Scand. J. Work. Environ. Health*, 25, 33, 1999.
14. **Thongsinthusak, T., Brodberg, R., Ross, J., Gibbons, D., and Krieger, R.**, Reduction of Pesticide Exposure by Using Protective Clothing and Enclosed Cabs (Poster Presentation - 199th National American Chemical Society, Boston, MA, 1990), Sacramento, CA, 1991, California Department of Food and Agriculture.

10. Susitaival, P. et al., *Scand. J. Work Environ. Health,* 1994; 20: 206-12.
11. Susitaival, P and Hannuksela, M., *Cont. Dermatitis,* 1995; 32: 233-237.
12. Smit, H.A. et al., *Cont. Dermatitis,* 1992; 26: 11-6.
13. Rea, J.N. et al., *Br. J. Prev. Soc. Med.,* 1976; 30: 107-14.
14. Meding, B., *Acta Derm. Venereol. (Stockholm),* 1990; (Suppl. 153): 1-43.
15. Berg, M. and Axelson, O., *Cont. Dermatitis,* 1990; 22: 71-7.
16. Meding, B. et al., *Am. J. Ind. Med.,* 1993; 23: 721-8.
17. Susitaival, P. et al., *Scand. J. Work Environ. Health,* 1995; 21: 30-35.
18. Lammintausta, K. and Kalimo, K., *Cont. Dermatitis,* 1981; 7: 301-8.
19. Kuokkanen, K. et al., *Suomen Lääkärilehti* 1982; 37: 22: 1948-1950, *(Finnish).*
20. Rystedt. I., *Acta Derm. Venereol. (Stockholm),* 1985; (Suppl. 117), Thesis.
21. Nilsson, E., *Acta Derm. Venereol. (Stockholm),* 1986; (Suppl. 128), Thesis.
22. Meding, B. and Swanbeck, G., *Cont. Dermatitis,* 1990; 23: 154-161.
23. Kristensen, O., *Cont. Dermatitis,* 1992; 26: 341-345.
24. Lammintausta, K., *Risk Factors for Hand Dermatitis in Wet Work,* Ph.D. Dissertation, Turku, 1982.
25. Diepgen, T.L., 4th Int. Course on Occupational Dermatoses, 7-11 June 1994, Naantali, Finland.
26. Peltonen, L., *Cont. Dermatitis,* 1979; 5: 27-32.
27. Menné, T. et al., *Acta Derm. Venereol. (Stockholm),* 1982; 62: 35-41.
28. Nilsson, E. and Bäck O., *Acta Derm. Venereol. (Stockholm),* 1986; 66: 45-50.
29. Schubert, H. et al., *Cont. Dermatitis,* 1988; 18: 237-9.
30. Susitaival, P. et al., *Cont. Dermatitis,* 1995; 32: 150-155.

December 15, 2000

Dear Customer:

Thank you for your purchase of *Pesticide Dermatoses,* edited by Homero Penagos, Michael O'Malley, and Howard I. Maibach.

In Appendix 1B, the questionnaire should be credited to: Susitaival, P., Kanerva, L., Hannuksela, M., Jolanki, R., and Estlander, T., *Tuohilampi Questionnaire for Epidemiological Studies of Contact Dermatitis and Atopy*, People and Work, Research Reports 10, Finnish Institute of Occupational Health, Helsinki, 1996. Below and on the reverse is the reference list for this appendix, which was inadvertently omitted.

We sincerely regret any inconvenience this may have caused you. Please let us know if we can be of any assistance regarding this title or any other titles that CRC Press publishes.

Best regards,

CRC Press LLC

Reference: appendix 1B

1. Cronin, E., *Cont. Dermatitis,* 1985; 13: 153-161.
2. Rystedt, I., *Acta Derm. Venereol. (Stockholm),* 1985; 65: 305-312.
3. de Boer, E. M. et al., *Cont. Dermatitis,* 1988; 19: 184-188.
4. Svensson, Å., *Acta Derm. Venereol. (Stockholm),* 1988; 68: 509-513.
5. Crippa, M. et al., *Cont. Dermatitis*, 1990; 23: 203-204.

6 Coenraads, P.J. et al., *Clin. Exp. Dermatol.*, 1983; 8: 495-503.

7. Coenraads, P.J. et al., *Clin. Exp. Dermatol.*, 1984; 9: 149-158.
8. Meding, B. and Swanbeck, G.S., *Br. J. Dermatol.*, 1987; 116: 627-634.

5 Work Activities and Patterns of Skin Exposure to Pesticides

Michael O'Malley

CONTENTS

5.1 INTRODUCTION

Exposures to pesticides occur in a wide variety of occupational and non-occupational settings (Table 5.1). Just as endogenous dermatoses demonstrate characteristic patterns of skin lesions (Figure 5.1),[1] exposures associated with particular occupational and non-occupational activities produce characteristic patterns of contact with the skin (for example, see Figures 5.2a and 5.2b).[2] In evaluation of systemic reactions to pesticides quantitative measurement of exposure allows estimation of absorbed dose and comparison with a known or estimated no-observed-effect level. For evaluation of contact reactions on the skin, it is perhaps more important to have an understanding of the qualitative pattern of skin exposure associated with the work tasks the patient performs. That qualitative comparison may be the only method of judging whether a patient's skin condition is related to a reported exposure.

A basic step in ascertaining the likely pattern of exposure is to examine the physical nature of the pesticide formulation. For example, pesticides used as fumigants are invariably gases or liberate gases on contact with soil, air, or water. Consequently, inhalation is usually more significant than dermal exposure for most of the fumigation-associated gases.

Most other active pesticidal compounds have extremely low vapor pressures (Table 5.2)[3] and skin contact usually accounts for more than 90-95% of the exposures. This is the case even when volatile contaminants (e.g., the mercaptan contaminants in formulations of organophosphates) or solvents present in a formulation are associated with noticeable odor.

1-56670-293-3/00/$0.00+$.50

TABLE 5.1
Principal Types of Skin Contact With Pesticides

Handlers

Mixing/loading: formulation diluted with water or other solvent; mixed with spreading agent. Transfer into spray tanks in planes or ground rigs, or into backpack or hand-held sprayers. Exposure may be substantially reduced by use of closed systems.

Application: Exposure is highly dependent upon the nature of spray equipment

Flagging: Standing at the beginning of rows to be sprayed to signal starting point of application for approaching crop-dusting aircraft. Flagging currently being replaced by automated flagging systems in some operations.

Residue exposures

Persistent residue following application is the principal means of exposure for pickers, sorters, packers, irrigatots snd other field workers. The degree of exposure is highly dependent upon the nature of the work and the nature of the crop. For example, hand labor in grapes results in markedly greater exposure than picking row crops such as lettuce or strawberries.

Indoor residue exposures

The nature of exposure depends upon whether broadcast application ("fogger") or "crack and crevice" application is employed.

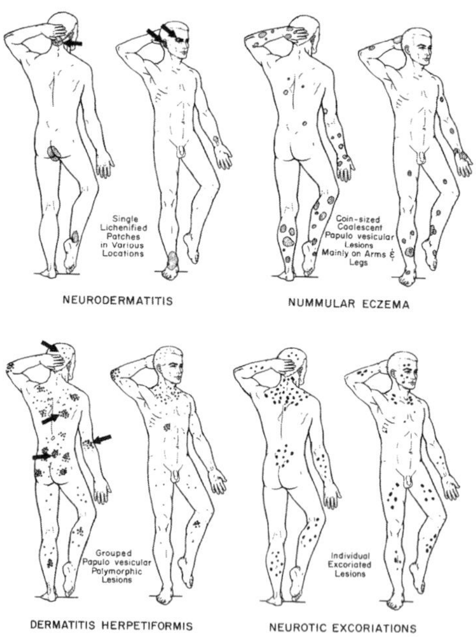

FIGURE 5.1 Pattern of lesions in the endogenous skin diseases, neurodermatitis, nummular eczema, dermatitis herpetiformis, and neurotic excoriations. (Reproduced from *Sauer's Manual of Skin Diseases*, 3rd Edition, with permission of Lippincott –Williams –Wilkins.)

TABLE 5.2
Vapor Pressures of Some Common Pesticides

Material	Vapor pressure
Fumigants/nematocides	***(mm Hg)***
Phosphine	23,369
Sulfuryl fluoride	>760
Chloropicrin	17
Methyl bromide	1,725
Metam-sodium	
Metam-sodium by-products	not meas
Methyl-isothiocyanate	16.0
Carbon disulfide	334.3
Hydrogen sulfide	15,981
Methylamine	2,324.3
Common solvents	
Water	24
Toluene	30
1, 1 Dichloroethane	234
Organophosphate contaminants	
Methyl mercaptan	1,261
Ethyl mercaptan	467
N-butyl mercaptan	83
Organophosphates	
Dichlorovos (DDVP)	1×10^{-2}
Mevinphos	2×10^{-3}
Malathion	1.2×10^{-4}
Chlorpyrifos	2.02×10^{-5}
Dimethoate	8×10^{-6}
Phosalone	$< 10^{-9}$
Carbamates	
Methomyl	5.0×10^{-5}
Organochlorines	
Chlordane	1×10^{-5}
Herbicides	
2,4-D	1.4×10^{-7}
Cyanazine	1.6×10^{-9}
Alachlor	2.2×10^{-5}
Glyphosate	$<7.5 \times 10$
Fungicides	
Hexachlorobenzene	1.09×10
Zinc bisdithiocarbamate (ziram)	$<7.5 \times 10$
Benomyl	$<3.75 \times 1$
Chlorothalonil	5.72×10
Triadimefon	1.5×10^{-7}

FIGURE 5.2 (a) Panamanian worker wrapping pesticide-coated plastic around banana plants. (Photo, H. Penagos); (b) Girdling trunk of grape vines, a high-contact work activity. (Photo, M. O'Malley)

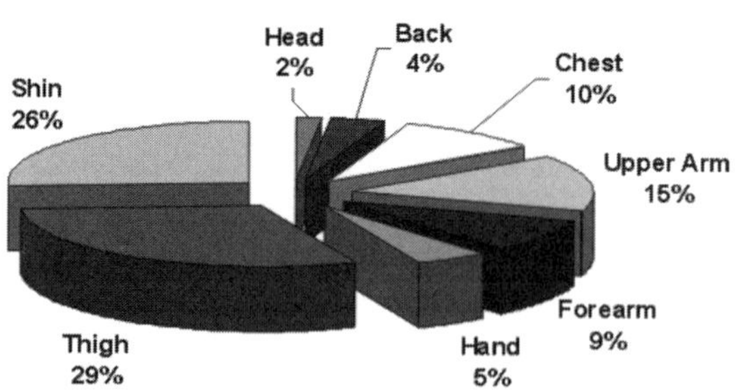

FIGURE 5.3 Distribution of estimated 344.9 mg exposure to outer protective clothing of nursery applicators using hand-held sprayers loaded with various materials that included abamectin, acephate, benomyl, captan, chorothalonil, and fluvalinate. Approximately 0.3% of pesticides handled are estimated to reach the skin through protective clothing. (From California Department of Pesticide Regulation, Spencer et al., 1996, HS-1455.)

This point is illustrated by data (Figure 5.3) from monitoring exposures of greenhouse workers applying abamectin, acephate, benomyl, captan, chlorothalonil, and fluvalinate. A total of 327.82 mg of pesticide was deposited on protective clothing worn by the applicators, 5.04 mg was deposited on the inner clothing. The total inhalation exposure was 0.12 mg, representing less than 0.1% of total pesticide deposited on exterior of the protective clothing, 2% of the pesticide deposited on the inner clothing, and 13% of the material penitrating through the inner clothing layer.

5.2 METHODS FOR MONITORING EXPOSURE

Visual (and photographic) observation of workers handling pesticides or doing hand labor tasks often provides good preliminary information about the anatomic distribution of potential exposure. This is illustrated by the photographs in Figure 5.2 which show (a) a worker wrapping pesticide coated plastic around banana plants on a Panamanian plantation and (b) a California worker girdling the trunk of a grape vine.

Information on the distribution of exposure may also be obtained from quantitative dermal exposure studies. Although these studies usually estimate total dermal exposure and the subsequent absorbed dose, they typically include data on regional as well as total exposure. Dermal patches made of absorbent material have traditionally been employed to monitor skin exposures to specific body regions, placed, for example, on the head, neck, shoulders, chest, back, upper arm, forearm, hip, thigh, calf, ankle, and wrist.[5] This method has been employed in a large number of studies, but has the disadvantage of producing wide variations in estimated exposures because exposures are not typically homogeneous (as the patch method assumes).

Whole garments have been used as an alternative method of monitoring skin exposure. Typically, investigators employ a long sleeved cotton shirt as a means of monitoring exposure to the upper extremities and torso. The lower extremity is monitored with a "long john" thermal undergarment and socks. Exposures to the uncovered areas of skin on the face and hands are monitored by skin wipes or skin washes.

Fluorescent markers added to pesticide mixtures have also been a useful means of providing qualitative information on exposure to pesticide applicators. Quantitative information from these studies has been derived from measuring the density of fluoresence with a video imaging system.[6] Information derived from these studies has underscored the limitations of the patch method by clearly demonstrating the nonuniform patterns of pesticide deposition on skin and clothing (Figure 5.4a).

5.3 EXPOSURE STUDIES OF PESTICIDE HANDLERS

The term "handler" describes workers who either mix or apply pesticides and also includes workers known as "flaggers" who direct aerial applications from the ground by signaling with a flag. The nature and distribution of exposure for handlers depends upon the type of mixing and loading operation, e.g., hand-pouring (Figure 5.4c) vs. closed mixing/loading (Figure 5.4d), and also upon the type of spray equipment they use. (It is obvious that exposures to a helicopter or biplane pilot will be markedly different than those experienced by a ground applicator using either a hand-held or

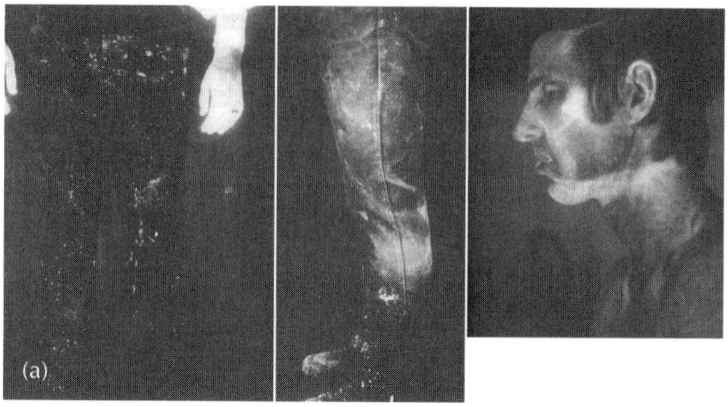

FIGURE 5.4a Nonuniform distribution of fluorescent material to nursery applicators. (Fenske, 1990, reproduced from the *Archives of Environmental Contamination and Toxicology*, Springer-Verlag. With permission.)

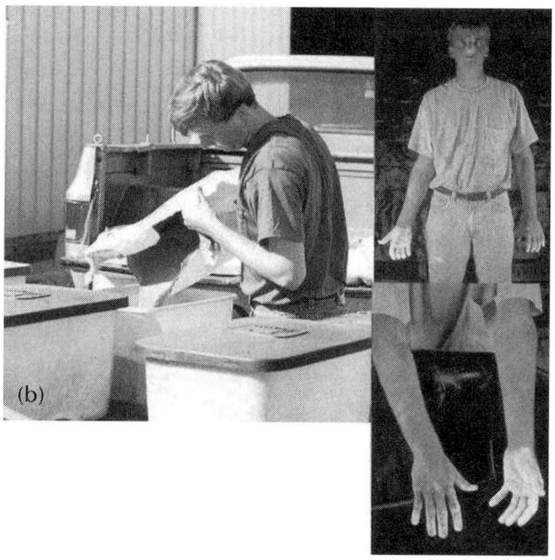

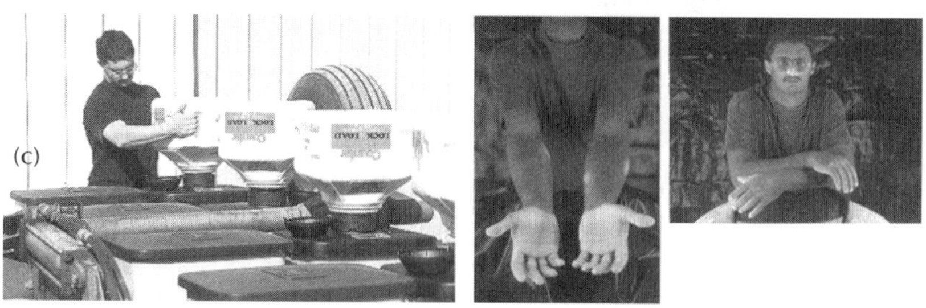

FIGURE 5.4 **(b)** Exposure from hand-pouring insecticide granules. **(c)** Fluorescent tracer exposure to hands occurs despite use of a lock-and-load system; facial exposure presumed to occur secondary to contamination of hands. (Reprinted with permission from *Successful Farming*, October 1992.)

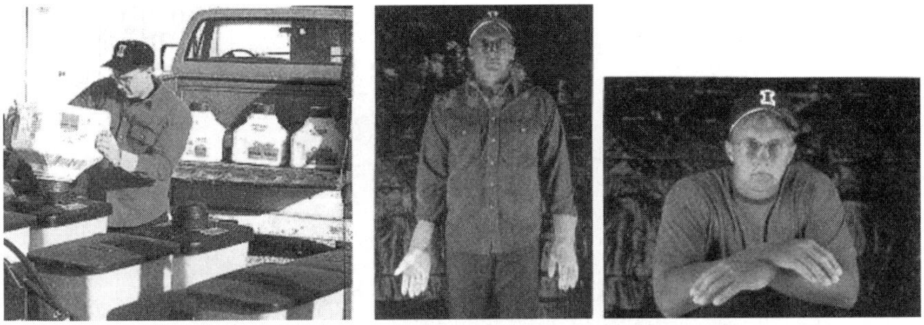

FIGURE 5.4d Contamination with fluorescent tracer minimized by combination of gloves and lock-and-load system. (Reprinted with permission from *Successful Farming*, October 1992.)

backpack sprayer.) Sprayers applying pesticides from within closed tractor cabs have less exposure than those applying the same materials from an open tractor seat. Applications are also made by adding pesticides to irrigation systems (chemigation), and exposure occurs only during the mixing and loading. Exposure to applicators is also minimized by use of enclosed tractor cabs. Examples of the distribution of exposure to pesticide handlers from quantitative studies are given below.

Figure 5.4 shows the distribution of exposure to nursery applicators, monitored using dermal patches in a study, who had diverse areas of contact with six different materials applied with a hand-held sprayer.[4] The exposure to external protective clothing was divided nearly equally between the head, upper extremities, and torso (43.5%) and the lower extremities (56.5%). The material reaching the skin was approximately 0.3% of the amount deposited on the external protective clothing.

The exposure distribution for two workers applying captan to grapes is shown in Figure 5.5. Clothing worn underneath protective gear was used for monitoring exposure rather than using patches. Both workers monitored showed unexpectedly more than 50% of the total pesticide measured from all areas of the body was found on the socks, used in the study to monitor exposures to the lower extremities.[7] In a study of Canadian farmers applying carbofuran, by contrast, nearly 90% of the

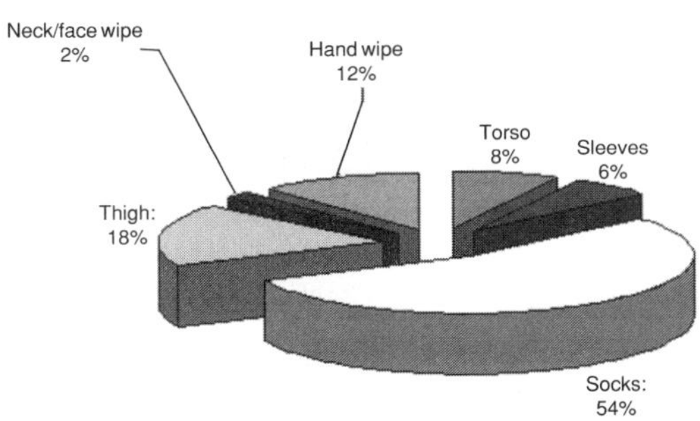

FIGURE 5.5 Distribution of exposure to two workers mixing/loading and applying captan to grapes with a ground rig. Mean total dermal exposure (beneath clothing) for both workers was 9.5 mg. (From California Department of Pesticide Regulation, O'Connell et al., 1990, HS-1511.)

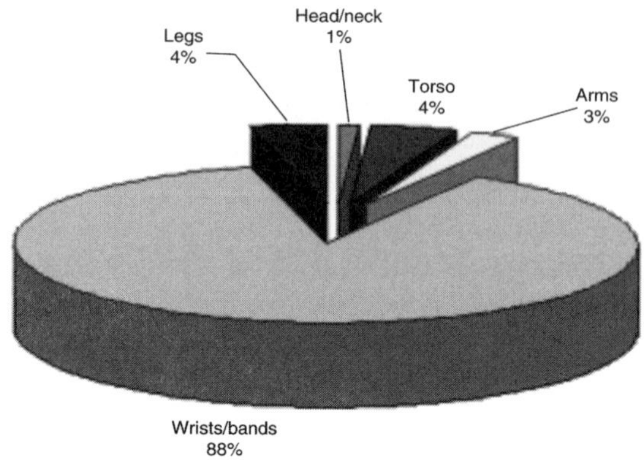

FIGURE 5.6 Distribution of exposure, measured with patches and hand washes in seven Canadian grain farmers wearing Tyvek coveralls and rubber gloves, mixing and spraying carbofuran. (From data in Hussain et al., Occupational exposure of grain farmers to carbofuran, *Arch. Environ. Contam. Toxicol.*, 19, 197, 1990.))

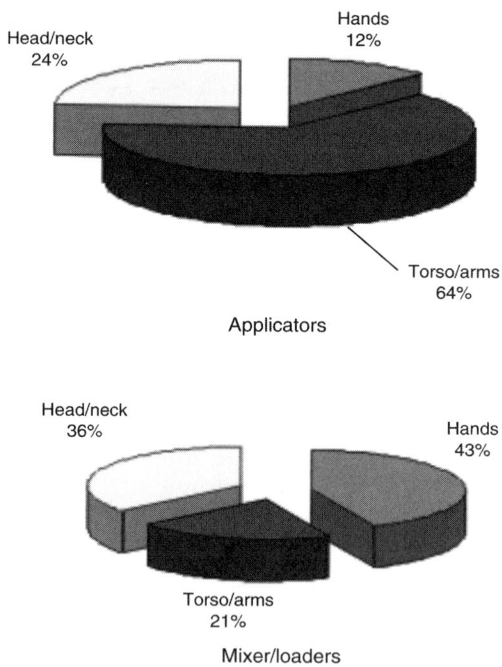

FIGURE 5.7 Distribution of exposure to malathion in California workers. (Based on data from Fenske, R., Correlation of fluorescent tracer measurement of dermal exposure and urinary metabolite excretion during occupation exposure to malathion, *Am. Ind. Hyg. Assoc.* J., 49, 438, 1988.)

dermal exposure was found on the wrist and hands. This was less unexpected, but could also have been an artifact since measurements from the skin of the hands were

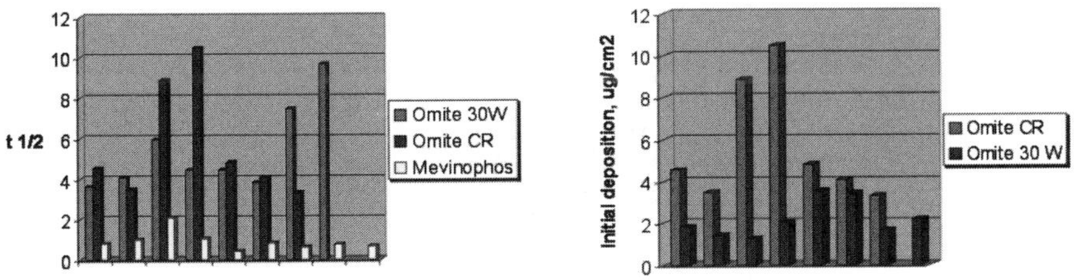

FIGURE 5.8 (a) Dissipation half-lives for mevinphos and two formulations of propargite. (Data from California Department of Pesticide Regulation.) (b) Variable deposition for two formulations of propargite (Omite).

made using washing techniques and were measured on other areas of the body with dermal patches (Figure 5.6).[5]

Figure 5.7 shows exposures to workers mixing/loading and applying malathion in California orchards with air blast sprayers. The mean total exposure was less for mixer/loaders than for applicators, but the range of exposure for the two groups overlapped considerably. Neither group had any lower extremity exposure recorded by either video imaging or using dermal patches.[8]

5.4 FIELD RESIDUE EXPOSURES

Exposure of non-handlers to pesticide residue varies with the work task and crop and with the timing of spraying relative to worker entry. Under ideal circumstances residue exposure can be eliminated completely by performing hand labor tasks before applications occur. Post-application tasks may also be safe if the pesticide residue has dissipated sufficiently by the time workers enter treated fields. The delay between application and safe re-entry depends upon the degree of variability in dissipation half-life. Relatively volatile compounds, such as mevinphos[9] dissipate largely by evaporation, typically with a half-life of 24 h. Low vapor pressure compounds such as propargite[10] depend upon chemical decomposition, and have markedly more variable half-lives (Figure 5.8a). Another important variable is the initial level of residue deposition. Both the initial deposition rate and the dissipation half-life may vary, even between two formulations of a given active ingredient. The sustained release formulation of propargite,[11] for example, is associated with initial residue deposition 2-3 times higher than the wettable powder formulation[12] of the same active ingredient (Figure 5.8b).

The safe residue level in turn depends upon the crop, the work activity, and the duration of contact with treated foliage. The interaction between crop and work activity can be crudely summarized in terms of residue transfer factors.[13] A fixed level of 1 μg/cm^2 of residue might be a danger for a high contact crop and work activity (e.g., hand harvesting grapes), but nevertheless be safe for a lower contact crop and work activity (mechanical harvesting and sorting of tomatoes). In terms of the model, harvesting grapes has a high dermal transfer factor (DTF) compared to work on mechanical tomato harvesters. The DTF has units of surface area per time (normally cm^2/h) and represents the amount of foliage contacted by

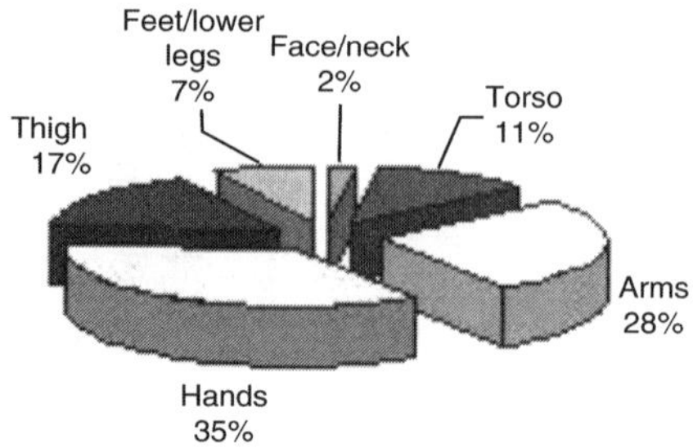

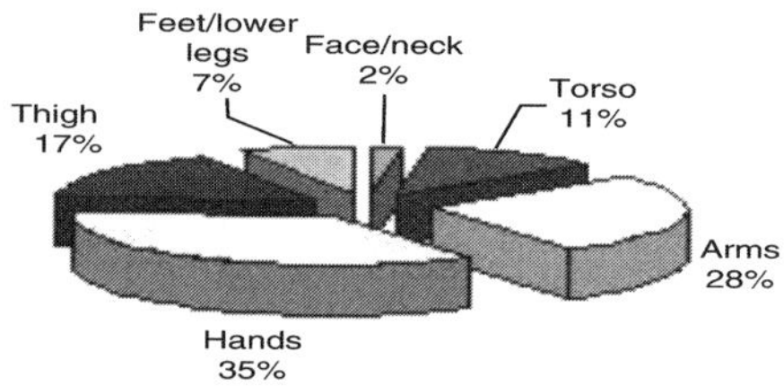

FIGURE 5.9 (a) Regional distribution of exposure in grape cane cutters. (b) Distribution of exposure in grape harvesters. (From California Department of Pesticide Regulation, O'Connell et al., 1993, HS–1522.)

a worker during a specific work period. The general form of the dermal transfer factor is as follows:

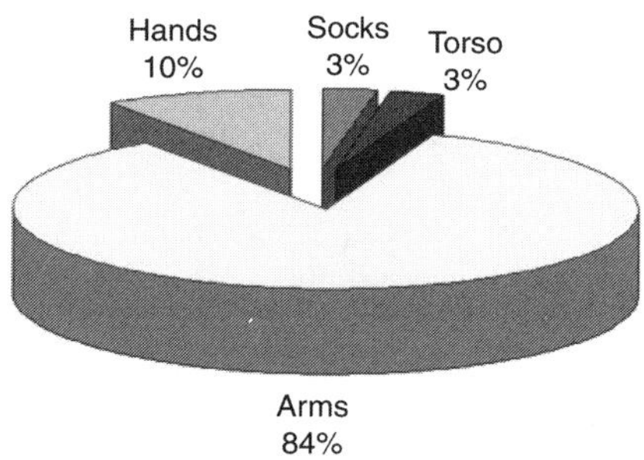

FIGURE 5.10 Regional distribution of exposure in workers picking bush tomatoes. (From California Department of Pesticide Regulation, Schneider et al., 1989, HS–1462.)

$$\text{DTF (cm}^2\ \text{h}^{-1}) = \text{Dermal Exposure } (\mu\text{g h}^{-1}) / \text{ Dislodgeable Residue } (\mu\text{g cm}^{-2})$$

Dermal transfer factors associated with mechanical harvesting and sorting range from 500–1000 cm^2/h. For hand labor activities in row crops, transfer factors generally range from 5,000–9,000 cm^2/h and for tree crops range from 10,000–20,000 cm^2/h. For grape harvesting transfer factors vary with specific work activity, but range as high as 93,000 cm^2/h.[14]

In addition to variations in total exposure, agricultural hand labor activities differ in associated regional distribution of exposure. For example, Figure 5.9 shows the broadly

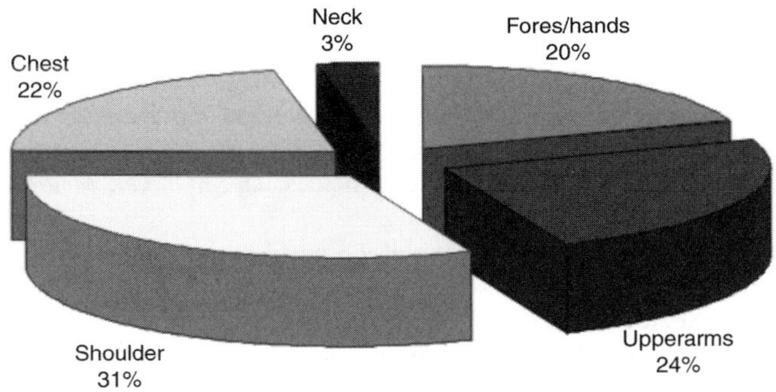

FIGURE 5.11 Distribution of chlorothalonil exposure in workers sorting tomatoes on mechanical harvesters. (From California Department of Pesticide Regulation, Spencer et al., 1988, HS–1466.)

distributed pattern of exposure to both the upper and lower extremities for workers doing hand labor tasks in grapes.[15] In contrast, exposure to workers picking bush tomatoes was principally to the hands and arms (94%) with minimal exposure to the torso and distal portion of the lower extremities (Figure 5.10). Workers sorting tomatoes on a mechanical harvester had pesticide contact largely confined to the upper body (Figure 5.11).

5.5 OTHER WORK CATEGORIES AND NON-OCCUPATIONAL EXPOSURES

Outside of the agricultural workplace, pesticide exposures have seldom been systematically studied with any quantitative methods. Some information can be inferred from cases of dermatitis reported in non-agricultural work settings and cases that follow non-occupational exposure — most cases occur in applicators and comparatively few are associated with any kind of residue exposure. This is to be expected from the prevalence of "crack and crevice" indoor applications and the low vapor pressure of pyrethrins and pyrethroids and organophosphates — the most commonly used indoor insecticides. More potential exposure would be expected with indoor broadcast applications or "bug bombs." For adults walking on treated surfaces, most of the exposure would be reasonably expected to occur on the feet, but exposure would be less predictable for children. Model studies of indoor exposure to pesticide residue have been conducted in order to identify a "worst-case" exposure for quantitative risk assessment,[16-18] but did not yield information about the distribution of skin contact expected from ordinary indoor exposures.

5.6 CONCLUSIONS

Knowledge of the pattern of exposure to pesticides is essential in making a clinical evaluation of the skin effects of pesticides. Exposures vary in pattern and in degree by job category and according to the physical properties of particular pesticides.

REFERENCES

1. **Sauer, G.,** *Manual of Skin Diseases*, 4th ed., J.B. Lippincott, Philadelphia, 1980.
2. **Marks, J. and DeLeo, V.,** *Contact and Occupational Dermatology*, Mosby Year Book, St. Louis, 1992.
3. **Rosenberg, J. and O'Malley, M.,** Pesticides, *Occupational Medicine*, LaDou, J., Ed., Lange-Appleton, 1997, chap. 32.
4. **Spencer, J., Edmiston, S., O'Connell, L., and Margetich, S.**, Greenhouse Pesticide Mixer/loader/applicator Exposure Study, Sacramento, CA, 1996, California Department of Pesticide Regulation, HS-1455.
5. **Hussain, M., Yoshida, K., Atiemo, M., and Johnston, D.,** Occupational exposure of grain farmers to carbofuran, *Arch. Environ. Contam. Toxicol.*,19, 197, 1990.
6. **Fenske, R.,** Nonuniform dermal deposition patterns during occupational exposure to pesticides, *Arch. Environ. Contam. Toxicol.*, 19, 332, 1990.
7. **O'Connell, L., Margetich, S., Bissell, S., and Graham, E.**, Worker Exposure During Captan Applications to Grapes, Sacramento, CA, 1990, California Department of Food and Agriculture, HS-1511.
8. **Fenske, R.,** Correlation of fluorescent tracer measurement of dermal exposure and urinary metabolite excretion during occupation exposure to malathion, *Am. Ind. Hyg. Assoc. J.*, 49, 438, 1988.
9. **Spencer, J., Hernandez, B., Schneider, F., Gonzales, M., Begum, S., and Krieger, R.**, Seasonal Phosdrin Degradation on Row Crops in Monterey County, 1990, Sacramento, CA, 1991, California Department of Pesticide Regulation, HS-1608.
10. **Smith, C. R.,** Dissipation of dislodgeable propargite residues on nectarine foliage, *Bull. Environ. Contam. Toxicol.*, 46, 507, 1991.
11. **Saiz, S. and Schneider, F.**, Degradation of Propargite Residues on Orange Foliage in Tulare County, CA, May–June 1986, Sacramento, CA, 1987, California Department of Food and Agriculture, 1408.
12. **Reeve, M., O'Connell, L., Fong, B., and Edmiston, S.**, Dissipation of Dislodgeable Foliar Residues of Propargite on Grape Foliage, 1989, Sacramento, CA, 1991, California Department of Pesticide Regulation, HS-1590.
13. **Fenske, R.,** Pesticide exposure assessment of workers and their families, *State Art Rev. Occup. Med.*, 12, 221, 1997.
14. **Ross, J., Thongsinthusak, T., and Dong, M.**, Factors Influencing Estimates of Reentry Worker Exposure to Pesticides (Poster Presentation at 8th IUPAC International Congress of Pesticide Chemistry, Washington, D.C., July 4-9, 1994), Sacramento, CA, 1994, California Department of Pesticide Regulation, HS-1707.
15. **O'Connell, L., Margetich, S., and Edmiston, S.**, Worker Exposure to Captan Residues While Performing Cultural Practices on Table Grapes, Sacramento, CA, 1993, California Department of Pesticide Regulation, HS-1522.
16. **Ross, J., Thongsinthusak, T., Fong, H., Margetich, S., and Krieger, R.,** Measuring potential dermal transfer of surface pesticide residue generated from indoor fogger use: An interim report, *Chemosphere*, 20, 349, 1990.
17. **Ross, J., Fong, H., Krieger, R., Thongsinthusak, T., and Margetich, S.**, Measuring Potential Dermal Transfer of Surface Pesticide Residue Generated from Indoor Fogger Use: Using the CDFA Roller Method, Interim Report II, Sacramento, CA, 1991, California Department of Food and Agriculture, HS-1603.
18. **Maddy, K., Goh, K., Edmiston, S., and Margetich, S.**, Dissipation and Consequent Possible Dermal Exposure Hazards of DDVP and Propoxur on Horizontal Surfaces Following the Release of Insecticide with An Indoor Fogger, Sacramento, CA, 1987, California Department of Food and Agriculture, HS-1334.

6 Regulatory Evaluation of the Skin Effects of Pesticides

Michael O'Malley

CONTENTS

6.1 INTRODUCTION

The requirements for skin testing of pesticides for the purposes of federal and state registration vary with the government jurisdiction. In the U.S., primary dermal irritation studies and sensitization studies are required for each manufacturing-use product and each end-use product.[1] The tests performed in this manner are considered part of the regulatory database and are not available in the public literature. The irritation testing requirments are similar to the standard Draize tests. Dermal sensitization tests may be done according to one of several standard protocols (Table 6.1). Details of the test protocols are given in Chapter 1.

In practice, registrants sometimes place strong warning language on pesticide labels in lieu of performing animal testing. Instead of submitting an animal sensitization study, the registrant simply identifies the compound as a potential sensitizer. This has often been done, for example, for compounds identified as sensitizers in the published scientific literature. Although the situations where this occurs vary tremendously, the practice is often described as "overlabeling."

Apart from tests required for registration, additional regulatory information is obtained from post-registration surveillance of pesticide illness reports and from public literature reports on adverse skin effects of pesticides.

The results of pre-registration animal testing are used to determine the content of precautionary statements on the pesticide product label. Materials showing corrosive effects are labeled as such and carry the signal word *Danger*, even if they cause minimal

1-56670-293-3/00/$0.00+$.50

TABLE 6.1
Pesticide Assesment Guidelines Subdivision F, Hazard : Human and Domestic Animals

Current Dermal Testing Requirements

Irritation

The substance to be tested (technical material with detailed characterization of contaminants, or an end-use product) is applied in a single dose to the skin of several experimental animals for 4 h. If severe irritation/corrosion is likely, a single animal may be tested in a stepwise procedure. The degree of irritation is read and scored at intervals until the irritation has resolved or is considered permanent. Based upon scores at 72 h and persistence of irritation more than 14 days, materials are categorized as corrosive (category I—72 h dermal irritation score >7), severe irritants (category II—72 h dermal irritation score 5-7), moderate irritants (category III—72 h dermal irritation score 2-5), or minimally irritant (category IV—72 h dermal irritation score 0-2).

Sensitization

Following initial exposure to a test substance by either intradermal or epidermal application, the animals are challenged, after an interval of at least 1 week, to establish whether the animal has developed hypersensitivity. This is evaluated by comparing scores during the induction period with those during the challenge period and with those of control animals who received the challenge.

Preferred test methods include:

- Buehler test
- Guinea-pig maximization test (GPMT)

Other acceptable tests include:

- Open epicutaneous test
- Maurer optimization test
- Split adjuvant technique
- Freund's complete adjuvant test
- Draize sensitization test

Source: Revised edition November 1984. [Title 40, Volume B, Parts 150 to 189]

acute systemic toxicity. Labels for products not found to be corrosive or irritating simply advise the user to minimize the degree of skin contact. Materials judged as sensitizers in animal tests, or reported as sensitizers in the public domain scientific literature, are required to indicate the possibility of skin sensitization on the product label.

A summary of information on irritation, sensitization, and post-registration illness surveillance information derived from the California Department of Pesticide Regulation (CDPR) is shown in Chapter 8 in the respective sections on insecticides, fungicides and antimicrobials, fumigants, and herbicides. The sensitization data are based on a review of study summary memoranda for 427 formulations or active ingredients reviewed between January 1989 and July 1997. The irritation data derive from a review of registration data and from the publicly available product label database[2] and relevant published literature. Probable or definite skin illness or injury cases involving the same products are also discussed. Because the pesticide illness database (1982-1995) contained more than 10,000 reports of skin injury or illness, the review was limited to 991 cases involving probable or definite skin effect to pesticide applicators from a single active ingredient (Appendix III). None of the 991 cases had patch testing performed.

6.2 REGULATORY CONTROLS OF SKIN EXPOSURE

Many pesticides are regulated exclusively through means of labeling, e.g., requirements for wearing goggles or glasses or protective clothing (personal protective equipment, PPE). The specific protections required depend upon the degree of hazard associated with the product. The most hazardous category I products require extensive PPE and less hazardous products signficantly less PPE (Table 6.2).

The recently inaugurated Worker Protection Standard (WPS) is intended to control all types of exposure through a uniform set of regulations for U.S. workers handling pesticides and performing hand labor operations in treated fields, forests, and greenhouses. Because the skin is the principal means of exposure for most pesticides, most of the WPS regulations aim to decrease contact with the skin — through safety training, restricted entry intervals, PPE, and worker notification — for a broad array of compounds.

Other controls on applicator exposure not included in WPS include closed systems (required in California, but not in other U.S. jurisdictions) for mixing/loading category I pesticides. Use of soluble packaging is another means of reducing exposure to dry product formulations. This was done specifically to reduce the irritation associated with mixing/loading powdered formulations of propargite (Omite®)[3] and also to minimize the risk of sensitization associated with mixing loading powdered formulations of anilazine (Dyrene®).

6.3 RESPONSE TO OUTBREAKS OF FIELD WORKER DERMATITIS

Under the WPS, the same re-entry intervals are specified for broad classes of materials based upon their acute toxicity categories, regardless of the time required for dissipation. The limitations of this approach are illustrated below.

Unlike pesticide handlers, field workers have indirect contact with pesticides, most commonly from contact with residues on foliage. The degree of exposure to pesticide residues depends upon the initial application rate, the half-life of residue dissipation, the nature of the crop, and the nature and timing of the work performed (see Chapter 5). The occurrence of skin reactions further depends upon the properties of the individual pesticides, to the extent these are established. For example, elemental sulfur is applied frequently to table grapes (every 2 weeks) and residues sometimes exceed 10 $\mu g/cm^2$. This means that a large percentage of skin injuries occurring in grape workers are associated with some exposure to sulfur.

Single case reports have several potential explanations that require some effort to differentiate. Some cases might be allergic reactions unique to a single crew member; some might be sentinel cases signifying the occurrence of an otherwise unrecognized outbreak; and some cases might prove to be non-work-related skin conditions not expected, in most circumstances, to occur in co-workers. In investigating outbreaks, it is usually obvious whether the problem is work related. The central question is whether the reported episode was related to pesticides and which material, among those reported, was principally responsible.

TABLE 6.2
Regulation Through Label Language

Product	EPA registration number	Handler protective clothing requirement
Nemacur (fenamifos)	3125-236	Chemical-resistant protective suit, waterproof gloves, chemical-resistant footwear plus socks, protective eyewear, chemical-resistant headgear for overhead exposure, for exposures in enclosed areas, a respirator with either an organic vapor- removing cartridge with a prefilter approved for pesticides (MSHA/NIOSH approval number prefix TC-23C) or a canister approved for pesticides (MSHA/NIOSH approval number prefix TC-14G) Engineering controls statements: when handlers use enclosed cabs in a manner that meets the requirements listed in the worker protection standard (WPS) for agricultural pesticides,[12] the handlers PPE requirements may be reduced or modified as specified in the WPS
Telone C-17 (dichloropropene)	62719-12	Handlers performing direct contact tasks must wear: coveralls over short sleeve shirt and short pants; chemical resistant gloves, such as barrier laminate (EVAL) or viton; chemical-resistant footwear plus socks; chemical resistant headgear for overhead exposure; chemical-resistant apron; full-face respirator with either an organic vapor removing cartridge with a prefilter approved for pesticides approved for pesticides (MSHA/NIOSH approval number prefix TC-23C) or a canister approved for pesticides (MSHA/NIOSH approval number prefix TC-14G)
Temik 15 G (aldicarb)	264-GA-01	Applicators and other handlers must wear coveralls over short sleeve shirts and short pants, waterproof gloves, chemical resistant footwear plus socks, chemical resistant headgear for overhead exposure, protective eyewear, chemical resistant apron and cleaning equipment, mixing or loading and a dust/mist filtering respirator (MSHA/NIOSH D/M approval number prefix TC-21C) Follow manufacturers instructions for cleaning/maintaining PPE and wash PPE separately from other laundry. Wash contaminated clothing in a strong washing soda solution and rinse thoroughly before reuse.
Chipco 26019 fungicide (iprodione)	264-481	Applicators and other handlers must wear long sleeve shirt and long pants, waterproof gloves, and shoes plus socks. Follow manufacturers instructions for cleaning/maintaining PPE and wash PPE separately from other laundry When handlers use close systems, enclosed cabs, or aircraft in a manner that meets the requirements listed in the worker protection standard (WPS) for agricultural pesticides,12 the handler PPE requirements maybe reduced or modified as specified in the WPS
Kelthane MF (dicofol)	707-202	Applicators and other handlers must wear: long sleeve shirt and long pants, chemical resistant gloves, such as laminate or nitrile rubber or neoprene rubber or Viton; chemical resistant footwear plus socks; chemical resistant headgear for overhead exposure; chemical resistant apron when cleaning equipment, mixing, or loading For exposures in enclosed areas, a respirator with either an organic vapor-removing cartridge with a prefilter approved for pesticides (MSHA/NIOSH approval number prefix TC-23C) or a canister approved for pesticides (MSHA/NIOSH approval number prefix TC-14G) must be worn
Confirm 2-F (tebufenozide)	707-238	Applicators and other handlers must wear: long sleeve shirt and long pants, waterproof gloves, shoes plus socks

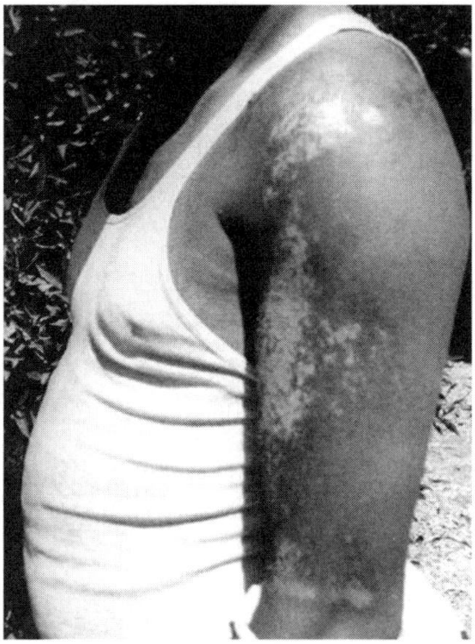

FIGURE 6.1 Postinflamatory hypopigmentation in worker affected by residue exposure to extended-release formulation of propargite in Tulare County, 1985. (Photo, Richard Ames)

The variation in residue dissipation is illustrated by data on propargite. Residue studies performed in a coastal area of California showed 1-2 day dissipation half-lives.[4,5] Residue studies in California's Central Valley typically showed half-lives of 5-7 days, but some fields showed half-lives up to 11 days.[6] Dissipation half-lives as long as 30 days have been measured in the context of outbreak investigations.[7,8] A delayed release formulation of propargite responsible for an extensive outbreak in 1985 (Figure 6.1) was associated with both long disspiation half-lives (Figure 5.8a) and very high levels of initial foliar deposition (Figure 5.8b).

In 1987, an outbreak occurred among workers turning cane in a vineyard near Bakersfield, California. Residue data revealed a history of applications of both sulfur and propargite. Residue data demonstrated foliar residues of sulfur ranging from 1.87–23.24 $\mu g/cm^2$ and foliar residues of propargite ranging from 0.08–0.87 $\mu g/cm^2$. Although a review of animal irritation studies demonstrated that the direct irritant capacity of propargite greatly exceeds that of sulfur, the environmental levels of sulfur found in the investigation greatly exceeded the levels of propargite. It was not possible to conclude which compound may have been the principal cause of the outbreak. No change was therefore made in the 7-day reentry interval on propargite or the "spray dries" interval in place for sulfur.

In 1988, an outbreak of dermatitis occurred among crews of nectarine harvesters in Tulare County (Figure 6.2). Examination of a comparison group of workers who had no dermatitis allowed an analysis of work history and residue history. This showed a strong correlation between cumulative exposure to propargite and the occurrence of dermatitis (Figure 6.3). A review of the work history for the group with no dermatitis showed their peak exposure was to propargite residues of 0.2 $\mu g/cm^2$. This value was used as an estimate for no-observed-effect-exposure-level

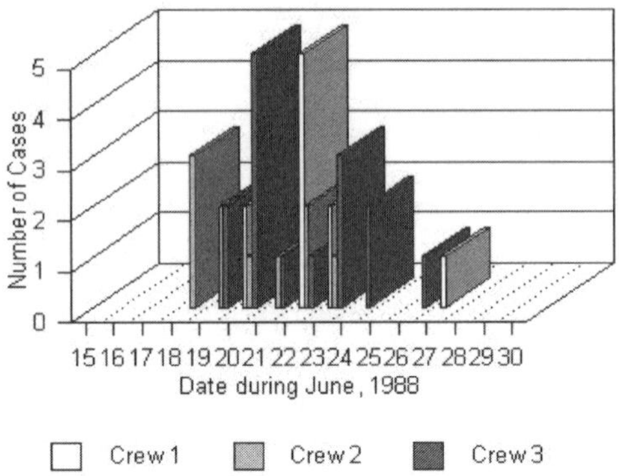

FIGURE 6.2 Dermatitis outbreak in nectarine harvesters in Tulare County, 1988. (Reprinted from O'Malley, M.A., Skin reactions to pesticides, *Occup. Med.*, 12, 327, 1997. With permission from Hanley and Belfus.)

for purposes of determining a safe reentry interval. The reentry interval for harvesting tree fruit was lengthened to 21 days following the episode.

In August 1995, the Department of Pesticide Regulation (DPR), Worker Health and Safety (WHS) Branch, received a report regarding an outbreak of dermatitis (sunburn-like erythema on the chest, neck, arms, and face) among crews of workers performing hand-labor activities on a table grape ranch in northern Fresno County near Kerman, CA. Of 202 workers (8 crews) lifting grape foliage over vine guide wires (turning cane), 65 (32.2%) sought treatment between 8/9/95 and 8/20/95. No rashes were reported among 54 workers (2 crews) working in the same vineyards pulling leaves. The incidence of workers seeking treatment varied from 3.7% in the crew apparently least affected to 87.5% in the crew with the largest number of reported cases. Because the crews left the Fresno area, it was not possible to determine whether the variation in reported cases was attributable to differences in the occurrence of rash or to differences in the likelihood of seeking treatment following onset of rash.

The crews affected by dermatitis worked between 8/7/95 and 8/11/95 in blocks of Thompson seedless and red globe grapes. Application records showed that these blocks had received treatments with propargite, Britz Buffer, glyphosate, methomyl, dichloronitroaniline, myclobutanil, Latron B1956, gibberelin, triflumizole, elemental sulfur, and iprodione. At site 304, 3 dislodgeable foliar residues (DFR) were taken 8/15/95 by DPR Pesticide Use Enforcement (PUE)/Fresno County Agricultural Commissioner (CAC). Single samples were also taken at site 301 and site 302. Propargite levels on 8/15/95 ranged from 0.37 to 0.66 $\mu g/cm^2$. Detectable residue was also found for elemental sulfur (0.16–1.1 $\mu g/cm^2$), iprodione (non-detect (nd)–0.16 $\mu g/cm^2$), myclobutanil (nd–0.16 $\mu g/cm^2$), and dichloronitroaniline (nd–0.045 $\mu g/cm^2$). At site 301, the sulfur residue (1.1 $\mu g/cm^2$) exceeded the residue of propargite (0.66 $\mu g/cm^2$), but at other sites the sulfur residue was <0.21 $\mu g/cm^2$.

Because animal studies demonstrated that propargite is a more severe dermal irritant than the other compounds identified in the residue samples, it was considered to be the probable cause of the outbreak. The levels of dislodgeable propargite

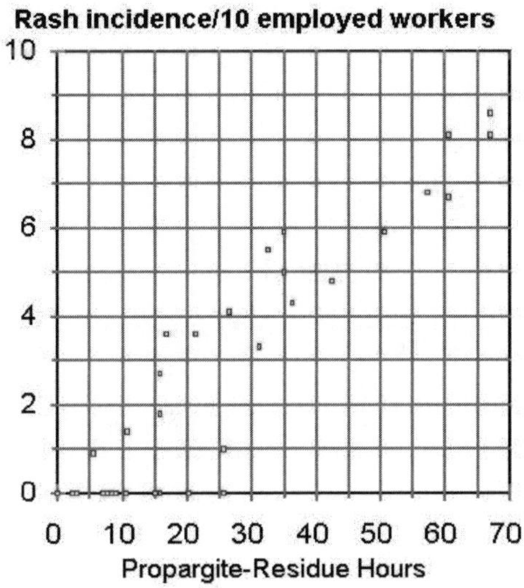

FIGURE 6.3 Dermatitis outbreak in nectarine harvesters in Tulare County, correlation between propargite exposure and cumulative incidence of derrmatitis. (Reprinted from O'Malley, M.A., Skin reactions to pesticides, *Occup. Med.*, 12, 327, 1997. With permission from Hanley and Belfus.)

residue were well above the estimated 0.2 μg/cm^2 no-observed-effect-level (NOEL) for repeated dermal exposure to propargite in nectarine harvesters. Propargite dissipation half-lives observed in follow-up monitoring studies showed residue half-lives of 9.4–14.4 days. Although some blocks of Thompson seedless grapes on the ranch received two applications of propargite within a 10-day interval, the residue levels on this site (0.55 μg/cm^2) did not exceed the highest levels (0.66 μg/cm^2) found on blocks receiving only a single treatment.

Slow dissipation of propargite, comparable to that seen in prior episodes of dermatitis associated with this compound, appeared to be the principal reason for this outbreak. However, an overapplication, or a reentry violation could not be definitely ruled out. Possible measures to prevent recurrent episodes include prohibition of propargite use prior to high contact work activities and routine monitoring of propargite dissipation prior to reentry. Additional studies might also evaluate whether dissipation of propargite is affected by spray adjuvants or other materials in the application tank mixes.

6.4 RESIDUE PROBLEMS IN OTHER JURISDICTIONS

Although the problems observed with propargite in California have not been reported extensively from other jurisdictions, the dry environment in the Central Valley of California is not unique. One must assume that the lack of reports is due to the lack of illness surveillance. A unique outbreak of residue related allergy, associated with the fungicide anilasine, was reported from Tennesee by Schuman in 1980. [9,10] Problems with irritant dermatitis have been reported in Washington related to residues of ziram on Bosc pears.[11] Prior to the outbreak, it was required that the ziram spray settle and dry prior to worker entry. Following the outbreak, this interval was extended to one week.

6.5 CONCLUSIONS

The principal means of regulating pesticide exposure is through product labeling. For applicators, exposure is controlled through specification of application equipment and PPE. For field workers, exposure is controlled by means of field reentry intervals. The degree of protection provided by reentry intervals depends heavily upon residue dissipation times. In the experience of both California and Washington, dermatitis outbreaks have required periodic adjustment of fieldworker re7474entry intervals.

REFERENCES

1. **U.S. EPA**, Pesticide assessment guidelines subdivision F, hazard evaluation: human and domestic animals, Washington, D.C., 1984.
2. **California Department of Pesticide Regulation,** Product label database, 1998.
3. **Maddy, K.**, Summary of Human Occupationally Acquired Ilinesses Reported in 1974 by Physicians in California as Due to Exposure of Omite or Comite, 1974, California Department of Food and Agriculture.
4. **Maddy, K., Kahn, C., and Rivera, L.**, A Study of the Decay Rate of Omite (Propargite) as a Foliar Spray on Strawberries in Ventura County, CA, April 1977, Sacramento, California, 1979, California Department of Food and Agriculture.
5. **Maddy, K., Riddle, L., Kahn, C., and Rivera, L.**, A Study of the Degradation of Propargite (Omite) on Strawberry Foliage and Fruit in Ventura County, CA, April 1977, Sacramento, California, 1977, California Department of Food and Agriculture.
6. **Reeve, M., O'Connell, L., Fong, B., and Edmiston, S.**, Dissipation of Dislodgeable Foliar Residues of Propargite on Grape Foliage, 1989, Sacramento, CA, 1991, California Department of Pesticide Regulation.
7. **Smith, C. R.,** Dissipation of dislodgeable propargite residues on nectarine foliage, *Bull. Environ. Contam. Toxicol.*, 46, 507, 1991.
8. **O'Malley, M., Smith, C., Krieger, R., and Margetich, S.,** Dermatitis among stone fruit harvesters, Tulare County, 1988, *Am. J. Contact Dermatitis*, 1, 100, 1990.
9. **Schuman, S. H., Dobson, R. L., and Fingar, J. R.,** Dyrene dermatitis [letter], *Lancet*, 2, 1252, 1980.
10. **Schuman, S. H. and Dobson, R. L.,** An outbreak of contact dermatitis in farm workers, *J. Am. Acad. Dermatol.*, 13, 220, 1985.
11. **O'Malley, M. A.,** Skin reactions to pesticides, *Occup. Med.*, 12, 327, 1997.
12. **Office of the Federal Registrar, N. A. o. R.**, Code of Federal Regulations, title 16, parts 1500.40-1500.42, 1985, General Services Administration.

7 Basic Epidemiology in the Investigation of Skin Disease

Michael O'Malley

CONTENTS

7.1 INTRODUCTION — BASIC CONCEPTS IN EPIDEMIOLOGY

To evaluate the information available about the effects pesticides on the skin of humans, it is necessary to understand some of the basic concepts of epidemiology. Particular attention will be paid to the quality of various kinds of information and how particular kinds of evidence can be used to infer cause-and-effect relationships.

1-56670-293-3/00/$0.00+$.50

This is similar to the necessity of understanding the design of basic animal tests used to test irritant and allergenic properties of pesticides. Information presented here is relevant to both discussion of the overall scientific literature in Chapter 8 and the discussion of the international experience with the skin effects of pesticides in Chapter 9.

7.2 CASE REPORTS AND CASE SERIES

Cases reports and case series often give clues to the distribution of cases within a population, and frequently give rise to follow-up studies that can evaluate causality. For example, the initial case reports of soft tissue sarcoma (STS) in Swedish workers exposed to phenoxy-acetic acids[1] gave rise to a series of case control studies showing a strong association between STS and exposure to phenoxy-acetic acids.[2]

The care with which both exposures and illnesses is documented is critical. For example, in many cases associating childhood aplastic anemia with exposure to organochlorines, there is minimal documentation of exposure.[3,4] The reports would be correspondingly more compelling if the cases were demonstrated to have elevated blood levels of chlordane or other organochlorine compounds. Laboratory evidence of exposure adds compelling weight to both case reports and to follow-up studies designed to evaluate cause-and-effect relationships, as evidenced by recent studies on blood levels of organochlorines and their relationships with breast cancer.[5]

Provocation testing used in the assessment of both immediate and delayed allergy is a unique kind of laboratory test in that positive findings may provide evidence of both exposure and disease. Cases of *de novo* or "active" sensitization do occur with some powerful antigens however, and this should be kept in mind when interpreting surveys employing patch batteries without specific information on prior exposure to each test antigen.

7.2.1 SPECIFICITY/SENSITIVITY/PREDICTIVE VALUE

Identification of cases in a population is called case ascertainment. Accurate diagnosis depends upon many elements including clinical history, examination, and laboratory tests or a combination of all three elements codified into a diagnostic protocol. The utility of a diagnostic protocol or test can be evaluated by reference to a comparison method or a gold standard. In the example below, the diagnostic protocol of concern is the clinical criteria for diagnosing chloracne based on skin examination alone (the presence of periorbital comedones, predominantly non-inflammatory lesions, and multiple small cysts). The reference method is the examination protocol plus a history of exposure. The data derive from an examination survey of 155 workers previously employed in a trichlorophenol manufacturing process heavily contaminated with dioxin and a group of 157 unexposed controls. Examinations were conducted in a blind fashion, so that examiners did not know the exposure status of the subjects. A surprising finding was that 48 (31%) of the unexposed controls were judged by the examiners to have chloracne. Like the general U.S. population, both cases (48%)and controls (53%) were found to have a high frequency of acne vulgaris. Another

TABLE 7.1
Clinical Judgment vs. NIOSH Criteria in Study of Workers Exposed to Trichlorophenol Manufacturing

		Chloracne pattern + exposure history		
		Yes	**No**	**Total**
Chloracne pattern acne	Yes	65	48	113
	No	0	199	199
	Total	65	247	312

acneiform condition, Favre-Racouchot syndrome (periorbital comedones secondary to sun exposure), was present in 10% of the cases and 5% of the controls.

The sensitivity is defined as the percentage of true cases (sum of cell a + cell c) actually recognized by the diagnostic protocol.

Sensitivity = a/(a + c) = 100% = 65/(65 + 0) = 100%

The specificity is the percentage of true negatives actually recognized by the protocol:

Specificity = d/(d + b) = 199/(199 + 48) = 80.5%

The predictive value of a positive test, defined as the percentage of all diagnosed cases that prove to be true cases = a/(a + b)

Predictive value of positive = 65/(65 + 48) = 57%

In this example, the predictive value is low, presumably because of the high prevalence of other acneiform lesions (acne vulgaris and Favre-Racouchot syndrome) in the study population.

The results of the above analysis may be usefully applied in evaluating a case reported in the literature (Figure 7.1), describing chloracne in a firefighter.[6] Although

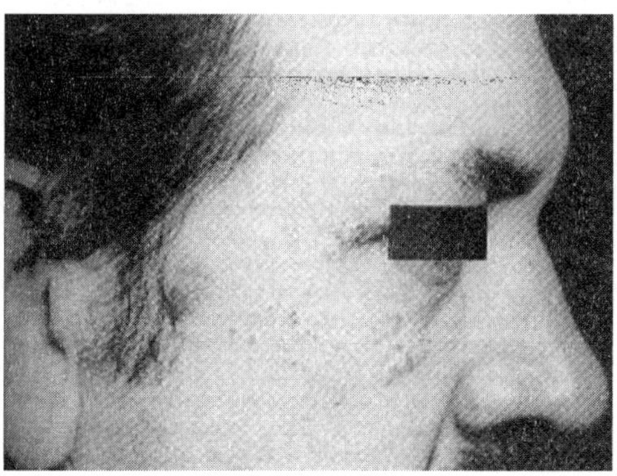

FIGURE 7.1 A reported case of chloracne in a Chicago firefighter. (From Orris, P. et al., Chloracne in firefighters [letter], 1986. With permission from *Lancet*.)

no exposure to either a novel or recognized chloracnegen was described, the authors apparently felt that the distribution of acne lesions was sufficiently characteristic of chloracne to rule out any other condition. Given the 57% predictive value of a simple dermatologic examination in detecting chloracne in the high prevalence group in the NIOSH study, the predictive value of the dermatologic findings in a population with a low prevalence of chloracne would be expected to be extremely low.

7.3 DISEASE FREQUENCY

Disease frequency measures depend both upon ascertaining cases and upon measuring the size of a study population. In large populations accurately ascertaining cases may be difficult and involve substantial tradeoffs between sensitivity and specificity. For example, a telephone or mail questionnaire might allow for rapid survey of a population, but a phone questionnaire would have a lower specificity than review of cases ascertained by physical examination or review of medical records.

In addition to case ascertainment, defining disease frequency depends upon identifying the number of individuals exposed to the agent of interest. This may not be possible to do with any accuracy. Many descriptive studies used proxy measures such as census or employment data.

7.3.1 INCIDENCE AND PREVALENCE

Incidence is defined as the number of new cases of illness or injury in a study population within a specified time period — which day, month, year, decade, etc. Prevalence, by contrast, is the number of cases present in the population at a specified point in time, e.g., the time of an examination survey. The two measures of disease frequency are related by the following simple equation:

$$\text{Prevalence} = \text{Incidence} \infty \text{Duration}$$

Both prevalence and incidence can be addressed in a single study. For example, in a survey of workers employed by a particular employer, all contact dermatitis cases identified by physical examination in the survey participants define the prevalence of that condition in the study group. In the hypothetical study of 100 workers, 4 were found to have contact dermatitis at the time of the exam (prevalence (1) = 4/100). A questionnaire administered on the day of the examination is also used to evaluate the occurrence of new rashes on the forearms since the beginning of the current agricultural season (5 months). This determines that 30% of the workers report having a new rash, of an unspecified type, since the onset of the current agricultural season. The estimated duration of the rash from Eq. 1:

Duration = incidence/prevalence = (4/100)*(100*5 months/30) = 2/3 month = **20 days**.

The above analysis assumes that all of the cases of contact dermatitis found upon examination are due to the exposure of interest. This may not be the case, as the hypothetical

TABLE 7.2
Comparison of Probable Occupational Disease Conditions Identified from Illness Survey with Employer's (OSHA-200) Log and Workers' Compensation Claims

Probable occupational disease	Survey only	Log only	Claims only	Survey and log	Log and claims	All three	Total
Anemia/other blood disease	7	0	0	0	0	0	7
Conjunctivitis/other eye conditions	30	0	2	1	7	0	40
Hearing Loss	156	0	1	0	0	0	157
Hypertension	1	0	0	0	0	0	1
Respiratory conditions	49	0	0	0	0	0	51
Skin conditions	58	6	3	1	3	5	76
Musculoskeletal	8	0	2	0	2	1	13
Chemical burns	0	1	3	0	0	2	6
Toxic and other adverse effects	22	3	2	0	3	0	30
Other conditions	30	0	0	0	0	2	32
Total	451	399	11	13	2	206	

Source: From Discher, D., Kleinman, L., and Foster, F., Pilot Study for Development of an Occupational Disease Surveillance Method, Cincinnati, OH, 1975, NIOSH, DHEW, (NIOSH) Publication No 75-162.

study described does not include a comparison group and examination surveys of the general U.S. population have shown a mean contact dermatitis prevalence of approximately 4%. A further premise is that all of the rashes reported on the questionnaire are due to the same hypothetical exposure, an assumption even less likely to be true.

7.4 ESTIMATING POPULATION INCIDENCE FROM SURVEILLANCE

Population incidence may also be estimated from illness registry data, ultimately derived from cases seeking medical treatment. The completeness of reporting depends upon: (1) the number of ill persons who seek treatment, (2) the number of persons seeking treatment who are properly diagnosed, and (3) the percentage of those who are diagnosed who are actually reported to the registry. These problems are illustrated by a survey conducted by Discher[7] who evaluated reporting from multiple sources on the same study population.

7.4.1 Classification of Surveillance Information

The quality of surveillance information in the surveillance system is obviously of critical importance. For dermatitis cases in pesticide handlers, it is frequently possible to ascertain whether there is a direct relationship between the pattern of a direct accidental exposure and the pattern of dermatitis. Dermatitis in field workers may be more difficult to classify because they have only indirect exposure to foliar pesticide residues. An isolated case of dermatitis in field workers could represent allergic contact dermatitis, unusual susceptibility to skin irritation, a non-work related condition, or possibly an index case of a disease outbreak. Careful evaluation of all such reported cases depends upon availability of sufficient resources to investigate individual cases and interview co-workers where this appears indicated by circumstances.

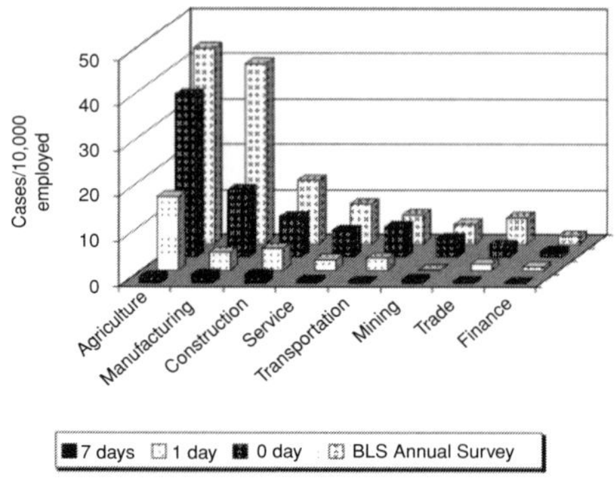

FIGURE 7.2 Identification of high-risk industry groups from Bureau of Labor Statistics (BLS) OSHA 200-log annual survey data from three separate groups reporting compensation cases — after 0 lost work days, 1 lost work day, and 7 lost work days — to the BLS Supplementary Data System, 1981. (Based on data from O'Malley, M.A. et al., 1988.)

7.4.2 Identification of Illness Patterns and High-Risk Groups

Identification of illness patterns does not necessarily depend upon complete reporting of cases. For example, surveillance of skin disease in the U.S. has shown that the same high-risk groups can be identified from a variety of surveillance data, despite substantial variation in the actual reported rate of illness. Among the eight major industry groups, agriculture shows the highest rate of skin disease in three of the four surveillance groupings. It is also apparent that at the very lowest level of reporting, differences in the rate of illness among the major industry groups become smaller and smaller (Figure 7.2).

7.5 OUTBREAK INVESTIGATIONS

Skin disease outbreaks present unique opportunities for evaluation of the dermal hazards of exposure to pesticides. Due to the organization of agricultural labor, outbreaks most often occur among groups of field workers. However, small clusters do occasionally occur among groups of pesticide handlers.

In contrast to the prevalence studies of healthy workers, acute contact dermatitis during an outbreak may affect 80-90% of exposed workers. Investigation may reveal unusual application histories, the use of newly registered products, or unusual persistence of compounds with normally rapid dissipation. For outbreaks of subacute irritation, there may be somewhat atypical work histories — crews working a week or more in a field where the residue has failed to dissipate — allowing expression of the cumulative effects of residues too low to cause acute irritation after a single day's exposure. An example of a brief questionnaire and field examination instrument used in California outbreak[8] investigations is included in Appendix IA.

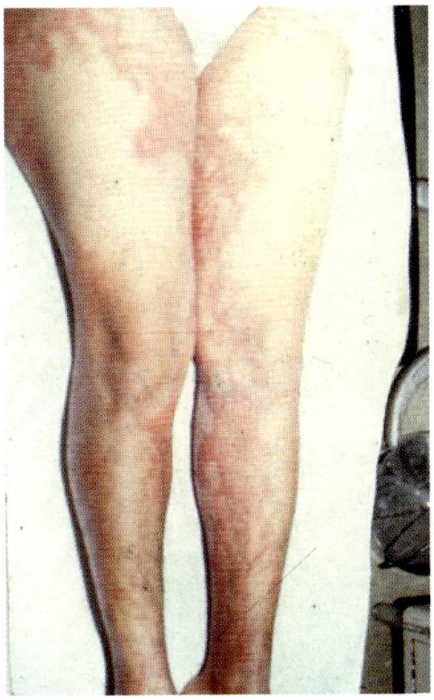

Figure 2.6 Dermal reaction following exposure to fenitrothion. (Photograph, H. Penagos.)

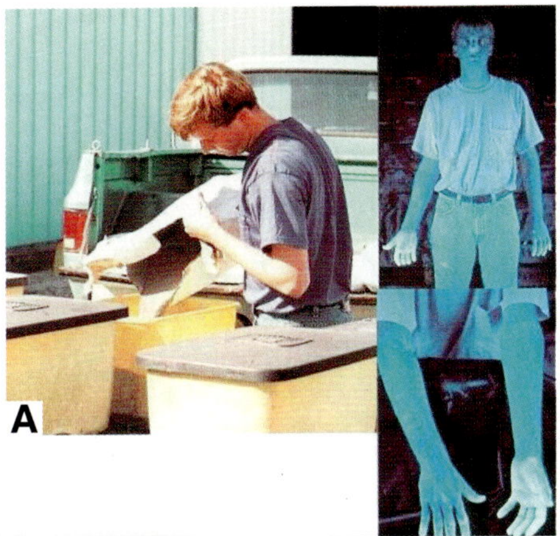

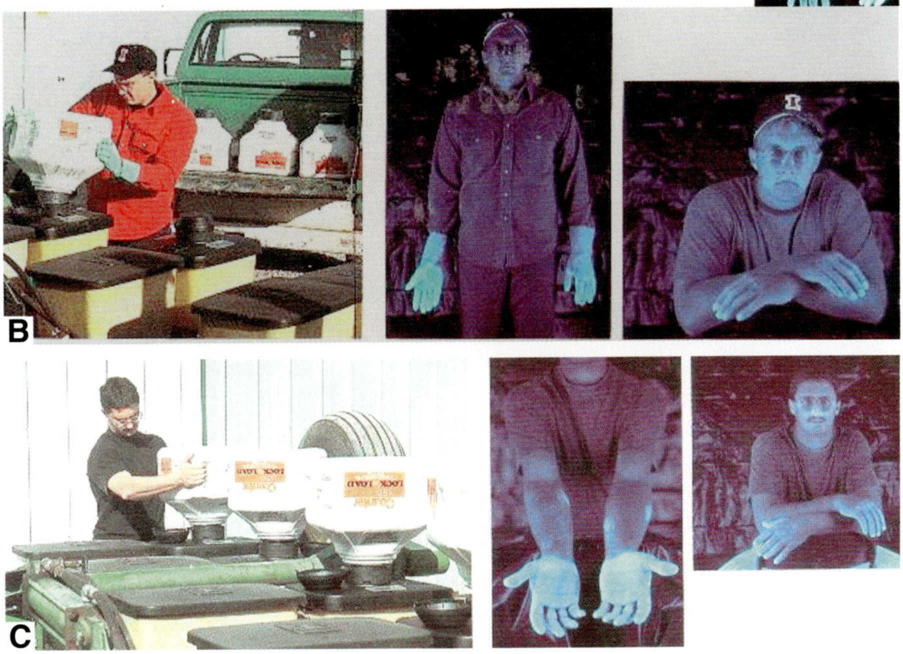

Figure 5.4 (A) Exposure from hand-pouring insecticide granules; **(B)** Fluorescent tracer exposure to hands occurs despite use of a lock-and-load system; facial exposure presumed to occur secondary to contamination of hands; **(C)** Contamination with fluorescent tracer minimized by combination of gloves and lock-and-load system. (Reprinted with permission from ***Successful Farming***, October 1992.)

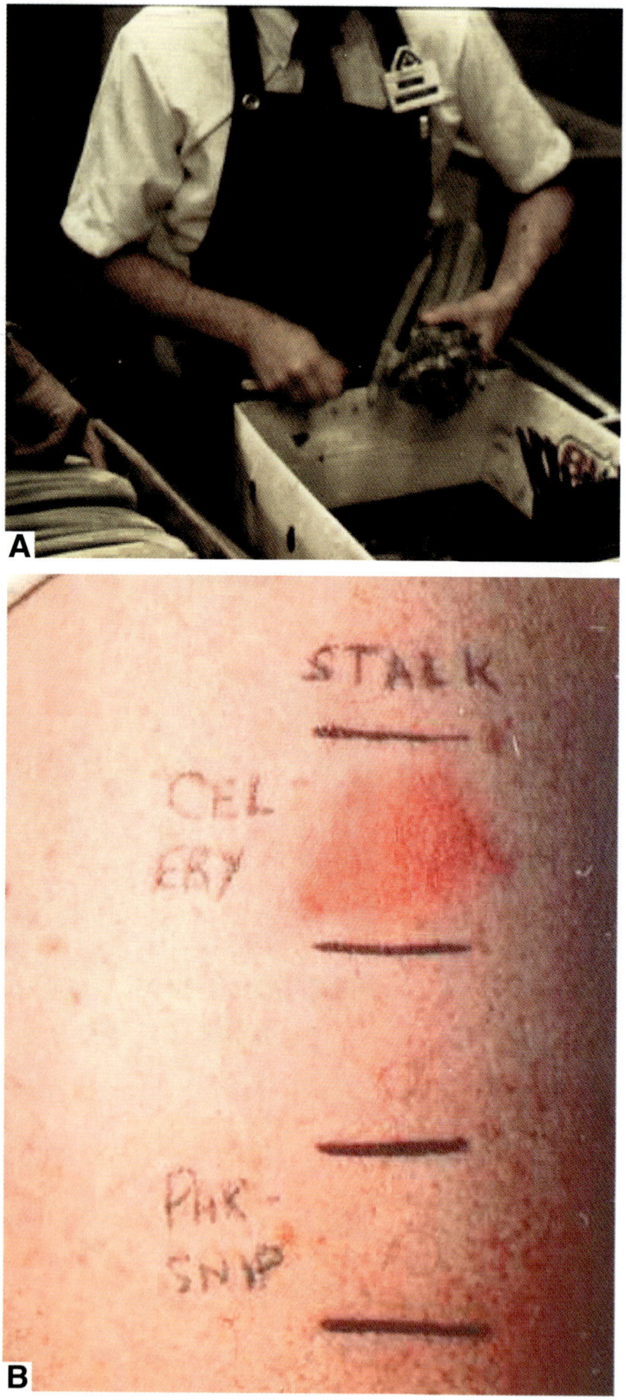

Figure 7.4 (A) Forearm contact while trimming celery by produce worker. **(B)** Results of exposure to cut end of celery stalk followed by exposure to UV light.

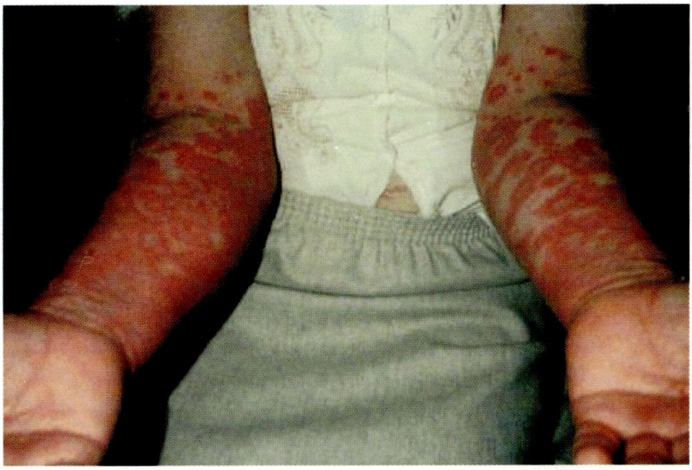

Figure 9.0 Dermatitis associated with pulling weeds in a sugar beet field — predominantly May weed, *Anthemis cotula*. (Reprinted from O'Malley and Barba, Bullous dermatitis in field workers associated with exposure to Mayweed, ***American Journal of Contact Dermatitis***, 1,34,1990. With permission from W.B. Saunders.)

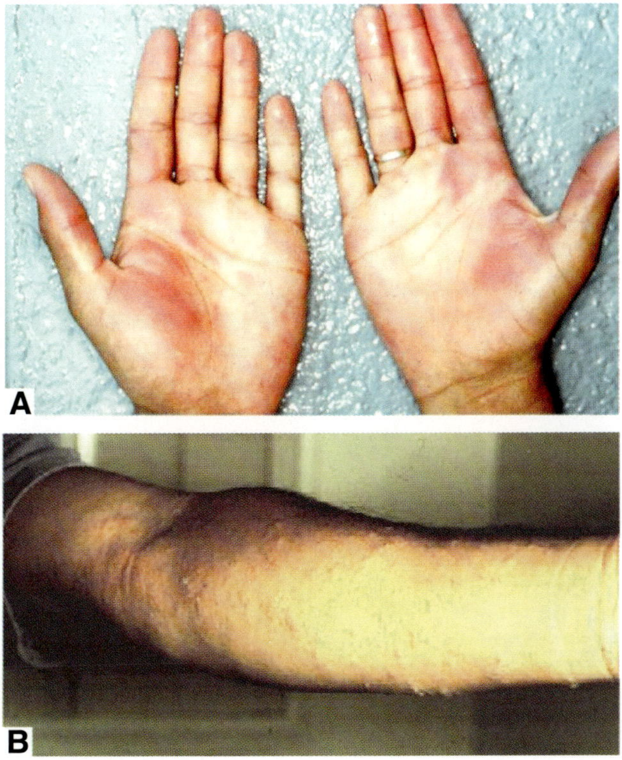

Figure 9.10 (A) Hand dermatitis associated with *Abutilon theophrasti* — velvetleaf. **(B)** Follicular irritant dermatitis associated with *Abutilon theophrasti*. (Photographs, Thomas Ferguson.)

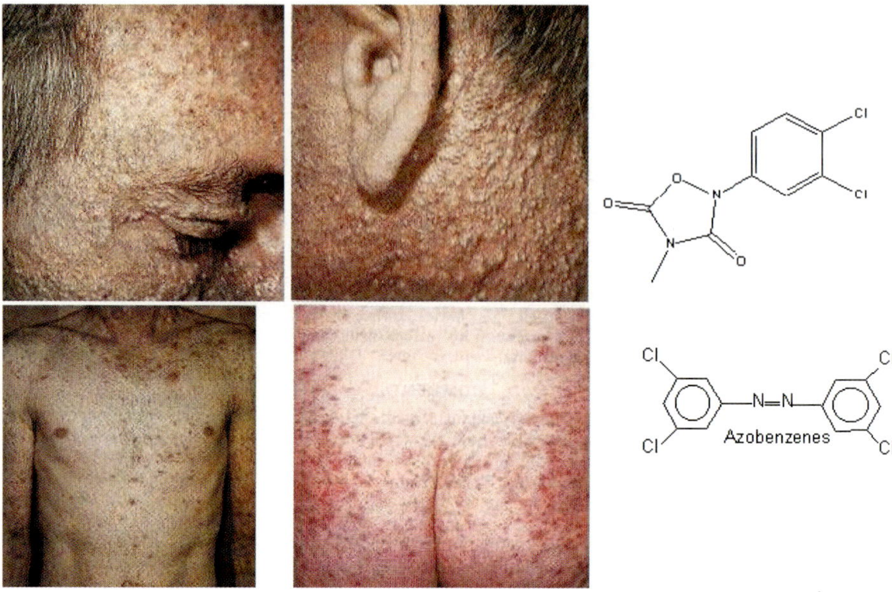

Figure 9.19 Chloracne from 3,3',4,4'–Tetrachloroazoxybenzene and 3,3',4,4'– Tetrachloroazobenzene, reported by Taylor, ***Archives of Dermatology***, 1977. (Reproduced with permission from the American Medical Association.)

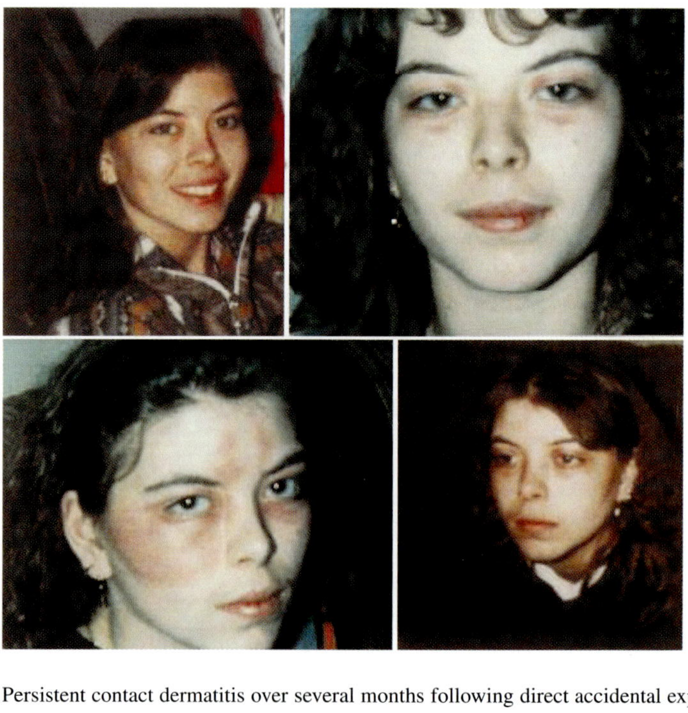

Figure 9.36 Persistent contact dermatitis over several months following direct accidental exposure to a mixture of pyrethrin and bioresmethin. (Muller-Mohnssen, ***Toxicology Letters***, 107, 161-175,1999. With permission from Scientific and Medical Division, Elsevier Science.)

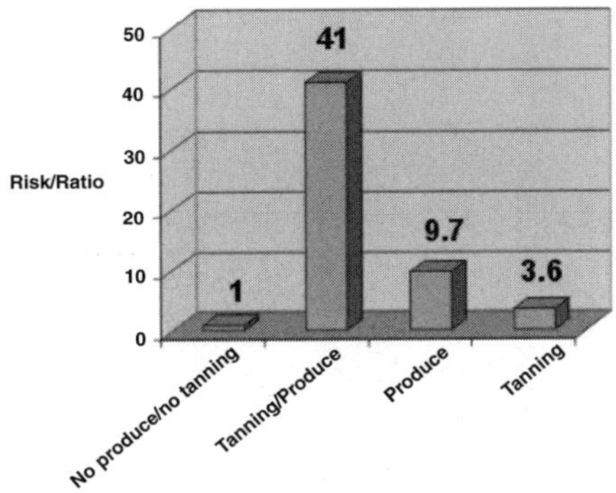

FIGURE 7.3 Risk factors for photodermatitis in grocery workers. (Data from NIOSH investigation reported in *MMWR*, 34, 11, 1985.)

The frequency of illness in investigation of an outbreak is important to establish by interviewing and/or examining as many of those potentially affected as possible. (A simple interview and examination form is shown in Appendix I.) When interview and examination demonstrate a frequency of contact dermatitis or other skin condition far beyond the background level (4%), this suggests exposure to a strong irritant or an extremely strong sensitizer. The nature of the reaction may also be indicated by clinical factors, such as the distribution limited to sun-exposed area or aggravation by exposure to UV light. These clinical details are extremely valuable when documented systematically during investigation of a large outbreak. For example, in an outbreak of dermatitis among grocery workers, the index case presented after visiting a tanning parlor, and this proved to be a clear risk factor (Figure 7.3) for the occurrence of a hyperpigmented rash on the forearms among other workers from the same store.[9] Additional valuable information can be obtained by correlating the pattern of dermatitis with observation of the work activities. In the grocery workers outbreak described above, exposures to various produce items (celery, limes, and parsnips) known to cause photodermatitis were highly concordant. Nevertheless, observation of the work activities revealed that handling celery was the only work activity that could explain the observed distribution of the rash on forearms (Figure 7.4). The pattern of exposure to pesticides may always not be obvious, but can be evaluated with phosphorescent tracers or by measurements of residue deposited on skin, clothing, or absorbent patches (see Chapter 5).

7.6 DESCRIPTIVE (ECOLOGIC) STUDIES

The term descriptive studies encompasses several different study designs. Case reports are also an important means of developing information about previously unrecognized connections between exposure and disease. These are frequently the only types of information about the occurrences of pesticide related skin diseases.

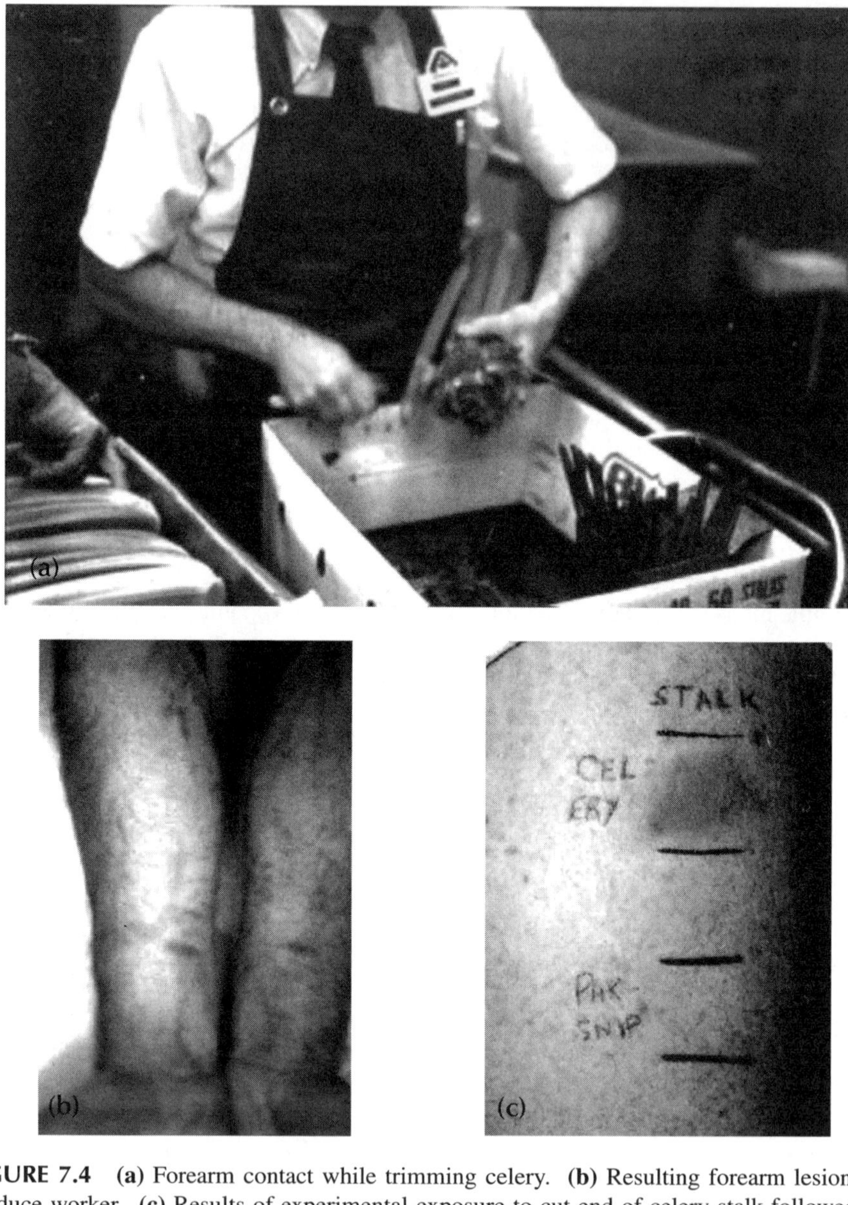

FIGURE 7.4 **(a)** Forearm contact while trimming celery. **(b)** Resulting forearm lesions in produce worker. **(c)** Results of experimental exposure to cut end of celery stalk followed by exposure to UV light.

Cross-sectional (survey) studies are a simple means of assessing the distribution of illness in the population. They may be relatively resource intensive if examinations and laboratory tests are conducted in addition to interviews. In the absence of a study control group, comparison of skin examination results may be made to the prevalence of skin conditions reported for the U.S. population from the National Health and Nutrition Examination Survey (NHANES II). Data from the 1971-1974 version of the survey are given in Appendix IV. As well as documenting the prevalence of numerous common skin conditions, the data also document that there was considerable variation among the 101 examiners in the frequency of identifying significant skin pathology.

Questionnaire studies are an alternative means of obtaining information about the distribution of skin problems. The most sophisticated questionnaires attempt to derive information about specific conditions, differentiating among possible alternative skin conditions by evaluating the type and distribution of lesions in a fashion akin to a systematic physical examination. A model of this type of questionnaire, evaluating possible causes of hand eczema, is show in Appendix IB (the Tuohilampi questionnaire).[10]

Results of cross-sectional studies are considered suitable for generating hypotheses and evaluating the distribution of skin conditions in a population of interest. The ecologic study is a unique type of descriptive study that seeks to evaluate underlying reasons for the distribution of disease within a population. Generally, ecologic studies correlate group characteristics or group exposures with the prevalence of illness in groups across a population. It is usually inappropriate to regard the associations identified in ecologic studies as cause-and-effect relationships.

7.7 EVALUATION OF CAUSE AND EFFECT

7.7.1 Experimental Studies

As in other types of science, the most desirable type of information derives from experimental studies of human subjects. In medicine, the model for this type of study is the randomized clinical trial. It is not usually possible to do this type of study with pesticides, but limited numbers of studies of experimental skin sensitization studies have been done using the repeated insult patch test. Patch tests on individual patients might be considered an experimental study with a large historical control group. Modeling done using the EPISTAT program for $2 \infty 2$ comparisons[11] indicates that a positive test on a single subject would not reach statistical significance unless there were at least 20 negative controls. For positive tests on two consecutive patients, fewer controls would be required to reach statistical significance (Figure 7.3).

Although experimental studies often have ideal conditions for evaluating the relationship between exposure and outcome, they do have predictable limitations. For example, a common question in interpreting data from experimental studies is how well the study group represents the population group of interest. Use of paid volunteers, for example, may make various systematic experimental studies possible, but still leaves a question about whether the results readily can be generalized to other populations.

7.7.2 Observational Studies

Approximation of the experimental method is sometimes possible by means of well-conducted epidemiologic research. This may be the case, for example, where accurate

Case Control Study Layout

	Exposed	Unexposed	Total
Subjects with illness to be studied	a	b	a+b
Healthy controls	c	d	c+d
Total	a+c	b+d	a+b+c+d

Measure of effect is called odds ratio = a*d/b*c

means of assessing both the exposure variables and the disease outcome are readily available and both the disease and the exposures of interest have a significant degree of variability in the study population. Because such ideal study conditions do not occur very frequently by chance, researchers, in some fields of medicine, endeavor to create them using clinical trials and other experimental studies. Observational studies vary in design depending upon the means and timing of identification.

7.7.2.1 Case Control Studies

Case control studies often serve as the only means of studying conditions that infrequently come to medical attention. They begin by identifying persons with a given illness and comparing the frequency of possible causes of illness between that group and a group demographically similar without the illness. The controls are ideally selected randomly from the general population group from which the cases are drawn, but sometimes are less randomly drawn from the clients of the same medical institution that identified the cases. An example of this study design, evaluating cases of pigmented dermatitis and an institutional, rather than a population-based control group, is discussed in Chapter 9. As illustrated in the example, multiple exposures can be evaluated in a case control study of patients with a particular illness.

7.7.2.2 Cohort Studies

In contrast with case control studies, cohort studies evaluate the frequency of illness in a group following an identified exposure compared to the frequency of illness in a group of unexposed controls. An example of this is discussed in Chapter 13, describing a cohort study used to evaluate an outbreak of irritant dermatitis. In contrast with a case control study (which allows evaluation of multiple exposures possibly linked to a given disease), the cohort study allows evaluation of multiple disease outcomes for a single exposure.

7.7.3 Rules for Evaluating Cause and Effect

Whether the data derive from experimental or observational studies, the rules for identifying causal associations are agreed upon by all epidemiologists, even though their application in practice is frequently the subject of great controversy. In brief, a causal relationship between an exposure factor and the subsequent occurrence of disease may be inferred to the degree that

1. The association is consistent from study to study
2. The association is statistically strong
3. There is a dose-response relationship

Cohort Study Layout

	Illness	Non-illness	Total
Exposure	a	b	a+b
No exposure	c	d	c+d
Total	a+c	b+d	a+b+c+d

Measure of effect is called risk ratio = (a/(a+b)/(c/(c+d))

4. The association is biologically plausible — in terms of pesticide related skin disease, this means there is reported association consistent with existing human literature and animal literature
5. The association is highly specific, e.g., as with vinyl chloride and angiosarcoma of the liver, or asbestos and pleural mesotheliomas

7.8 LIMITATIONS OF EPIDEMIOLOGIC STUDIES

7.8.1 Exposure Evaluation

The limiting problem with most epidemiologic studies is an inability to accurately assess the level of exposure. The reliability of the exposure information in epidemiologic studies may vary from extremely reliable information generated from biologic markers measured systematically in a study population to possibly inaccurate information generated from questionnaires. The greatest difficulty arises from questionnaires attempting to identify the nature and extent of exposures that occurred many years previously. Bias from selective recall of ill individuals or relatives of deceased patients may be extremely difficult to control.

7.8.2 Control Selection

Careful selection of control populations is a principal means of minimizing bias. In a randomized clinical trial, this is achieved by randomly assigning subjects from a uniform group to treatment or non-treatment status. In a cohort study the comparison group may simply be the population-at-large, if data exist for the distribution of the outcome of interest within the overall population. The population-at-large is commonly the comparison group for the cohort mortality study. For disease outbreak investigations, the comparison group is usually an at large population. Unfortunately, there is no objective means of measuring the extent of bias, except by comparing the difference in odds ratios obtained using different control populations. For example, in a study of childhood brain cancer and chemical exposures, controls were selected from the general population by random digit dialing and also from a list of children with non-brain cancers. Even when this is done, there is no *a priori* means of knowing which control population should be considered the "gold standard" in terms of recall bias. When controls are selected from among patients with another type of illness, the studies may be prone to another type of negative statistical bias, if there is an unrecognized association between the exposure of interest and the "control" illness.

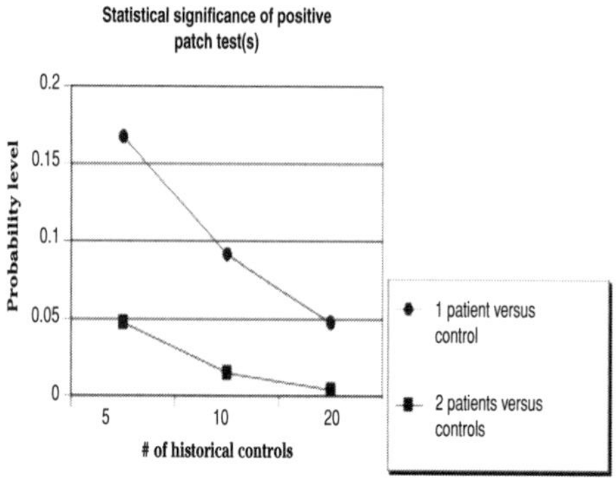

FIGURE 7.5 Significance level of positive patch test(s) for a single patient vs. a variable number of historical controls — modeled using the Fisher's exact test from the STATCALC module of the CDC EPISTAT software.

7.8.3 Multiple Comparisons

A final methodologic consideration is the problem of multiple comparisons. In case control studies, it is possible to record on a questionnaire many exposures and evaluate their statistical association with the illness in question. In this situation it is possible that some exposures will be statistically associated with the occurrence of illness purely by chance. It is therefore important in evaluating case control studies to consider which exposures have truly been identified *a priori* as concerns. For cohort studies, the problem of multiple comparisons arises from the possibility of evaluating the difference in incidence between the exposed and control populations for multiple illness outcomes.

7.9 CONCLUSIONS

Reproducibility is the basic criterion for evaluation cause and effect in science. Epidemiologic information can contribute to the evaluation of cause and effect in situations where experimental studies are not possible for ethical or logistical reasons.

REFERENCES

1. **Hardell, L.,** Malignant mesenchymal tumors and exposure to phenoxy acids — a clinical observation, *Lakartidningen*, 74, 2753, 1977.
2. **Eriksson, M., Hardell, L., Berg, N. O., Moller, T., and Axelson, O.,** Soft-tissue sarcomas and exposure to chemical substances: A case-referent study, *Br. J. Ind. Med.*, 38, 27, 1981.
3. **Infante, P. F., Epstein, S. S., and Newton, W. A., Jr.,** Blood dyscrasias and childhood tumors and exposure to chlordane and heptachlor, *Scand. J. Work. Environ. Health*, 4, 137, 1978.

4. **Epstein, S. and Ozonoff, D.,** Leukemias and blood dyscrasias following exposure to chlordane and heptachlor, *Teratog. Carcinog. Mutagen*, 7, 527, 1987.
5. **Wolff, M., Toniolo, P., Lee, E., Rivera, M., and Dubin, N.,** Blood levels of organochlorine residues and risk of breast cancer, *J. Natl. Cancer Inst.*, 85, 648, 1993.
6. **Orris, P., Worobec, S., Kahn, G., Hryhorczuk, D., and Hessl, S.,** Chloracne in firefighters [letter], *Lancet*, 1, 210, 1986.
7. **Discher, D., Kleinman, G., and Foster, F.**, Pilot Study for Development of an Occupational Disease Surveillance Method., Cincinnati, OH, 1975, NIOSH, DHEW, (NIOSH) Publication No. 75-162.
8. **O'Malley, M., Smith, C., Krieger, R., and Margetich, S.,** Dermatitis among stone fruit harvesters, Tulare County, 1988, *Am. J. Contact Dermatitis*, 1, 100, 1990.
9. **NIOSH,** Phytophotodermatitis among grocery workers — Ohio, *MMWR,* 34, 11, 1985.
10. **Susitaival, P., Husman, L., Hollmen, A., and Horsmanheimo, M.,** Dermatoses determined in a population of farmers in a questionnaire-based clinical study including methodology validation, *Scand. J. Work Environ. Health*, 21, 30, 1995.
11. **Centers for Disease Control**, EPISTAT software, 1997.

8 The Reported Skin Effects of Pesticides by Use and Structural Category

Homero Penagos, Woraphong Manuskiatti, Kenneth Abrams, Daniel J. Hogan, Howard I. Maibach, and Michael O'Malley

1-56670-293-3/00/$0.00+$.50

8.1 INTRODUCTION

In this section, the published literature is organized and reviewed by use and chemical categories. Where possible, the emphasis is placed upon structure-activity relationships to allow the reader to understand the relationships between the chemical properties of and the dermal effects of similar products. Animal irritation and allergy tests, from product registration studies, are reviewed, where available, and relevant case examples (involving, in most cases, handlers with isolated exposures to single compounds) are included from the California pesticide illness registry (referred to as part of the "handler database" put together for purpose of this review). The illness registry data provide basic information about the circumstances of exposure for the reported cases and usually contain sufficient description of the resulting skin effects to infer a relationship between the exposure and the dermatitis. Background information about these cases is given in Appendix III. Few cases reported to the registry include information about the results of patch testing because this procedure is so seldom done in clinical practice. Recommended patch test concentrations are summarized in Appendix II, but are also mentioned in the text below in the discussions of the clinical literature.

The basic design of the animal tests and the quantitative structure activity relationship (QSAR) terminology discussed below are described by Dr. Bashir and Dr. Maibach in Chapter 1. Although animal data were not available for every compound reviewed, they were discussed where available, for the technical formulation of each compound, or a formulation with the maximum concentration of active ingredient. Basic physical data on active ingredients relevant to QSAR, such as molecular weight, and log P (partition coefficient between octanol and water), and vapor pressure,* a physical parameter relevant to exposure (see Chapter 5), are included where available. Except where indicated, these physical and chemical parameters derive from the U.S. Department of Agriculture's ARS pesticide properties database.[1]

* These differ slighty for some compounds from the pressures for the sane compounds listed in Chapter 5.

It was not possible to review the data for the thousands of different formulations for the entire list of active ingredients reviewed. The focus on active ingredients is inevitable, but unfortunate. As illustrated in Section 8.5.7, the skin effects associated with some pesticides may be due to surfactants or preservatives in the formulated product, rather then their active ingredients.

8.2 INSECTICIDES AND INSECT REPELLENTS

8.2.1 ORGANOPHOSPHATES

Many organophosphates, such as chlorpyrifos,[2] have high octanol/water partition coefficients (for example, the log P for chlorpyrifos = 5.0) and penetrate the skin readily. (They are differentiated from other highly lipophilic compound groups, see Section 8.2.3, by their susceptibility to rapid metabolism and often water soluble transformation products.) Nevertheless, organophosphates have very specific activity rather than diffuse chemical reactivity, unless there is a reactive chemical moiety, or structural element, in the non-phosphate portion of the compound, sometimes termed the leaving group. Organophosphates are thus thought to cause minimal skin irritation[3] and this can be confirmed by the results of predictive Draize tests submitted for purposes of pesticide registration. Compounds that show only transient irritation in the Draize assay, according to California registration data, include acephate, diazinon, dichlorvos, dimethoate, malathion, methamidophos, methidathion, oxydemeton-methyl, phosmet, and sulfotep.[4] Matsushita[5] found many of the organophosphates cause some degree of primary irritation when evaluated for use in epidermal application in the challenge (epicutaneous) phase of the guinea pig maximization test.

Sensitization studies conducted for pesticide registration, principally conducted by the Buehler method, show negative findings for acephate, chlorpyrifos, dimethoate, malathion, methamidophos, methidathion, and phosmet.[4] Matsushita, in his review of the skin experience with the organophosphates in Japan, found that several were strong sensitizers in the maximization test. These included diazinon, fenitrothion, and methidathion. In the clinical series of slightly more than 202 cases of organophosphate associated dermatitis described in conjunction with the predictive animal tests, he noted that approximately one fourth had at least mild coincident symptoms of systemic poisoning.[5]

8.2.1.1 Irritant or Sensitizing Organophosphate Compounds

Although the phosphates as a group cause few skin effects, individual compounds with reactive substituents attached to the phosphate moiety can cause either irritation or sensitization. These are discussed below.

Chlorpyrifos (Lorsban®, CAS# 2921-88-2, MW 350.62, vp 1.88 ∞ 10-5 mmHg, log P 5.0). Chlorpyrifos is one of the most extensively used organophosphates, having applications in agriculture, household sprays, and in structural termite control. Data on the irritant effects of chlorpyrifos are somewhat ambiguous. A 40.7% formulation of chlorpyrifos for agricultural use is labeled as a skin irritant,[6] but data for a technical formulation of 98% chlorpyrifos show only transient irritation in the Draize test (category IV).[7] Some cases reported to the California pesticide illness registry suggest that formulations of chlorpyrifos can occasionally cause mild to moderate irritation when direct exposure occurs among applicators:

92-1560 — A hose broke during treatment of nursery stock with chlorpyrifos, soaking upper half of applicator's body with insecticide. He immediately removed coveralls and shirt and washed. He finished the application wearing fresh coveralls. Examination showed faint, mottled, nearly confluent erythematous macular rash on chest.

Naled (Dibrom®, CAS # 300-76-5, MW 381, vp 0.00195 mmHg, log P NA). Although no animal data were available for review, naled is labeled as a corrosive,[8] presumably based upon the use experience with the compound described below. In its native form, the reactive element in naled is the dibromodichloropropanol moiety (QSAR descriptor IX). Naled loses bromine readily in the presence of sunlight and forms dichlorvos. Surface residues after application on grapes, for example, contain about 30% dichlorvos 6 h after application. (The irritant effects of dichlorvos on the skin are described below.) Both compounds have relatively high vapor pressures compared to typical organophosphate compounds and dissipate rapidly after initial application.[9]

Mick et al.[10] described irritation due to naled in an agricultural pilot who was exposed to naled through a hole in protective gloves worn during a loading and cleaning operation. The area became red and edematous, then blistered. This 1970 case is comparable to a 1988 case reported to the California pesticide illness registry:

88-924 — A worker hand poured naled (Dibrom®), and then spilled some on his leather boots, which he continued to wear the rest of the day. He developed a severe blister on the exposed foot, but did not seek medical treatment for several days.

Cases of possible contact sensitivity to residue of naled while cutting chrysanthemums were reported by Edmundson and Davies in 1967. Captafol, captan, and naled were applied 2 h prior and the foliage was apparently still wet when 12 harvesters entered the treated area to cut flowers; 9 of the 12 women developed dermatitis on the face, neck, hands, and arms and, 4 were interviewed and examined by the investigators. All reported the onset of symptoms within a few hours of exposure. After treatment with triamcinolone cream, the affected workers tolerated re-exposure without suffering recurrent skin problems. The 24-h provocation tests conducted were negative to the two fungicides, but were positive in 3 of the 4 women to 60% naled in xylene and the same material caused a 1+ reaction in 1 of 8 control subjects tested.[11] The difference between the test reactions in the controls could have been attributable to differences in circumstances of testing between the exposed patients and the controls, but could also have been due to chance. The high prevalence of complaints among the work force, the short intreval between exposure and onset of symptoms, the nature of the exposure, and the absence of recurrent dermatitis on re-exposure, all suggest that the primary problem in this small scale outbreak of dermatitis may have been irritation.

Malathion (Carbophos® and other brand names, CAS# 121-75-5, MW 330.4, vp 3.38×10^{-6} mmHg, log P 2.7). California registration data show that technical malathion causes transient dermal irritation (clearing two days after removal of patches) in the Draize test. However, valid animal dermal sensitization data for malathion were not available for review. Milby and Epstein[12] found that nearly half of 87 volunteers developed contact sensitization following a single exposure to 10% malathion and that many of them reacted to dilutions as low as 1 ppm. Similar results were reported by Kligman.[13] However, it was found as part of the first study that only about 3% of people with occupational exposure to malathion had a positive patch test reaction to 1% malathion, and no worker had to change work because of malathion allergy. Only one positive patch test reaction to 0.5% malathion was obtained in 455 routine eczematous patients patch tested by the International Contact Dermatitis Research Group (ICDRG). This one reaction was of unknown relevance. In practice, malathion appears to be a weak allergic contact sensitizer. Kligman has suggested that the usage concentration of malathion is too low to provoke sensitization. Milby and Epstein determined that the sensitizer in malathion is diethyl fumarate, which is used in the manufacture of malathion. Diethyl fumarate was present at a 3% concentration in technical grade malathion.[13] An experimental study reported by Maibach demonstrated that diethyl fumarate can produce non-immunologic contact urticaria.[14] In QSAR terms, the compound (Figure 8.1) contains a reactive conjugated olefin similar to the ICONJ structural element described in Table 1.4.

Schanker et al.[15] evaluated 1874 illnesses reported in conjunction with the aerial spraying of malathion to eliminate the Mediterranean fruit fly in Los Angeles. These included 47 reports of urticaria, 38 reports of angioedema, and 213 reports of a non-specific skin rash. Although epidemiologic conclusions from the study were limited, because of the low rate of participation, some objective evidence for the occurrence of bait related allergy was found. Of ten subjects who underwent provocation testing, one child showed an immediate allergic reaction to the bait protein used as a carrier for malathion, but none showed delayed hypersenstivity. One child had an irritant reaction to the bait protein and to malathion.

Diazinon (Spectracide® and other trade names, CAS# 333-41-5, MW 304.35, vp 0.00014 mmHg, log P 3.02-3.81). Although diazinon causes only slight irritation in the Draize (rabbit) assay,[16] it was among the more irritating organophosphates in the guinea pig maximization tests conducted by Matsushita. In the epicutaneous challenge phase, the maximum non-irritant concentration of diazinon identified was 0.5% (the same as for DDVP), compared to 2% for naled, malathion, and leptophos, and to 5% for fenitrothion, methidathion, and the organophosphate fungicide kitazin. Cases reported to the California illness registry indicate that, in humans, formulations of diazinon can cause irritant response on direct exposure. A typical case (87-2537), involved a nursery worker who was applying diazinon on ornamentals when a hose line ruptured. The material got on his shirt, which he changed, but he did not shower. He then developed a rash on his arms and chest, corresponding to the areas of exposure. In the clinical series of cases reported by Matsushita, diazinon was among the most frequently reported organophosphates associated with dermatitis, but the clinical report did not distinguish between cases due to irritation and those due to sensitization.[5]

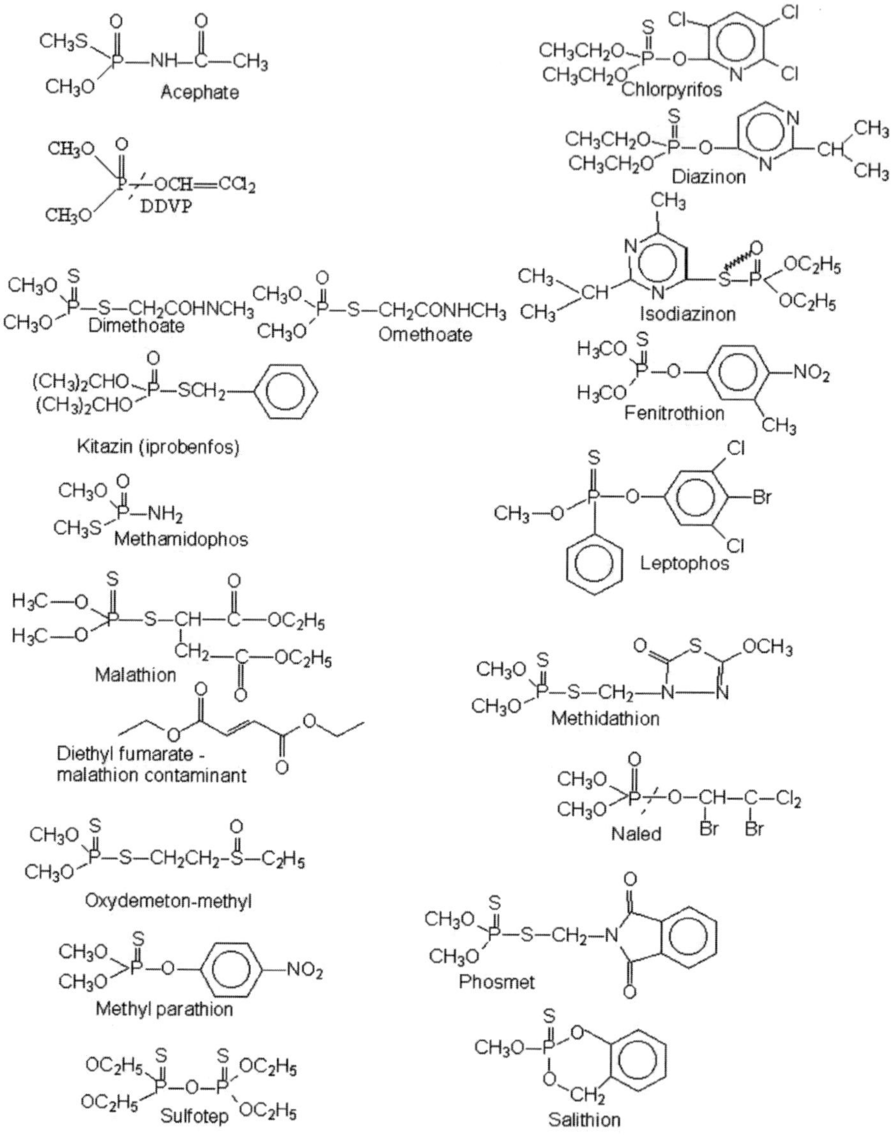

FIGURE 8.1 Some organophosphorous compounds with reported skin effects.

California registration data show negative sensitization studies conducted by the Buehler method for formulations (including wettable powders, granules, and emulsifiable concentrates) ranging from 0.5 to 87.5% diazinon.[17] A single study conducted by the Buehler method on a 2% dust formulation showed a very minimal degree of edema (0.2 average score at 48 h post challenge), and a mild-to-moderate degree of sensitization response (average score for erythema of 1.25 at 48 h).[18] In the guinea pig maximization test, Matsushita identified diazinon as a sensitizer at a challenge concentration of 0.5%, with a grade V (most potent) score at both 24 and 48 h.

A case report by Collins from Australia identified an isomer of diazinon called isodiazinon as a cause of porphyria cutanea tarda. The affected worker was a sheep rancher who developed typical lesions of porphyria after handing diazinon-based insecticide dips. Urine porphyrin levels were elevated. The authors also investigated the effects of diazinon and isodiazinon in a tissue culture model, and in an animal study using rats treated topically with isodiazinon in xylene. Isodiazinon showed greater effect on porphyrin synthesis in both the tissue culture model and in the rats treated topically.[19] A subsequent study showed that isodiazinon acts by inhibiting the ferrochelatase enzyme in the rat liver's porphyrin metabolic pathway.[20]

Parathion (Phoskil® and other trade names, CAS# 56-38-2, MW 291.27, vp 9.75 x 10^{-6} mmHg, log P 3.83) and Methyl Parathion (Penncap® CAS# 298-00-0, MW 263.21, vp 9.75 x 10^{-6} mmHg, log P NA). Parathion was removed from the U.S. market in 1992 by the U.S. EPA because of its long history of causing systemic poisoning in applicators and field workers.[21,22] Evidence that parathion is an experimental sensitizer is based upon two reports. In a guinea pig study reported by Dikshith,[23] repeated dermal application of parathion caused hyperkeratinization of the epidermal layer, thickening of the stratum corneum, and scattered dermal infiltration with mononuclear cells, but did not produce frank dermatitis. Palmintieri[24] studied the effects of repeated parathion applications on human volunteers and noted that some had an eczematous response on simultaneous abrasion of the skin and application of parathion.

Pevny[25] reported eczema in a winemaker/vineyard operator, associated with a positive parathion provocation test and accompanied by cross-reactions to other organophosphates, including methyl parathion, azinphos methyl, metathion, and metasystox. A positive reaction to the organochlorine compound fentichlor in the same case was felt to represent *de novo* sensitization.[25] Svindland[26] reported a case involving a gardener who used a concentrated spray containing parathion and emulsifiers on begonias and also had a cut finger. A swollen and bluish-red discoloratiom developed at the site of the cut finger a day later. The finger was cool and painless. The author believed the patient had an unusual topical reaction to parathion, resembling erysipeloid. Although the patient was treated with penicillin, bacterial cultures eventually proved negative.[26] Bhargava[27] reported a case of erythema multiforme, occurring in a woman who lived in a dwelling treated with methyl parathion to control household insects, and whose skin conditin was found to recur upon re-exposure to that compound.

DDVP, dichlorvos (Duravos®, Marvex®, No-Pest®, CAS# 62-73-7, MW 221, vp 0.031503 mmHg, log P 1.40-2.29). DDVP contains a vinyl group with two reactive elements (IX and ICONS). Strong positive reactions to flea collars were reported in two patients who had dermatitis attributed to contact with dogs wearing DDVP-impregnated flea collars.[28] Cronce and Alden[29] reported primary irritant contact dermatitis to DDVP flea collars in four patients. Patch testing produced bullous primary irritant reactions in four patients and five controls. Mathias[30] noted negative patch tests to 1% and 0.1% DDVP in petrolatum in a truck driver who developed prolonged contact dermatitis to DDVP. This man also had systemic symptoms of organophosphate toxicity following his exposure to DDVP from a spill of this

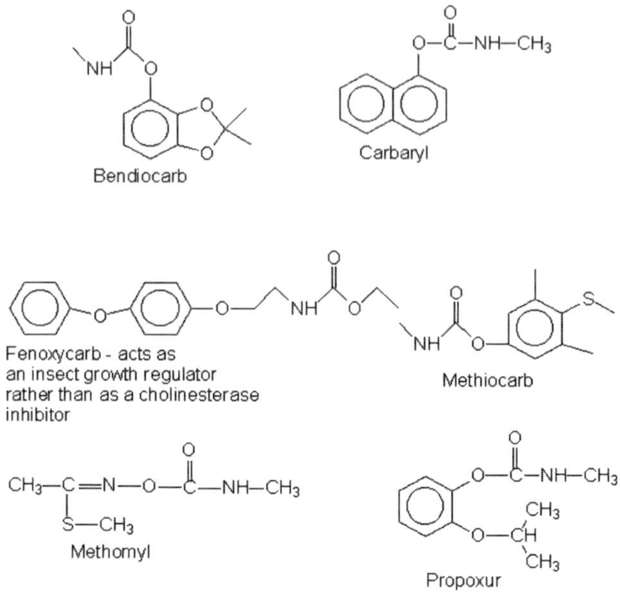

FIGURE 8.2 Carbamate insecticides.

chemical in his truck. Mathias emphasized that individuals who develop dermatitis from skin contact with organophosphates should be carefully questioned regarding possible symptoms of systemic poisoning.

Thiometon (Ekatin®, CAS# 640-15-3, MW 246.3, vp 0.0002 mmHg,[31]log P NA). A case of contact dermatitis has been attributed to thiometon. The patient was a 17-year-old female working in a sprayed field. She developed a rash on the back, trunk, and arms, as well as non-specific symptoms (headaches and nausea) compatible with mild systemic poisoning. It appears, from the limited information about this case available, no provocation testing was conducted.[32]

Omethoate (Folimat®, CAS# 1113-02-6) and Dimethoate (Cygon®, De-fend®, CAS# 60-51-5, MW 229.3, vp 8.25 x 10^{-6}mm Hg, log P NA). California registration data show that technical dimethoate (96-98%) causes transient dermal irritation (clearing within 24 h) in the Draize test.[33] No data from the Draize test were available for the structurally related compound omethoate. Sensitization studies conducted by the Buehler method for a dimethoate emulsifiable concentrate and a wettable powder were both negative. Allergic contact dermatitis caused from omethoate and dimethoate have been reported.[34]

8.2.2 Carbamates

None of the carbamate insecticides (Figure 8.2) appeared to be markedly irritant or to be sensitizers in animal test models (see Table 8.1). Cases associated with carbamates in the California illness registry appeared to be principally transient irritation following direct contact. Examples included case 84-512, involving a worker cleaning methomyl from sprayer nozzles without wearing gloves, who subsequently

TABLE 8.1
Summary of Carbmate Predictive Dermal Irritation and Sensitization Tests in Rodents

Compound (brand name, CAS#), physical data	Dermal irritation study	Dermal sensitization studies
Bendiocarb (Ficam®, 22781-23-3), MW 223.23, vp 3.45 x 10–5 mmHg, log P 1.7	None reviewed	Negative study by Buehler method conducted on mixture of bendiocarb and piperonyl butoxide
Carbaryl (Sevin®, 63-25-2), MW 201.23,vp 0.005026 mmHg, log P 2.29	No irritation observed on technical formulation	13% and 90% formulations tested by Buehler method; dermal scores all 0 at 24 and 48 h
Fenoxycarb (Torus®, 72490-01-8), MW 301.3, vp 6.49 x 10–9 mmHg, log P 4.07 (this compound acts as an insect growth regulator, rather than as a cholinesterase inhibitor)	No study on file for dermal irritation; observations from a 21-day dermal toxicity study indicate that technical fenoxycarb is a minimal irritant	A 24% fenoxycarb formulation was tested by the maximization method (24.4%); the reported scores were 0 for both erythema and edema
Methiocarb (Mesurol®, 2032-65-7), MW 225.31, vp 2.1 x 10-7 mmHg, log P 2.97-3.34	No erythema or edema observed in study with 75% methiocarb	Methiocarb (1%) tested for sensitization with the Buehler showed no erythema or edema in challenge phase of the assay
Methomyl (Lannate®, 16752-77-5), MW 162.2, vp 5.63 x 10–6 mmHg, log P 0.57	Study on 98.35% methomyl showed no erythema, eschar, or edema noted at 1, 24, 48, or 72 h post exposure	Buehler assay conducted on a bait mixture of 1% methomyl and 0.25% muscalure
Propoxur (Baygon®, 114-26-1), MW 209.25, vp 9.68 x 10–6 mmHg, log P 1.56	Study on 99.6% technical formulation showed no erythema or edema or other signs of irritation	Buehler assays conducted on a 70% formulation of propoxur and a 0.25% formulation for a flea and tick spray showed no evidence of sensitization

developed hand dermatitis. Case 88-297 involved a worker spraying propoxur who wiped his hands on his shirt. He subsequently developed rash on his chest where he wiped his hands. The transient irritation reported in these two cases may have been due to non-pesticidal ingredients of the formulation or spray mix since the Draize tests on methomyl and propoxur did not show even transient irritation. The cases reported from California were similar to a report from Nigeria by Vandekar,[35] who described a case of dermatitis in an applicator whose back was sprayed with the material while a sprayer was being filled.

Cases of suspected allergic contact dermatitis associated with application of carbamates have also been reported. Among 30 farmers with contact dermatitis evaluated by Sharma,[36] several reacted to the carbamate insecticides carbofuran and carbaryl, but the precise clinical relevance of the positive tests was difficult to determine. Willems[37] described a case of hand eczema in a carnation grower. The case was described as severe and it had a very acute onset. Patch testing was negative for a standard pesticide series, several additional pesticides used in the operation, and for carnation leaves and stems. Only the provocation test to methiocarb was positive.

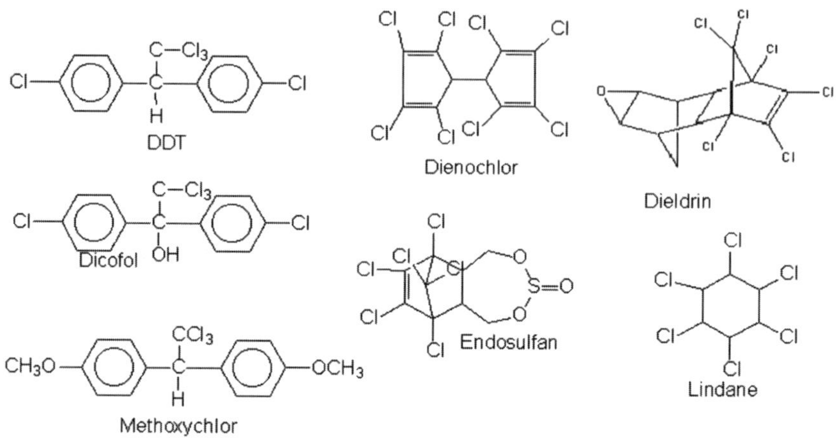

FIGURE 8.3 Organochlorine compounds.

Bruynzeel[38] also reported a case of hand eczema in a nursery worker with a history of nickel and cobalt sensitivity who regularly handled pelargonium, primula, and cyclamen. Patch testing with leaves of pelargonium and cyclamen, plant materials, and the standard European allergen series was negative. She reacted to methomyl and found that her skin cleared upon avoiding contact with the material. A second nursery woman had intermittent hand dermatitis, associated with 15 years employment work in a gerbera (a composite herb) nursery. Upon patch testing she reacted to fragrance mix, methomyl, and gerbera leaf and root. Her condition improved after she began working in a fern growing operation and avoided contact with methomyl.[38] A similar case, involving a nursery worker with simultaneous sensitivity to methomyl and gerbera, was recently reported by Paulsen.[39]

8.2.3 Organochlorines

DDT (Anofex® and other trade names, CAS# 50-29-3, MW 354.5, vp 1.88 x 10^{-7} mmHg, log P NA; for DDE metabolite DDE, log P 5.69-6.96). DDT is the model organochlorine compound, but it is not used legally anywhere in the world, except in public health vector control campaigns. As noted, its principal breakdown product, DDE, retains a very high degree of fat solubility (log P 5.7-7.0). Since 1973, DDT has been banned in the U.S. for all except emergency public health measures. Organochlorines still registered in the U.S. are those with the shortest environmental and biological half-lives. Dicofol, for example, has a soil half-life of 45-68 days compared to 2-15 years for DDT.[1] Compound structures are shown in Figure 8.3.

Given the wide use of DDT, it is not surprising that a variety of skin problems have been described in association with its use. A cluster of chloracne associated with production of DDT in a Texas chemical plant is discussed in Section 9.2.7.[40] Case reports have also described pemphigus,[41] alopecia,[42] and a variety of eczematous eruptions.[43] Behrbohm[44] described a case of dermatitis in a German chemical worker, who reacted on patch testing to technical, but not purified DDT.[44] Nevertheless, as emphasized by Abrams, allergic contact dermatitis due to DDT has not been a widespread problem.[45] Positive patch test reactions to 3% DDT were reported in cotton workers with dermatitis of exposed areas. No positive patch test reactions to 1% DDT were obtained among 665 routine eczematous patients patch tested by the ICDRG, as described by Cronin.[46]

Dicofol (Kelthane®, CAS# 115-32-2, MW 370.51, vp 3.75 x 10^{-7} mmHg, log P 4.28). Dicofol is an agricultural insecticide with 11 product labels currently registered in the U.S. It is structurally similar to DDT, but contains a central hydroxyl group that allows more rapid environmental and metabolic degradation. A 50% formulation of dicofol caused transient post-application erythema in the Draize test in 2/6 test animals, but cleared completely by 7 days. Some dicofol products are labeled as sensitizers, but the and is also labled as a sensitizer. The handler database contains reports of irritant contact dermatitis following direct contact with dicofol:

84-954 — An employee was mixing dicofol wearing gloves and face shield when some material splashed up on his neck. He developed symptoms and informed his supervisor who gave him some ointment. He saw a doctor three days later when his symptoms did not improve and was found to have dermatitis on neck akin to a second degree burn.

84-1454 — A mixer/loader/applicator was splashed on the arms while mixing dicofol. It soaked through his protective clothing and a rash developed on his skin. When he saw a doctor three days later he was noted to have residual erythema on the forearm

Dieldrin (Dieldrex®, CAS# 60-57-1, MW 380.9, vp 3 x 10^{-6} mmHg, log P 3.7-6.2). Dieldrin is an insecticide prohibited in the U.S., but was formerly used for termite control. Dieldrin probably caused dermatitis of the lower legs in 200 of 1209 police recruits exercised to sweating who wore socks mothproofed with dieldrin.[47]

Dienochlor (Pentac®, CAS# 2227-17-0, MW 474.6, vp 2.18 x 10^{-6} mmHg, log P 3.5). Dienochlor is an organochlorine miticide used principally in nurseries, field grown ornamental plants, and in landscaping. Typical cases reported to the California illness registry include:

83-239 — A worker developed a rash after spraying with Pentac®, mixed in excess of the label recommendation.

83-2688 — A worker exposed to pesticide used for spider mites on indoor plants later developed a rash.

A 1% concentration of dienochlor was non-irritating in control subjects tested as part of the California Nursery Workers' study by O'Malley, Rodriguez, and Maibach.[48,49] However, several control subjects, as well as occupationally exposed subjects developed hyperpigmentation, similar to that produced by the standard series test antigen, balsam of Peru (Figure 8.4a and b).

Lindane (Isotox®, CAS# 58-89-9, MW 290.85, vp 4.2 x 10^{-5} mmHg, log P NA). Lindane is used for control of ectoparasites on cattle and for insects on a variety of commericial crops. A 20% formulation used for control of borers and leaf miners is labeled as a skin irritant. It is also a sensitizer in the Buehler assay. The handler database contained two cases of contact dermatitis following direct accidental exposure to lindane, both consistent with irritant reaction. Dermatitis has also

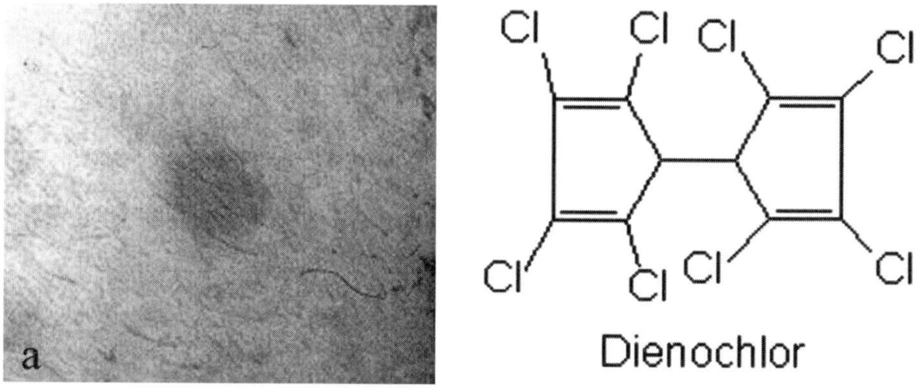

FIGURE 8.4 (**a**) Hyperpigmentation patch reaction to dienchlor observed in California Nursery Workers' study. (**b**) Dienchlor.

been reported among workers in lindane manufacturing operations, but the reported cases were possibly attributable to precursors and byproducts not typically found in commercial formulations of lindane.[50] Although the agricultural products may contain as much as 40% lindane, post-treatment dermatitis also occasionally occurs in patients treated for scabies with 1% formulations of lindane. The extensive series reported by Farkas also contained cases reacting to a 20% scabicidal formulation of sulfur.[51] No positive patch reactions to 1% lindane were obtained in 665 routine eczematous patients patch tested by the ICDRG.[47]

Methoxychlor (Marlate®, CAS# 72-43-5, MW 345.65, vp NA, log P NA). There are 2 products containing methoxychlor currently registered in California (mixtures with captan and malathion) and 134 inactive labels listed in the product label database (4/98). Its range of applications is similar to that for DDT, but it has a markedly shorter environmental and biological half-life.[52] No animal data are on file for either sensitization or irritation studies. There were no cases associated with methoxychlor included in the handler database.

Endosulfan (Thionex®, CAS# 115-29-7, MW 406.9, vp 1.73E-07, log P 2.23-3.62). There are 14 products containing endosulfan currently registered in California. No data were reviewed for either dermal sensitization or dermal irritation, but the product is not labeled as either a sensitizer or an irritant. No cases associated with endosulfan were included in the handler database.

8.2.4 Biological Insecticides, Repellents, and Related Synthetic Materials

8.2.4.1 Pyrethrum, Pyrethrins, and Pyrethroids

Pyrethrum and pyrethrins (Black Flag®, Raid®, and other trade names — a total of 899 registered products in California, 1999 — for pyrethrum oleoresin, CAS# 121-21-1 for pyrethrin I, MW 328.43, vp 7.5E-09, log P NA). Pyrethrum is an extract of the dried flowers of *Chrysanthemum cinerariaefolium*, imported from

Kenya and Ecuador, containing not less than 1% pyrethrins — the active compounds pyrethrins I and II, cinerins I and II, and jasmolin I and II. These esters function by inactivating the sodium channel in the insect nervous system. Mammals are relatively immune to their action because of rapid detoxification by the liver enzymes. They have a low order of systemic toxicity to man and produce no harmful residues on food crops.

There are more than 1000 pyrethrin formulations registered in the U.S.,[6] and the California illness registry cases associated with pyrethrins reflect their broad scale use — in pet sprays, ready-to-use household sprays, aerosols, and food protection in warehouses. Pyrethrins are commonly formulated in mixtures with piperonyl butoxide and with pyrethroids, carbamates, organophosphates, and other materials.

Dermal irritation studies with technical pyrethrins (57% concentration) showed mild erythema at 24-h evaluation, which was reduced substantially by 72 h, and cleared completely 7 days after application.[53] These observations are compatible with cases reported to the illness registry.

86-385 — A worker set off a fogger in the kitchen area of workplace, but failed to exit the area for about 10 min and subsequently developed skin irritation on his face.

The reaction reported in a second case, indicates that pyrethrin products formulated in petroleum distillates may cause more intense irritation than that suggested by the findings of the Draize test on technical pyrethrin.

90-2621 — A fairgrounds employee suffered a reported chemical burn to his right leg while applying an insecticide to livestock barns. "The fogger" machine he was using had a loose cap on the reservoir tank causing insecticide concentrate to come in contact with his leg.

Many individual pyrethrin formulations are positive in the Buehler test and the few tested by the intradermal method are also positive. These positive findings are appropriately reflected on product labels.[54] However, no well-documented report of sensitization to purified pyrethrins (i.e., not containing the lactone sensitizer described by Mitchell, see below) has been published.

Clinical reports of sensitization include bullous reactions, reactions in 1-2% of tested patients, [55] reports of cross-reactions occuring among pyrethrum, chrysanthemum, shasta daisy, and ragweed oleo resin,[56,57] and asthma and urticaria reported as reactions to natural pyrethrin.[58] In 1972, Mitchell et al.[59] found that a sesquiterpene lactone, pyrethrosin, was the chief allergen in pyrethrin.

8.2.4.1.1 Pyrethrin synergists

Piperonyl butoxide (CAS# 51-03-6, MW 338.443, vp NA, log P NA) is a synergist used with formulations of pyrethrins and pyrethroids, but is not chemically related to either group. The technical formulation (92%) is an irritant in the Draize test, but the concentration in ready-to-use formulations is less than 1%. The only dermal sensitization studies reported involve mixtures with pyrethroids. Cases reported in the handler database also had simultaneous exposure to pyrethrins or pyrethroids.

***N*-octyl-bicyloheptene dicarboximide (NOBD, CAS# 113-48-4, MW 275.39, vp NA, log P NA).** This is also used as a pyrethrin synergist in hundreds of formulations. The only irritation study available for review involved a mixture with diethyl toluamide, isochromyl cinchonerate. This caused only transient irritation and was scored as category IV. There is no study on the capacity of NOBD to cause sensitization. Cases reported to the handler database all involved mixtures with pyrethrins and other compounds.

3.2.4.1.2 Synthetic pyrethroids

Pyrethroids are synthetic compounds produced to duplicate the biologic activity of the active principals of pyrethrum. Pyrethroids have a longer duration of activity against insects than pyrethrum. These compounds include allethrin, alphametrin, barthrin, bioresmethrin, biopermethrin, cismethrin, cyclethrin, cytluthrin, cypermethrin, decamethrin, deltamethrin, dimethrin, fenothrin, fenpropanate, fenvalerate, flucythrinate, fluvalinate, furethrin, indothrin, permethrin, phthalthrin, resmethrin, and tetramethrin. Allergic contact dermatitis due to pyrethroids has not been reported. Temporary paresthesias manifested by numbness, itching, burning, tingling, and warmth have been reported following cutaneous exposure to the synthetic pyrethroid fenvalerate. Fenvalerate, produced in the U.S., produces more paresthesias following topical exposure than pyrethrin and other pyrethroids. It has been suggested that topical vitamin E acetate is highly effective in treating paresthesias induced by pyrethroids.[60] The pyrethroid compounds are generally characterized by relatively high molecular weights, high octanol/water partition coefficients, and extremely low vapor pressures.

Effects of synthetic pyrethroids on the sodium channels of cutaneous nerve endings may cause paresthesias at levels of exposure that do not provoke visible erythema.[61 62] The standard Draize irritation study is thus a poor means for evaluating such purely symptomatic endpoints. An alternative animal test developed by Cagen evaluates the sensory effect of pyrethroids through observations of grooming behavior focused on the site(s) of applied test material. The behavioral test demonstrates direct effect on grooming behavior for 4 h after pyrethroid application and increased response to other chemical irritants (oil of mustard in the test model) for 24 h after application.[63]

Type I pyrethroids (Figure 8.5) contain two cis/trans isomeric sites (up to four isomers each), and have demonstated moderate ability to stimulate cutaneous nerves in human studies.[64,65] The results of standard animal individual type I compounds are described in Table 8.2.

Noteworthy cases from the California illness registry involving type I pyrethroids include

Case 94-1052 — An applicator bumped his right arm against the nozzle of a sprayer containing bifenthrin for use on cotton and got some of the spray solution on the arm. He suffered numbness and tingling at the site of exposure.

Case 94-138 — A worker spilled a mixture of allethrin and piperonyl butoxide on the outside of a backpack sprayer and on his arms while mixing a tank load. He wiped off the sprayer with a paper towel, but after spraying, he

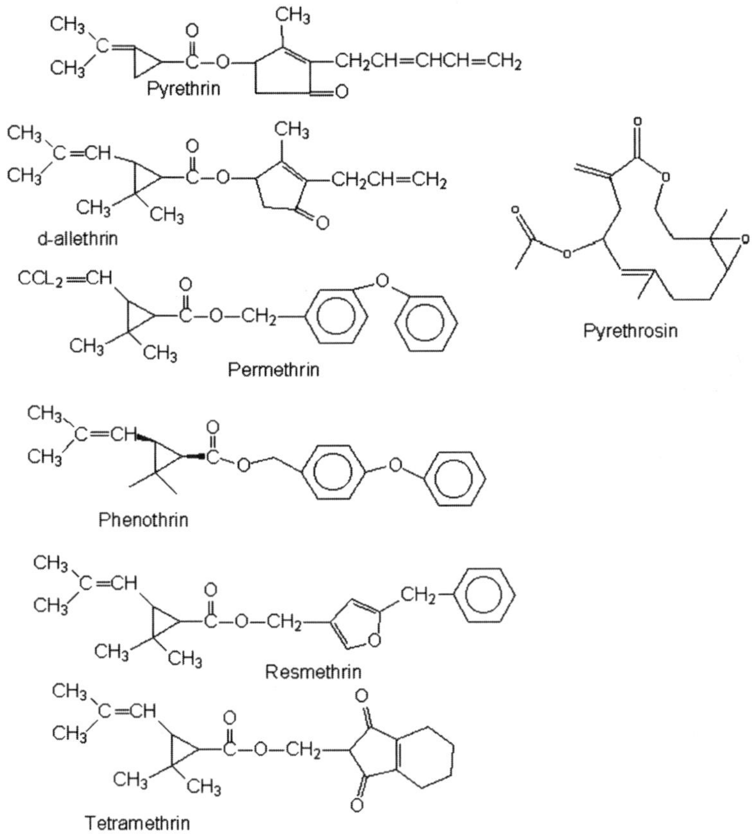

FIGURE 8.5 Pyrethrin and Type I synthetic pyrethroids (non-cyano compounds).

developed itching, redness, and swelling on the skin of the lower back and arms.

Case 92-1381 — A worker was mixing permethrin for an aerial application; small amounts of the material kept getting under gloves and shirt. As a result, he developed pain, swelling, and blisters on hands and forearms.

8.2.4.1.3 Type II pyrethroids

These contain three isomeric sites and most contain a cyano group attached near the ester linkage (Figure 8.6). They are relatively more potent systemic toxins than the type I pyrethroids and cause a greater degree of paresthesia in experimental studies on human volunteers.

Noteworthy cases reported to the California illness registry involving type II pyrethroids include

Case 88-2388 — A mixer/loader developed dermatitis associated with accidental transfer of cypermethrin from the hands to the genitalia.

TABLE 8.2
Summary of Predictive Dermal Irritation and Sensitization Tests in Rodents

Compound (brand name, CAS#)	Dermal irritation study	Dermal sensitization studies
Allethrin (Banish® and other trade names, 42534-61-2, MW 302.4126, vp NA, log P NA)	Technical d-allethrin (95.6% concentration of a cis-trans mixture) shows transient irritation, clearing by day 3 in the Draize test[66]	A Buehler sensitization study was negative on a dilute end use product (a mixture of allethrin, cypermethrin, piperonyl butoxide and petroleum distillates)
Bifenthrin (Capture®, Talstar®, 82657-04-3, MW 422.88, vp 1.8E-07, log P 6)	Draize assay on technical bifenthrin (88.4%) showed no signs of erythema or edema	
Permethrin (Ambush®, 52645-53-1, MW 391.28, vp 3.38E-07, log P 6.1).	Technical permethrin shows minimal irritation in the Draize test, with only very slight erythema persisting 72 h after initial application	Several formulations appeared to be non-sensitizers in the Buehler assay, but a mixture of 1% permethrin and 1% piperonyl butoxide showed mild sensitization reaction on the rechallenge portion of the assay
Phenothrin (Raid®, Black Flag®, Enforcer®, Ortho®, and multiple household products, 26002-80-2, MW 350.4566, vp NA, log p NA)	A Draize irritation test (97.7% phenothrin) demonstrated that the material did not cause any sign of erythema or edema	A mixture of tetramethrin (.05%) and phenothrin (.096%) showed marginally increased reaction on re-challenge (erythema graded as 0.2 at 24 h on challenge and as 0.4 at 24 h on re-challenge), the dermal scores for erythema and edema were 0 in the Buehler assay conducted on a 1% formulation of phenothrin
Resmethrin (Raid®, Black Flag®, Bug Boss®, Enforcer®, Ortho®, Sunbugger®, and multiple household products, 10453-86-8, MW 338.4, vp <7.50075E-08, log P 5.42)	A Draize irritation test on technical (84.5%) resmethrin showed moderate erythema and mild edema present at 48 h post dosing	Mild to moderate erythema, accompanied by slight edema found in all test animals in challenge phase of a Buehler assay done with 3% resmethrin (erythrema score =1.75 vs. 0.6 for control animals) indicating the product was a sensitizer under test conditions
Tetramethrin (Black Flag®, Bug Boss®, Enforcer®, Good Riddance, Green Thumb®, Ortho®, Petscription, Raid®, Spectracide®, Sunbugger®, Terminator and multiple household products 7696-12-0, MW 331.4108, vp NA, log P NA).	Animals treated with a 20.8% formulation of tetramethrin showed transient erythema which cleared in all treated animals by day 4 post application	See study above on phenothrin which showed marginally increased response; a test on a ready-to-use mixture of piperonil butoxide (1%), NOBD (1%), fenoxycarb (0.15%), pyrethrins (0.14%), and tetramethrin (0.063%) was negative

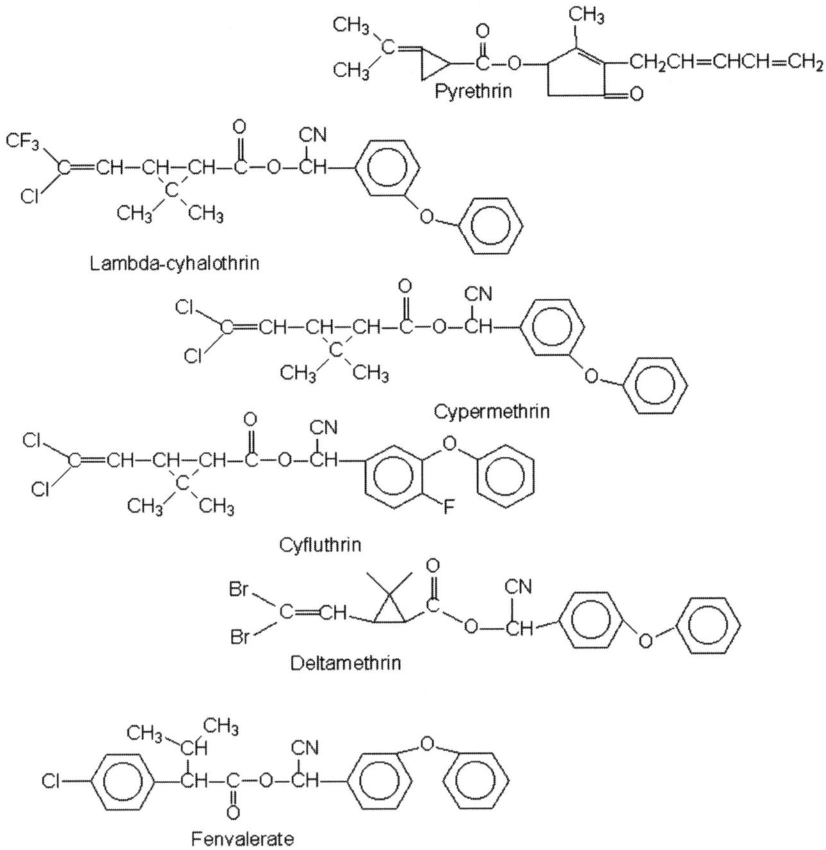

FIGURE 8.6 Pyrethrin and cyano-pyrethroid (Type II compounds).

Case 86-1191 — A mixer/loader suffered from a burn on the arm after he spilled fenvalerate on himself and did not promptly decontaminate his clothing.

8.2.4.2 Nicotine and Synthetic Analogs

Nicotine (Black leaf 40®, Nicotine Smoke Generator®, CAS# 54-11-5, MW 162.234, vp NA, log P NA). Nicotine has largely been replaced by organophosphate insecticides and other newer synthetic products, including nicotine analogs like imidacloprid and related compounds (Figure 8.7). For example, slightly more than 200 applications of nicotine products were reportedly used in California during 1995, compared to more than 11,000 for naled, a typical moderate toxicity organophosphate, in the same year.[67] The only remaining registered product is a nicotine smoke generator used for control of nursery and greenhouse pests.[6]

Vero and Genovese[68] patch tested workers with dermatitis who made cigars from various tobacco leaves; no positive reactions were found. Transdermally absorbed nicotine induced predominantly sudorific and rubiform reactions which may be accompanied by subtle piloerection, hyperalgesia, and pruritus.[69] Nicotine also may be absorbed through the skin of people harvesting tobacco leaves, particularly in the presence of dermal abrasions. Exposed workers may develop nausea, vomiting,

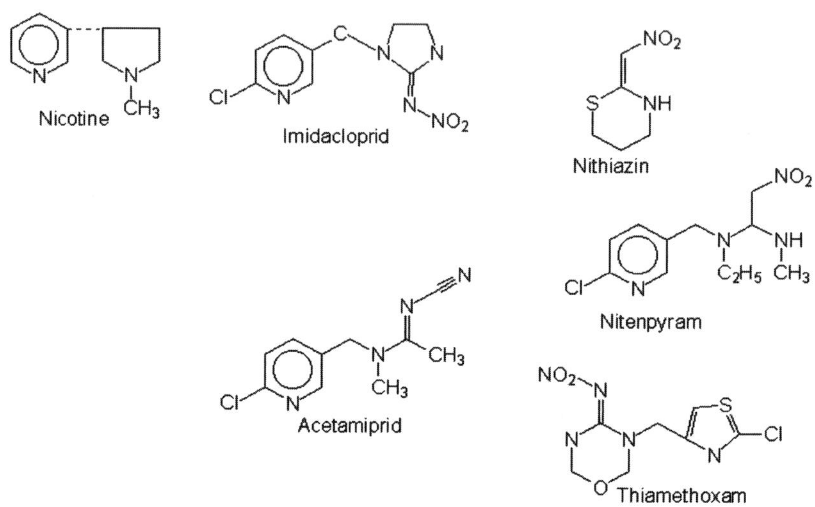

FIGURE 8.7 Nicotine and some related analogs.

dizziness, prostration, and weakness very similar to the symptoms of organophosphate poisoning.[70-72] The use of work gloves significantly decreases nicotine absorption in workers.[73]

Imidacloprid (Admire®, Gaucho®, and Marathon®, 105827-78-9, MW 255.6633, vp NA, log P NA). Imidacloprid is a nicotine analog, sold as a wettable powder, and in flowable and granular formulations. It is used as a soil, seed, or foliar insecticide on vegetable, orchard, and nursery crops. Registration data show imidacloprid caused only transient irritation in the Draize test and was a non-sensitizer in the Buehler assay. There were no cases associated with the compound in the handler database.

8.2.4.3 *Bacillus Thuringiensis*

Bacillus thuringiensis (Dipel®, Thuricide®) and related toxins. CAS# 68038-71-1). Two active and 132 inactive products registered in California contain *B. thuringiensis* (BT; variant Berliner). Related products include BT variant San Diego (and other) and BT derived endotoxins. A 15% formulation of BT (Berliner), subsp. *Kurstaki*, strain EG 2348 caused no dermal irritation. A formulation of BT variant San Diego showed a slight reaction in the Buehler assay (score = 0.2 at 48 h), but the reaction was markedly less than the reaction to the 0.05% dinitrochlorobenzene (DNCB) used in the study as a positive control. A formulation of BT endotoxin mixed with killed *Pseudomonas fluorescens* is labeled as a dermal sensitizer, but was not among the products for which dermal sensitization studies were available for review.

TABLE 8.3

Cyfluthrin (Tempo®, CAS# 68359-27-5, MW 434.3, vp 3.3E- 08, log P 5.91-6.00) and cyhalothrin (Demand®, CAS# 91465-08-6, MW 449.8562, vp NA, log P NA)	Both caused minimal irritation in the Draize test	Both were non-sensitizers in the Buehler assay
Cypermethrin (Ammo®, CAS# 52315-07-8, MW 449.8562, vp 1.42514E-09, log P 6.60)	Technical cypermethrin causes minimal irritation in the Draize test	Buehler assay on the 18% formulation of cypermethrin showed no evidence of sensitization
Fenvalerate (Tribute®, CAS# 51630-58-1, MW 419.9, vp 1.5E-09, log P >4)	The 24% formulation is irritant (category II) in the Draize test	Also a sensitizer in the Buehler assay.

A recent study by Bernstein et al.[74] evaluated the occurrence of allergy symptoms and immune response in workers harvesting BT treated field crops (celery, parsley, cabbage, kale, spinach, and strawberries). The study identified a clear increase in skin test reactivity to BT extracts and IgE anti-bodies over the course of a growing season in workers with high potential exposure to BT compared to those with low exposure. Cases of contact dermatitis occurred in the group, but were attributed by the authors to plant contact dermatitis. Allergic rhinitis possibly associated with exposure to BT was also reported in conjunction with its use in gypsy moth eradication, but the etiology of the reported cases was not established by specific provocation tests.[75]

8.2.4.4 Other Biological Insecticides

The non-pyrethroid biological insecticides (Figure 8.8) are complex molecules with high molecular weights. Although octanol/water partition coefficients are not available for most of the compounds, most are soluble in organic solvents and sparingly soluble in water. Many contain lactone-like reactive elements akin to those present in plant antigens such as tuliposide A or olefins conjugated with amino groups (ICONJ, as indicated in Table 1.4).

Avermectin[76] (Agri-mek®, CAS# 71751-41-2, MW 873.11 **vp** negligible**, log P NA).** Avermectin is a mixture of avermectin B1a and B1b used as an ant and cockroach bait and sometimes for nursery and agricultural pest control. The 1.9% formulation used as an agricultural miticide is labeled as a dermal sensitizer, but the bait formulations are not. Avermectin produced no irritation in the Draize test reported for registration, but a case reported in the California handler database (92-520) was compatible with an irritant mechanism.

Azadirachtin[77] (Azatin®, CAS# 11141-17-6, MW 646.687**, vp** >2 mmHg, **log P** 12.3**).** Azadirachtin is a triterpenoid derived from Neem oil, used on a large variety of vegetable and nursery crops. California registration data demonstrate that 10%

technical material was completely non-irritating in the Draize test. A dermal sensitization conducted by an unspecified method was reported for a product containing 0.25% azardirachtin, but was not adequate to allow judgment of its sensitization status. None of the azadirachtin products listed in the label database are reported as dermal sensitizers and there were no cases associated with its use listed in the California handler database.

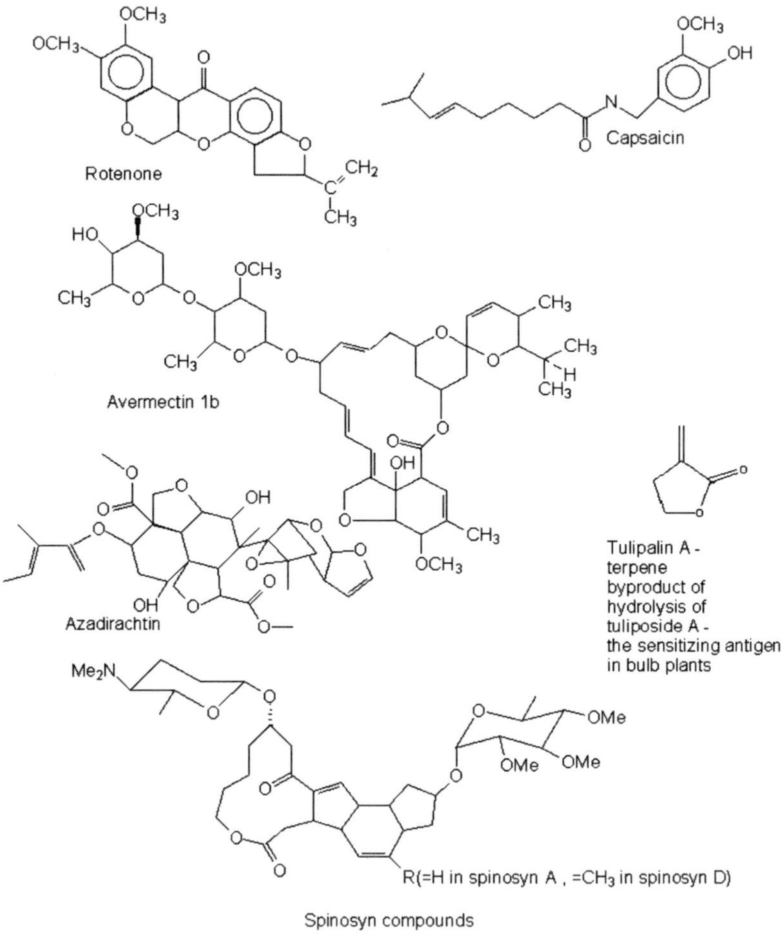

FIGURE 8.8 Non-pyrethroid biological insecticides.

Capsicum oleoresin-capsaicin (Insect Control® -For Ants, Counter Assault Bear Deterrent®, Dog Shield®, Get Away Squirrel and Raccoon Repellent®, Hot Sauce Animal Repellent®, CAS# 404-86-4 for capsaicin, MW 305.4162, vp NA, log P NA). Capsaicin is a material extracted from pepper plants, formerly registered for ant control, but now used principally as an animal repellant. No Draize test was available, but the concentration of resin in a typical animal repellent formulation is below the 1% identified as non-irritating for the purposes of patch testing.[78] Although the formulation reviewed was negative in the Buehler sensitization assay, it was labeled as a sensitizer. [79] Garlic, per se, is a well-recognized sensitizing agent. [78, 80-83]

Rotenone (Green Light Tomato & Vegetable Garden Dust®, Noxfish®, Farnam Flys-off Insect Repellent for Dogs®, Brittle Extract of Cube Root®, CAS# 83-79-4, MW 394.4, vp <7.50075E-06 mmHg, log P NA). Rotenone is a non-selective mitochondrial poison extracted from the CubÈ root (or from derris or timbo). It is often used for control of unwanted fish species,[84] but is also used as an insecticide on a large number of nursery, orchard, and vegetable crops. Rotenone may also be used in combination with pyrethrin and piperonyl butoxide (a synergist) for control of a wide variety of insects on food crops. Skin irritation has been reported among workers in rotenone- processing plants in South America. Skin inflammation was most notable in intertriginous areas or where the powder had been trapped by perspiration on the skin. A similar outbreak was reported among workers in France but improved ventilation and dust masks diminished the occurrence of dermatitis in these workers.[85]

Spinosad #1 (Conserve®, Success®, Precise®, CAS# 131929-60-7, MW 731, vp NA, log P NA). Spinosyn compounds are a new category of insecticide, derived from the actinomycete organism *Saccharopolyspora spinosa,* active against thrips, beetles, moths, flies, ants, and mites. They are currently used on turf and ornamental crops.[86] Although the compound contains one conjugated olefin site (ICONJ) similar to those present in sensitizing lactone compounds, its relatively high molecular weight suggests it will penetrate skin to only a limited degree. EPA registration data indicate that spinosad does not irritate or sensitize the skin in animal tests,[87] but experience with the compounds is too limited to determine whether spinosad or related compounds will prove to sensitize humans.

8.2.5 Other Insecticides (Figure 8.9)

Boric acid and related borate salts (Enforcer ant baits®, Roach Stop®, CAS# 11113-50-1, MW 61.84, vp NA, log P NA). Borates (including borax and hydrated octaborates) are non-sensitizers and non-irritants in animal tests. Case 84-197 in the handler database described a transient skin reaction associated with handling borates.

Amitraz (Mitac®, Ovasyn®, CAS# 33089-61-1, MW 293.4, vp 2.55E-06 mmHg, log P 6.5). The aniline derivative amitraz is used for ectoparasites in veterinary practice, for psylla control on pears and for whitefly on cotton. There were no dermal irritation data on file, but a formulation of 50% wettable was labeled as corrosive. Amitraz also proved to be a sensitizer in the guinea pig maximization test. There were no cases associated with isolated exposure to amitraz in the California handler database.

Benzyl benzoate (Acarosan Moist Powder®, Aid Mange Lotion®, CAS# 120-51-4, MW 212.2476, vp NA, log P NA). This benzoic acid derivative is used for control of mange in dogs. The technical material produced transient erythema in the Draize test. No adequate dermal sensitization study was reviewed. The technical material is labeled as a sensitizer, but the derivative formulations are not.

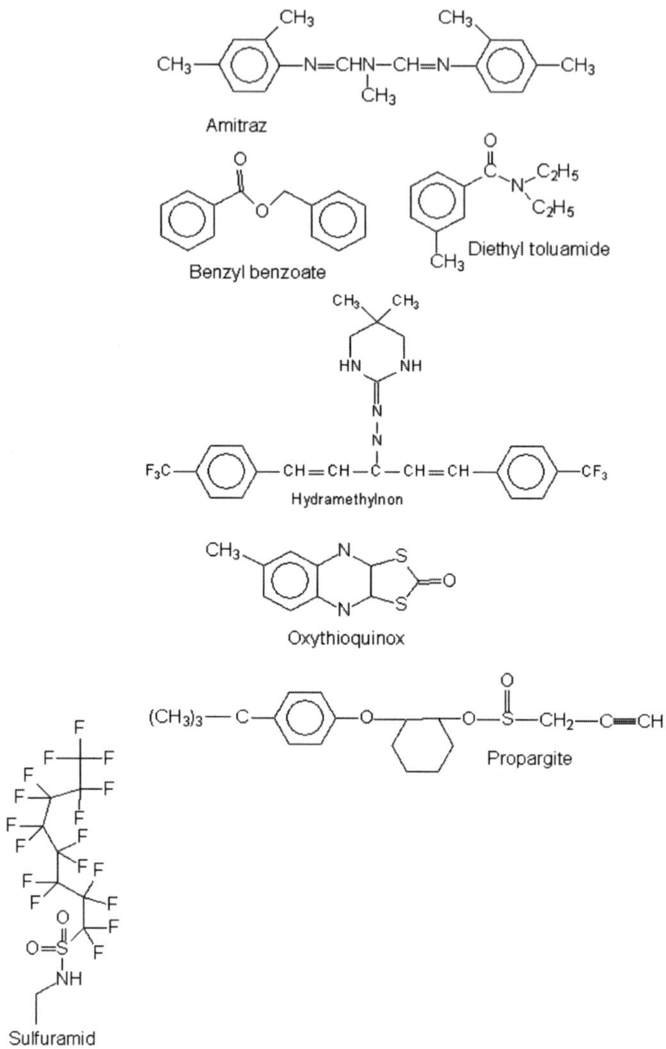

FIGURE 8.9 Miscellaneous insecticides.

Butoxy polypropylene glycol (Crag fly repellent®, CAS# 9003-13-8, MW NA, vp NA, log P NA). Butoxy polypropylene glycol is a fly repellent used for dogs and cattle. Registration data show that a mixture with piperonyl butoxide and pyrethrins (both non-irritants) caused mild irritation in the Draize test, but no dermal sensitization study was available for review. No cases were reported in the handler database.

Diethyltoluamide or DEET (Deep Woods Off®, Cutter Insect Repellent®, Repel®, CAS# 134-62-3, 191.2724, **vp NA, log P NA**). DEET is an insect repellent sold in many over-the-counter formulations that range from 20 to 100% active ingredient. Registration data show that DEET caused transient irritation in the Draize test and was a non-sensitizer in both the Buehler and maximization tests. Nevertheless, some formulations are reported as sensitizers on the product label because of published reports, beginning with Maibach and Johnson in 1975,[88] indicating that DEET can cause contact urticaria. Two cases reported to the California illness registry (82-1871 and 93-1422) involved apparent immediate allergic reactions similar to those reported in the published literature.

Hydramethylnon (Wipeout®, CAS# 67485-29-4, MW 494.5, vp 2.03E-08, log P 2.31). Hydramethylnon is an insecticide used in roach, ant, and termite baits. It is a non-sensitizer in the Buehler assay, but no dermal irritation studies were available for review. No cases associated with its use were included in the California handler database.

Insectidal soaps — potassium salts of fatty acids, potash soap (Safer®, Concern®, CAS# 61790-44-1, MW NA, log P NA). Insecticidal soaps are used to control aphids and spider mites on plants and vegetables in both household gardens and commercial nurseries. Registration data show that a 49% formulation was reported as corrosive in the Draize test. Dermal sensitization data were not reported and there were no cases associated with insecticidal soaps in the handler database.

Oxythioquinox (Morestan®, CAS# 2634-33-5 63, MW, vp, log P). Oxythioquinox is a miticide, insecticide, and fungicide used for control of mite eggs and mildew on deciduous fruit. Registration data indicate that oxythioquinox is a minimal irritant in the Draize test, but the 40% formulation is a sensitizer in the Buehler test.[89] A case (82- 869) from the handler database involved apparent irritant dermatitis.

> 82-869 — While filling a nursery tank, some of the material overflowed, landing on the mixer/loader's face, neck, and arms. The subsequent clinical examination showed marked irritation of the skin in the exposed areas.

Propargite (Omite®, Comite®, CAS# 2312-35-8, MW 350.5, vp 4.48E-08, log P 3.73). Propargite is a miticide with broad agricultural uses. The technical material is corrosive in the Draize assay. A study compared the dermal sensitization potential of propargite with that of iprodione following a California field worker's dermatitis that involved exposure to both compounds. It used a modified version of the Buehler method. While the study had some technical deficiencies (i.e., lack of a positive control group and the use of the same animals to test both products), a number of the findings were significant. In the range-finding portion of the study it was determined that iprodione could be applied during the challenge tests at the maximum concentration allowed by the protocol (5%); propargite could only be applied at concentrations of 0.1%. During the challenge portion of the study both materials produced less reaction than during the induction phase, indicating neither material was a sensitizer under the conditions tested. In QSAR terms, the structural element most likely responsible for propargite's irritant capacity is the terminal carbon-carbon triple bond (propargyl group) and its moderate degree of octanol solubility, indicating its capacity to at least partially penetrate the skin.

The handler database contained 55 cases associated with propargite between 1982 and 1995. Typical cases shown involved accidental direct exposures.

82-1667 — A mixer/loader splashed a few drops of concentrated propargite formulation (for use on corn) on his neck while opening a can for a closed system. He did not promptly decontaminate and over 24 h, he developed a rash that persisted for a week until he got it treated.

85-11642 — During a mix/load operation, a nursery tank overflowed, dousing the worker's arm with propargite. He subsequently developed chemical burns on his arm.

Many irritation cases occurred following initial registration of propargite, prior to control of the hazard of mixing powdered forms of propargite by using water soluble bags.[90] The issues involved in regulating the hazards of propargite residues on crops harvested, or cultivated, with manual labor are discussed in Chapter 6 and in the international context in Chapter 9.

Sulfluramid (Chemsico®, Raid Max®, Volcano®, Ultrabait®,CAS# 4151-50-2, MW 527.2, vp NA, log P NA). Sulfluramid is a sulfonamide derivative used as a cockroach and ant control bait. It is a non-irritant and non-sensitizer. There were no cases associated with its use listed in the California handler database.

Arsenic pentoxide, arsenic acid compounds (Bardec®, Osmose®, CAS# 7778-39-4, MW 459.6804, vp NA, log P NA) and cacodylic acid (Green light weed killers®, CAS# 75-60-5, MW 137.9977, vp NA, log P NA). The earliest insecticides against chewing insects were the arsenicals, chiefly copper acetoarsenite (Paris green) lead arsenate, and calcium arsenate. Sodium arsenite has been used as a sterilant herbicide and a potato-vine killer. Inorganic arsenicals have been superseded because of their hazard to man and animals. Sodium arsenate was formerly the toxicant in many ant syrups for household use, but this application also has been discontinued. Organic arsenicals are, at present, of the most interest because of their value as selective herbicides.

Skin effects associated with inorganic arsenic include both cutaneous irritation and sensitization,[46] and hyperkeratosis, hyperhidrosis, and hypermelanosis occur following chronic systemic exposure. The hyperpigmentation is most marked on surfaces exposed to light. There may be a speckled depigmentation of pigmented areas giving a "raindrop" appearance. Compounds similar or identical with those used as pesticides have caused skin cancer in man.[91] Agricultural workers are also exposed to chronic ultraviolet light, and it had been believed that ultraviolet light exposure was a more significant factor for these workers than their exposure to inorganic arsenic.[92]

8.3 FUNGICIDES AND ANTIMICROBIALS

Compounds used to control fungi on plants overlap to some degree with the broader category of antimicrobial compounds. The fungicidal activity of many compounds depends upon the presence of electrophilic groups that allow the inhibition of fungal sterol synthesis enzymes, 14-demethylase, or affect mitochondrial electron transport.[93] This same chemical property gives rise to many interactions with skin proteins that cause many fungicides to be skin sensitizers (element IUNIQ in the QSAR model shown in Table 1.4). The same property is generally associated with carcinogencity in animal bioassays,[94,95] and many fungicides have been identified as animal carcinogens.[96]

8.3.1 Phthalimido Compounds

The phthalimido compounds have a high degree of strucutural similarity (Figure 8.10), with the core ring structure of each compound resembling that of the prototype compound, captan, and varying principally in the side chains attached to the dicarboximide ring.

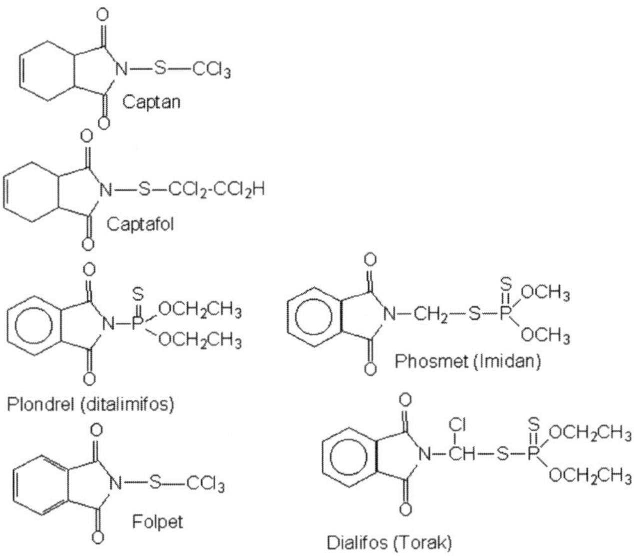

FIGURE 8.10 Phthalimido fungicides.

Captan (Orthocide® and other product names, CAS# 133-06-2, MW 300.61, vp 9E-08, log P 2.8). Captan is a fungicide commonly used on grapes, apples, almonds, and other crops. It is formulated as a wettable powder, dust, and flowable powders and used by itself or in combination with other fungicides and insecticides. In addition to use as a pesticide, it has been used successfully as a treatment for pityriasis versicolor.[97]

A technical formulation of captan tested in 1984 caused no irritation in a rabbit dermal irritaton study test reported as part of the California registration database,[98] but a currently registered technical formulation (88% captan) is recognized as a sensitizer on the product label[6] and the compound has been documented to cause sensitization in experimental human studies, as described below.

Marzulli and Maibach[99] demonstrated that captan in a concentration of 1% is a significant contact allergic sensitizer. Out of 205 human test subjects, 9 were sensitized. The International and North American Contact Dermatitis Groups later found captan in a 1% concentration to be irritant on patch testing; and 0.25% in petrolatum is now preferred for patch testing.[46] Relevant allergic reactions to captan nevertheless appear to be rare.[46] In another experimental study, Jordan and King[100] demonstrated a 5% sensitization rate for captan using a modified Draize test on volunteer subjects and a 10% sensitization rate on volunteers using captan in the human maximization test. Women appeared to become sensitized more frequently than men.[100]

In case reports, urticaria due to captan has been documented in a gardener who reacted to captan and to captan-treated plants.[101] Fregert[102] reported a fruit farmer who developed dermatitis of his hands and face after three weeks exposure to captan. Patch testing with captan and the related compound folpet (Phaltan), 1% in petrolatum, gave positive reactions. Surveillance data demonstrate that dermatitis associated with captan exposure has been a frequently reported problem among all groups of California agricultural workers.[103] A series of 178 patients at the Nagoya City University Medical School who were routinely tested between 1977 and 1980 with the North American Contact Dermatitis Research Group of standard allergens, 5.6% had significant positive reactions to captan. No clinical details were given in the report, but the surprisingly high percentage reacting to captan, presumably, an uncommon exposure, raises the possibility that the material cross reacts with other allergens in the standard series or causes active sensitization.[104] Rudner[105] observed a similar high percentage of captan reactors in the North American Contact Dermatitis Group results in 1976 and speculated that results might be due to cross reaction with thiurams. Nevertheless, current recommended patch test concentrations are 0.1 or 0.25% (see Appendix II) compared to the 1% concentration in the test series reported by Rudner.

Captafol (Dilfolitan®, CAS# 2939-80-2, MW 349.1, vp 9.98E-09 mmHg, log P NA). Captafol has a chemical structure nearly identical to that of captan, with the exception of a chlorinated carbon (IX element in QSAR terms) in its side chain. It has many similar uses as a fungicide, including prevention of blight on potatoes and fungal diseases on fruit and cereal crops.

In a case report, Groundwater[106] described the sudden appearance of wheezing, vesiculation, and edema of the face and hands of a welder, employed by a maintenance firm which serviced captafol distribution plants, associated with contact with bags of captafol. Subsequent re-exposure led to recurrences. Cottel[107] also reported two farmers with dermatitis of exposed areas who had positive patch tests to 0.1% captafol in water.

In an outbreak investigation, Camarasa reported severe pruritus, morbiliform urticarial eruptions, and asthma among 7 out of 41 workers in a company that packed captafol (Difolatan).[108] Four of these patients had strong patch test reactions. In Japan, among workers employed in tangerine and other citrus crops, captafol was associated with several hundred cases of contact dermatitis between 1966 and 1980.[109-112] Lee[113] also found that reactions to captafol were prevalent among Korean farmers, affecting 18.2% of the 121 subjects tested in that series. Stoke[114] conducted a questionnaire survey of 133 New Zealand workers from several different wood treatment plants at a time when captafol was being introduced to replace pentachlorphenol as an antisapstain agent. He identified responses consistent with contact dermatitis in 23% of those completing the survey. He felt that most cases were attributable to irritant, rather than allergic dermatitis, but provocation was not conducted as part of the study.[114] An outbreak of dermatitis due to captafol sensitivity was also seen among a group of 36 workers on a Kenyan coffee plantation.[115]

Captafol thus is capable of causing both delayed and immediate types of hypersensitivity as well as irritant dermatitis, and is labeled accordingly.[116]

Folpet (Phaltan®, CAS# 133-07-3, MW 296.555, vp NA, log P NA). Folpet differs from captan only in having a full six-carbon aromatic ring in place of a partially saturated six member ring in its core structure. It is a protective fungicide used as a 50% wettable powder and in various dusts on fruits, berries, vegetables, flowers, and ornamentals. In the series reported by Lisi,[117] six agricultural workers had positive

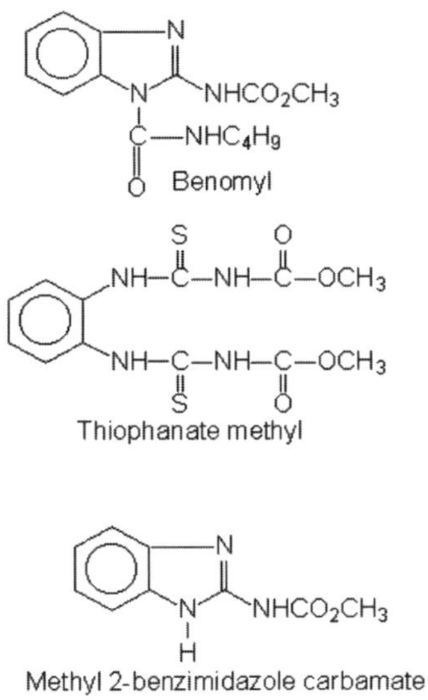

FIGURE 8.11 Carbamate fungicides.

patch test reactions to 0.1% folpet, but irritant reactions were seen in some controls. California registration data indicate that the 88% technical material is a minimal skin irritant in the Draize test, and is a sensitizer in the maximization assay.[118]

Ditalimfos (Plondrel 50W®, CAS# 5131-24-8, MW 299.28066, vp NA, log P NA). Ditalimifos is a fungicide that structurally resembles captan and captafol, but can also be classified as an organophosphate because the sidechain attached to the dicarboximide group contains a phosphothioate group. In 1975, van Ketel[119] reported the cases of two culturists and two florists who sprayed Plondrel on roses and subsequently developed dermatitis. All four reacted to 0.1% Plondrel in petrolatum, but no reactions occurred in 20 control subjects tested with the same material.[119] van Ketel[120] subsequently reported a third case of hand eczema in a 21-year-old florist who had a 3+ reaction to 0.1% plondrel. A case of occupational dermatitis in the foreman of an English chemical plant making Plondrel was seen at the St. John's Hospital dermatology service and also reported by Cronin.[46]

8.3.2 Carbamates

Benomyl (Benlate®, CAS# 17804-35-2, MW 223.23, vp 3.75E-08 mmHg, log P 1.37). Benomyl is a systemic carbamate fungicide (Figure 8.11) with extremely low potency as a cholinesterase inhibitor, used in the control of many diseases of fruits, nuts, vegetables, and ornamental plants. Guinea pig tests of benomyl for irritancy conducted at 12.5 and 25% aqueous dilutions were reported to be negative, but the maximization test reported in the same publication showed 2% benomyl to be a potent experimental allergen.[121]

In 1972, allergic contact dermatitis in areas exposed to benomyl was reported in seven Japanese women working in a hot, humid California greenhouse where the compound was being sprayed on carnations. Ten Hispanic co-workers were unaffected. Undiluted benomyl and benomyl diluted 1:5 in olive oil produced negative patch tests in 3 controls, but a medical assistant developed a 2+ reaction when these patch test sites were exposed to 30 s of ultraviolet light. The 7 Japanese workers had positive patch test reactions to benomyl 1:10 in olive oil.[122] Fregert[123] reported one case with a positive patch-test reaction to benomyl 0.1% in water. van Ketel[124] also reported a case of benomyl sensitivity, confirmed by patch testing (with a 1% preparation which elicited no reaction from 10 controls), in a begonia grower exposed principally through contact with residue. Patch testing with 1% in petrolatum was positive. A report from The Netherlands, by van Joost,[125] also highlighted the occurrence of benomyl hypersensitivity in nursery workers and florists.

The above cases illustrate the capacity of foliar residues of benomyl to cause allergic contact dermatitis in nursery workers. Zweig et al.[126] demonstrated that exposure up to 5.4 mg per person per hour to benomyl is also a potential problem in strawberry harvesting. Everhart[127] studied benomyl applicators and noted a maximum total exposure of 26 mg of benomyl in mixing/loading operations, and markedly lower total exposures associated with field residue exposure (12 mg), and home use of the material (<1 mg). Hargreave[128] noted the possibility of exposures from handling treated commodities, which demonstrated persistent benomyl residues on litchi nuts up to 15 days after postharvest treatment in a dipping process consisting of 20 ppm of benomyl in skin and 1.3 ppm in the flesh of the nut.

Thiophanate methyl (TOPS®, CAS# 23564-05-8, MW 342.4, vp 7.13E-08 mmHg, log P 1.5). Thiophanate methyl is a systemic fungicide used on vegetables, beans, nuts, and turf. An 85% formulation is a minimal irritant in the Draize assay A formulation containing both mancozeb (63%) and thiophanate methyl (15%) was a sensitizer in the Buehler assay, but a formulation containing only thiophanate methyl (2.5%) did not provoke a sensitization response.[129] Many cases of apparent irritant contact dermatitis but few cases of allergic contact dermatitis to thiophanate methyl have been reported.

8.3.3 Thiocarbamates

The thiocarbamate group of fungicides structurally resembles the rubber accelerator disulfiram (also employed as the pharmaceutical Antabuse®), a common sensitizer present in the U.S. standard patch test series (Figure 8.12). Disulfiram differs from the fungicide thiram,[130] the prototype thiocarbamate compound, only in having an N-ethyl group in place of the N-methyl group present in disulfiram.

Thiram (Arasan®, Thylate®, Tersan 75®, Vancide Tm-95®, and other trade names, CAS# 137-26-8, MW 240.44, vp NA, log P 1.593 -1.73). Thiram, thirame, and TMTD are common names used for tetramethylthiuram, variously used as an animal repellent, fungicide, and seed protectant. Commonly treated crops include vegetables, fruit, grain, and ornamentals.[67]

California registration data indicate that a technical formulation of thiram caused no residual irritation at 72 h in the Draize test. However, a thiram case reported to the handler database was consistent with an irritant mechanism, involving an

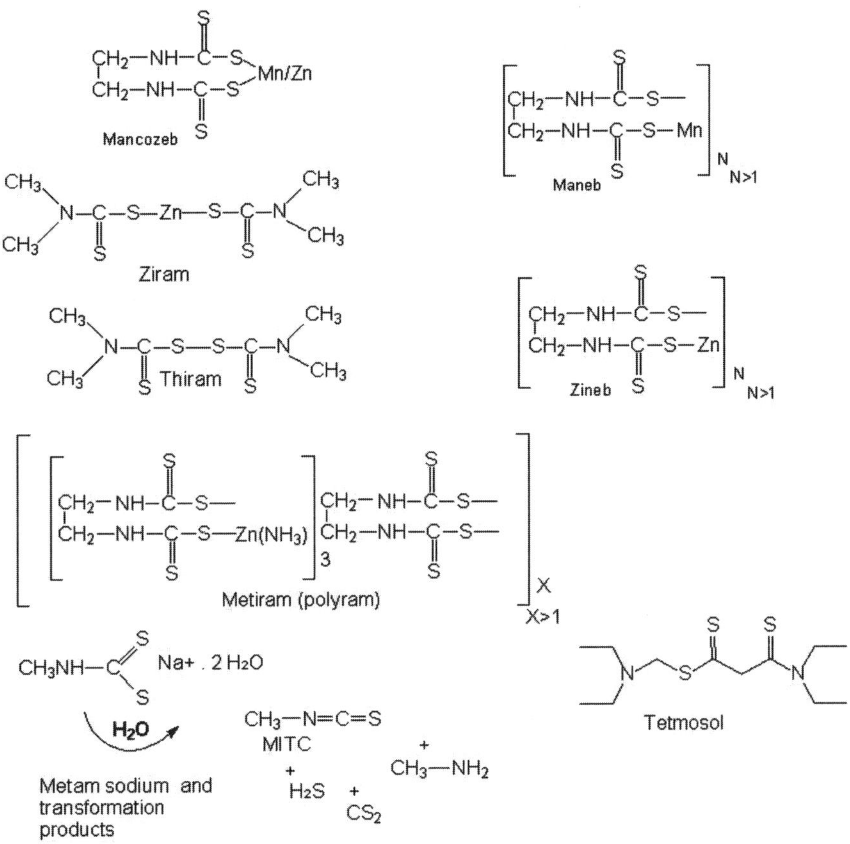

FIGURE 8.12 Thiocarbamate fungicides.

applicator treating seeds with thiram dust, who developed dermatitis around the respirator line (84-1488). Most thiram formulations are appropriately labeled as potential sensitizers, presumably based upon numerous case reports in humans. Some of the many fungicides and animal repellents that contain thiuram are described in a review by Fisher.[131]

Shultz and Hermann[132] described dermatitis in five dock laborers exposed to thiram in the course of unloading bananas treated with the fungicide. Shelley[133] described dermatitis from use of thiram as a golf course fungicide. Cronin[46] related dermatitis in a man who had applied thiram to his garden, and Rudzki and Napiorkowska[134] descirbed thiram-related allergic contact dermatitis in a Polish flower vendor. Riordan and Nahass[135] described occupational vitiligo in a firefighter as a consequence of allergic contact dermatitis associated with wearing protective clothing containing thiram and mercaptobenzothiazole.

Ziram (Fuklasin®, Zerlate®, Milam®, Vancide MZ-96®, and other trade names, CAS# 137-30-4, MW 305.8,vp 7.5E-08 mmHg, log P 1.086). The structure of ziram is very similar to that of thiram, but the compound contains a zinc atom between the two atoms of sulfur. It is also similar to zineb, which is a zinc/thiocarbamate polymer and to the manganese/thiocarbamate polymer, maneb. A 96% formulation of ziram used as an industrial biocide was labeled as both an irritant and as a sensitizer, but no animal data were available for review. Clinical reports regarding ziram are far less numerous than those involving maneb and zineb (see below), but include a positive patch test reaction in an agricultural worker in the series described by Lisi and colleagues.[117] Of two ziram associated cases reported to the handler database, one (84-518) involved accidental direct exposure and the other (83-298) failure to wear protective gloves. Both cases were consistent with an irritant mechanism, but neither was patch tested.

Maneb (Dithane M-22® and other trade names, CAS# 12427-38-2, MW 265.29, vp 7.5E-08 mmHg, log P NA) and Zineb (Lonacol®, Dithane Z-78®, CAS# 12122-67-7, MW 275.8, vp 7.5E-08 mmHg, log P <1.3). Maneb and zineb are used to treat many plant diseases. Maneb may be used in combination with other pesticides including lindane, hexachlorobenzene, captan, and zineb. Matsushita[136] tested maneb and zineb with the guinea pig maximization procedure and found both to be potent sensitizers with a high degree of mutual cross reactivity. Concentrations of 5% or more were found irritating upon intradermal injection.

Six patients with dermatitis from zineb were reported by Sarpa and Ippolito in 1959.[137] Cases of allergic contact dermatitis in workers spraying tobacco were reported in 1964 by Laborie.[138] In 1979, Nater[139] described two cases in office workers who had purchased maneb spray to care for the plants in their office, and a third in a 51-year-old woman who worked as an assistant in a flower shop. Patch testing was done at 1, 2, and 5%, but studies of cross reactions to zineb and other dithiocarbamates were not done.[139] The 1980 review by Cronin[46] described a case of dermatitis in a rose gardener who had a strong patch test reaction to maneb rose spray and also to 1% zineb. In 1982, Adams and Manchester[140] reported a case of allergic contact dermatitis to maneb in the wife of a residential gardener. Several severe episodes of dermatitis occurred before the cause was discovered. The patient's husband had stored a large bag of maneb next to the washing machine in the garage. Patch test to maneb 1% and thiram 1% in petrolatum were strongly positive, whereas testing with zineb was negative.[140] Peluso et al.[141] described a female agricultural worker with a two-month history of finger and hand eczema, which improved when she stopped work. Patch testings gave positive results to carbamate-mix 1% pet., bis-dithiocarbamate (maneb) and thiophthalimide (captan 0.1% pet., captafol 0.1% pet. and folpet 0.1% pet.) at 2 and 3 days (+/++).[141] Piraccini et al.[142] reported a case of allergic contact dermatitis from pesticide maneb in an agricultural worker with a seven-year history of recurrent dermatitis of the face and arms each summer. Patch testing was positive to maneb 1% pet. (D2+/D3++). The patient had no recurrence of dermatitis the following summer, after discontinuing maneb exposure.[142]

Cronin[46] described results of testing by members of the ICDRG with zineb 1% and maneb 1% in petrolatum on 655 eczematous patients. Three patients had positive reactions to zineb but none had clinical relevance and 35 patients had positive reactions to maneb. In only 1 case, involving hand dermatitis in a florist, did the reaction have some possible clinical relevance.[46] The frequency of reactions without obvious history of pesticide exposure suggests that sensitization resulted from exposure to thiocarbamate rubber accelerators.

International experience with reactions to zineb and maneb are discussed in Chapter 9. These include case reports, surveillance, and epidemiologic investigations from the U.K. and Ireland (Section 9.4.1), France, Belgium, and The Netherlands (Section 9.4.2), Austria, Germany, and Switzerland (Section 9.4.4), southern Europe (Section 9.4.5), Japan (Section 9.5.2), and Latin America (Section 9.8). Dermatitis associated with the manufacture of zineb[143] in eastern Europe proved to be due to the precursor compound ethylenediamine, rather than to the active ingredient.

Mancozeb (Dithane M-45®, Policar S®, Vondozeb Plus®, CAS# 8018-01-7, MW 265.2828 for thiocarbamate portion of molecule, vp 1.32E-10 mmHg, log P 1.33). Mancozeb is a compound similar in structure to maneb, but contains zinc coordinated with the manganese-thiocarbamate complex. The Draize irritation study reported to the registration database for the 80% formulation of mancozeb showed a minimal irritant reaction and a combined formulation of mancozeb and thiophanate methyl was a sensitizer in the Buehler assay.

Burry reported allergic contact dermatitis from mancozeb in South Australia in a worker treating barley and wheat seeds and in a farmer planting the seeds.[144] Patch testing was performed at 0.5% concentration and 10 controls were negative. Testing for cross reactivity to thiram was not done. Allergic contact dermatitis due to mancozeb in an agricultural worker has also been reported by Lisi and Carafinni.[145] This patient was also allergic to maneb. Allergic contact dermatitis associated with the manufacture of Rondo-M® fungicide (containing pyrifenox and mancozeb) and the insecticide alfacron (azamethiphos) has also been reported.[146] Testing with dilutions of the individual components of Rondo-M demonstrated a positive reaction to mancozeb and no reaction to pyrifenox. Testing with azamethiphos was also positive.

Photoallergic contact dermatitis from mancozeb has been reported in a patient presented with eczema on sun-exposed areas,[147] consistent with the absorption spectrum of mancozeb is in the UVA and UVB range. Photopatch test with mancozeb was positive while testing in 20 control cases was negative.

Tetmosol (Sarcocide B®, CAS# 95-05-6, MW 264.4614, vp NA, log P NA). Tetmosol (tetraethylthiuram monosulfide) is used to treat parasitic infestations in dogs and cats and is used as a scabicide and a fungicide. Cronin[46] has reported a 58-year-old woman seen at the St. John's Hospital dermatology service with chronic eczema whose dermatitis flared after treating her cat with tetmosol. Cross reaction with thiocarbamate rubber accelerators was suggested by aggravation associated with wearing rubber gloves. She had positive patch tests to tetmosol solution and to both 1% tetramethylthiuramdisulfide and 1% tetramethylthiuramdisulfide in petrolatum.

8.3.4 Copper Fungicides

Copper compounds are used as both fungicides and antimicrobial agents. Although formulations of copper compounds appear to be uniformly non-sensitizers in the Buehler test, based upon the California registration database,[148] copper is recognized as a human sensitizer in the public domain literature.[115,149-151] Most formulations of the copper fungicides are consequently labeled as potential sensitizers, despite the negative animal sensitization studies.[6] Technical formulations (with concentrations as indicated) of elemental copper (84.8% — category III), copper naphthenate (80% — category II), and cupric oxide (85.7% — category III) are irritant, but not corrosive to rabbit skin in the Draize test (see Chapter 1 and Chapter 6 for details of scoring procedure).[152]

Technical formulations of copper ammonium carbonate (24%), copper hydroxide (90%), copper oxide (97%), copper sulfate (50 and 99%), and copper triethanolamine complex (3.5% as elemental copper) cause only minimal irritation in animal tests. Noteworthy cases reported to the California illness registry for copper compounds appeared to result principally from irritation.

87-1724 — A wood worker was painting a copper naphthenate wood preservative to the cut end of wood and developed a chemical burn on his arms.

94- 561 — A worker applied copper hydroxide to walnuts using a high volume sprayer. His face began to itch and burn a few hours after he finished the application. The affected area was on the unprotected portion of his face.

90-2588 — A worker was wearing gloves, goggles, and coveralls; during mixing, some powder got inside the glove causing irritation to his right forearm.

93-1839 — A worker applied copper sulfate granules to canal water using a piece of equipment he called the "sandblaster," which is designed to air blast the application material on canal water. Dust landed on his neck and chest, but he did not wash until his skin began itching and burning. Examination identified large erythematous patches on the neck and chest.

8.3.5 Amine, Nitro, and Nitrile Chlorinated Benzene Derivatives

The amine, nitro, and nitrile benzene derivatives (Figure 8.13) range from intermediate (metalaxyl) to high values of log P (PCNB) and have in common their ability to act as electrophiles (the IUNIQ QSAR structural element described in Table 1.4).

Metalaxyl (Ridomil®, Apron®, Subdue®, and other trade names, CAS# 57837-19-1, MW 279.34, vp 5.63E-06 mmHg, log P 1.5-1.75) Metalaxyl is an analog of the amino acid phenylalanine that inhibits fungal protein synthesis by interference with the synthesis of ribosomal RNA.[93] In the Draize test, a 50% wettable powder formulation of metalaxyl causes mild irritation present at 24 h, but not persisting beyond 48 h in any of the animals tested. A formulation containing 25% active ingredient was negative in the guinea pig maximization protocol and a flowable formulation containing 33% metalaxyl was negative in the Buehler test. A case of direct accidental facial exposure (91-787) to metalaxyl and iprodione (Chipco 2619®) resulted in facial swelling, slight conjunctival injection, photophobia, a painful crusted lesion at base of right nares, and periorbital and diffuse facial edema accompanied by slight erythema. Iprodione (see below) does not even cause transient irritation in Draize test, but has been associated with occasional episodes of dermatitis, usually involving mixed exposures. An applicator case involving isolated, direct accidental exposure to a dilute solution of metalaxyl was reported in 1985, in that

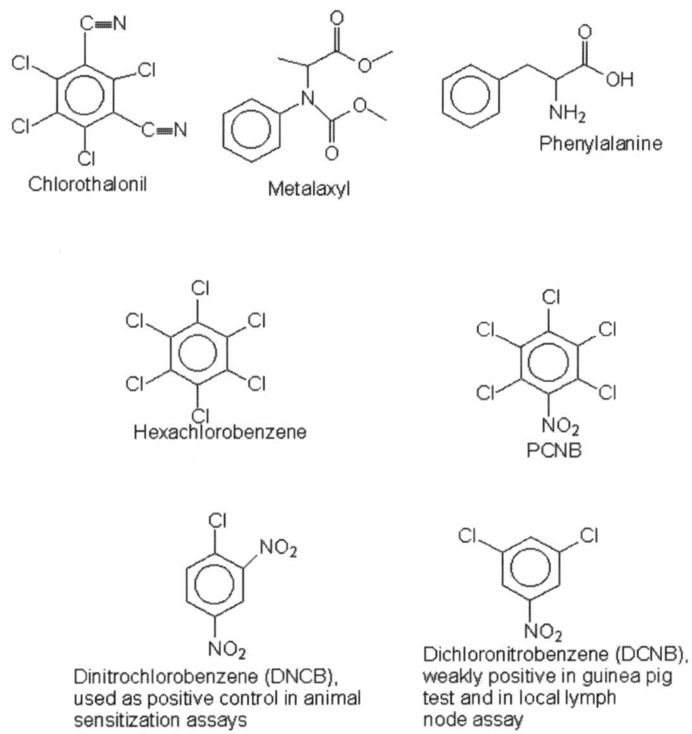

FIGURE 8.13 Amine, nitro, and nitrile fungicides.

instance provoking a subjective sensation of facial irritation, but no visible erythema on the skin (85-1312).

Chlorothalonil (Daconil®, Bravo®, CAS# 1897-45-6, MW 265.92, vp 5.85E-07 mmHg, log P 2.88). Chlorothalonil is a broad-spectrum fungicide approved for use on vegetables, fruits, flowers, and trees. Stable upon exposure to ultraviolet light, it has a soil half-life of about two months.[1]

Experimental studies with New Zealand white rabbits demonstrated that chlorothalonil 0.1% in acetone is a moderate cutaneous irritant; 0.1% chlorothalonil in petrolatum is much less irritating; and 0.1% chlorothalonil in saline is non-irritating to the skin of New Zealand white rabbits.[60] Contact dermatitis has been reported in vegetable growers, woodworkers, and flower growers.[153] Patch testing is performed with 0.01% chlorothalonil in petrolatum. This is a marginally irritant concentration of chlorothalonil. Chlorothalonil is a strong cutaneous irritant at the concentrations used in spraying. Penagos et al.[154] recently reported chlorothalonil as a possible cause of erythema dyschromicum perstans (ashy dermatitis). Positive patch test reactions to chlorothalonil (0.001% in acetone) were observed in 34 of 39 banana farm workers presented with erythema dyschromicum perstans-like dermatitis. Biopsies from all patients were compatible with a chronic pigmented dermatitis or erythema dyschromicum perstans-like dermatitis.

Chlorothalonil is used as a wood preservative in northern Europe. Johnson has reported an epidemic of contact dermatitis in a Norwegian woodenware factory where 14 out of the 20 workers had work-related skin complaints, and 7 workers had contact dermatitis. Bach and Pederson reported contact dermatitis to tetrachloroisophthalonitrile in a cabinet maker.[155] Spindeldreier and Deichmann[156] have also reported three cases of contact dermatitis to tetrachloroisophthalonitrile.

Fatal toxic epidermal necrolysis (TEN) has been attributed to chlorothalonil. A 30-year-old navy pilot had played 81 holes of golf in the week prior to developing TEN. One golf course had been sprayed with chlorothalonil. The authors state that special photographic techniques using ultraviolet light demonstrated chlorothalonil on the deceased's golf clubs, balls, and shoes.[157] In a case reported by Dannaker,[158] chlorothalonil was associated with contact urticaria in a nursery worker and an anaphylactoid reaction on provocation testing with 1% aqueous chlorothalonil.

Dinitrochlorobenzene-DNCB (Nalpac 8343 Microorganism Control Chemical®, CAS# 97-00-7, MW 202.5537, **vp NA, log P NA).** DNCB is a potent sensitizer used as a positive control in most animal sensitization assays (see Chapter 1). Although DNCB is no longer registered in the U.S., during its use as an algicide it was associated with cases of allergic contact dermatitis.[159] Zimmerman et al.[160] emphasized that this potent sensitizer should only be used in completely enclosed systems not allowing skin contact. The dual nitro groups are apparently key elements in its allergenic potency, because the structurally related compound dichloronitrobenzene is a very weak sensitizer in both the guinea pig maximization test and in the local lymph node assay.[161]

Pentachloronitrobenzene, PCNB (Terraclor®, Tritisan®, Quintozene®,[162] CAS# 82-68-8, MW 295.3365, **vp** 4.9505E-05 **mmHg, log P** 4.46**).** PCNB is used as a soil fungicide and seed-treatment chemical. It is especially useful for Brussels sprouts, broccoli, and artichokes. Cronin[46] described a 46-year-old man packing pesticide powders who developed dermatitis of his arms, legs, forehead, trunk, and scrotum. Patch testing to PCNB, 1% in petrolatum, was positive. The patient's shoes were heavily contaminated with PCNB dust.[46] O'Malley and Rodriguez observed positive PCNB patch tests in a series of tests on nursery workers, but could not establish from work records or the study questionnaire data a definite history of prior exposure. The observed reactions could have resulted from active sensitization induced by the provocation procedure.[48,49,163]

Hexachlorobenzene (Amatin®, Anticarie®, Bunt-cure®, and other trade names, CAS# 118-74-1, MW 284.8, vp 1.09E-05 mmHg, log P 3.93-6.42). Hexachlorobenzene is a seed protectant used on wheat. Contact with hexachlorobenzene may cause slight skin irritation. Hexachlorobenzene produced an epidemic of severe porphyria cutanea tarda in Turkey from 1955-1959. The disease occurred almost exclusively in persons who admitted eating hexachlorobenzene wheat distributed for seed rather than for food. Many infants died from exposure to hexachlorobenzene in this epidemic.[164] This episode is discussed further in Sector 9.5.4.2.

Dinobuton (Dessin®, Dinofen® CAS# 973-21-7, MW 326.3054, vp NA, log P NA). Dinobuton is a fungicide and acaricide previously registered to control the mites of deciduous fruits, citrus, cotton, cucumbers, and other vegetables and for controlling powdery mildew on apples, cucumbers, hops, and other crops. In 1974, Wahlberg[165] reported yellow staining of the hair and nails in workers with allergic contact dermatitis from dinobuton who were manufacturing the compound in a

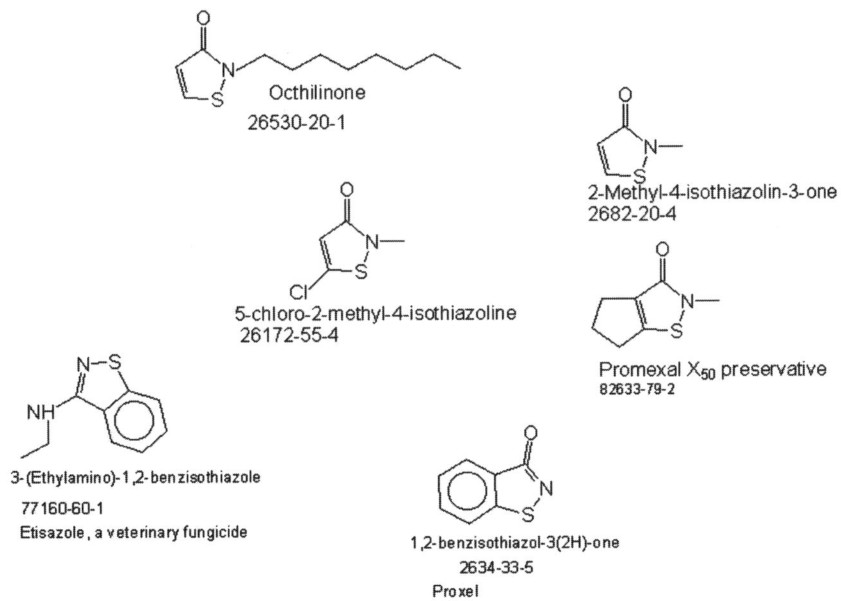

FIGURE 8.14 Isothiazolin (Kathon) compounds.

factory in northern Sweden. Wahlberg noted that dinobuton did not appear to be a strong primary irritant; patch testing with 40% dinobuton caused spontaneous flare reactions (sensitization) in 2 workers 9-10 days after application. The chemical relationship to picric acid was considered significant.

8.3.6 Isothiazolin Compounds (Figure 8.14)

Isothiazolin fungicide/preservative compounds are structurally related to the thiadizine compounds dazomet and fosthiozate formulated as solid pro-fumigants (see Section 8.4.2.2). The compounds discussed below include 1,2-benzisothiazolin-3-one (Proxel®, 82633-79-2), octhilinone (Kathon 893®), methylisothiazolinone (2038 Acticide®, 2682-20-4), 2-methyl-4,5-trimethylene-4-isothiazolin-3-one (MTI), Kathon FP-15(2039), and methylchloroisothiazolinone (2038; 26172-55-4). In addition to these compounds, allergic reactions to the chemically related veterinary antifungal agent and preservative etisazol (1,2 benzisothiazolin-3-1 and 3-ethylamino-1,2 benzisothiazolin hydrochloride-etisazol) have also been reported.[166] A complete review of the literature on these compounds is beyond the scope of this discussion and only illustrative reports for each compound are cited. In QSAR terms, the isothiazolins contain both an electrophilic sulfur atom (IUNIQ) and a conjugated olefin moiety (ICONJ).

1,2-Benzisothiazolin-3-one (Proxel®, Nipacide®, and other trade names, CAS# 2634-33-5, MW 151.1826**, vp NA, log P NA).** 1,2-Benzisothianolin-3-one is formulated in 9-20% concentration for use as an industrial fungicide. In the Draize test, it caused severe immediate irritation that diminished but did not disappear during 14 days of post-application observation. It has been identified as a sensitizer in the local lymph node assay and also in the repeated insult patch test.[167] It has been reported as a human sensitizer under a variety of use conditions.[168,169]

2-Methyl-4-isothiazolin-3-one (CAS# 2682-20-4, MW MW 115.1496**, vp NA, log P NA) and 5-Chloro-2-methyl-4-isothiazilone-3-one (Acticide®, CAS# 26172-55-4, MW** 149.5947**, vp NA, log P NA).** Like other isothiazolin compounds, 5-chloro-2-methyl-4-isothiazilone-3-one is used as an industrial biocide and fungicide. In the Draize test of a 10.3% formulation mixed with a 4% concentration of the related isothiazolin compound 2-methyl-4-isothiazolin-3-one, a single animal was tested. Severe immediate irritation was noted, persisting with minimal change over 14 days of observation. Allergy associated with the use of 5-chloro-2-methyl-4-isothiazilone-3-one as a cosmetic preservative has been documented,[170] and also with the use of 2-methyl-4-isothiazolin-3-one in the paper industry.[171] Irritant and allergic dermatitis associated with a mixture of the two compounds is illustrated in Figure 8.15.

Octhilinone (Kathon 893® and other brand names, CAS# 26530-20-1, MW MW 213.3372**, vp NA, log P NA).** Octhilinone is a thiazolone biocide used in cooling water plants, cutting oils, cosmetics, shampoos, and as a leather preservative. Although no dermal irritation study was available for review, it was identified as an irritant on the product label.[172] It is a sensitizer in the guinea pig maximization test.[173] Two cases of occupational allergic contact dermatitis to octhilinone used as a fungicide for surface paint of roof sheets were reported by Thormann.[174] Both workers had severe allergic contact dermatitis. Mathias[173] also reported a case of poison oak-like contact allergy in a worker manufacturing latex paint containing octhilinone as a preservative.

8.3.7 Other Heterocyclic and Multicyclic Fungicides

Many of the heterocyclic fungicides are benzimidazole derivatives related to the compounds used as antifungal drugs such as ketoconazole (Figure 8.16).

Flusilazole (Nustar®, CAS# 85509-19-9, MW 315.3974**, vp NA, log P NA)**. Flusilazole is used for control of ascomycetes and other fungi on cereals, fruits, and vegetables, and is formulated as dry granules and as an emulsifiable concentrate. Registration data indicate that a 60.7% formulation of the compound produces prolonged, but low grade irritation in the Draize test. A 20% formulation was a non-sensitizer in the Buehler assay. There were no cases associated with flusilazole in the California handler database.

Imazalil (Clinaform®, CAS# 3554-44-0, MW 297.18, vp 1.19E-06, log P 3.82). Imazalil is a systemic fungicide active against benzimidazole resistant strains of fungi, formulated as an emulsifiable concentrate, water soluble powder, and soluble liquid. In the Draize test it was a non-irritant and the Buehler assay was negative on the 13.5% formulation. Nevertheless, the technical material is labeled as a potential sensitizer. There were no cases associated with imazalil in the handler database.

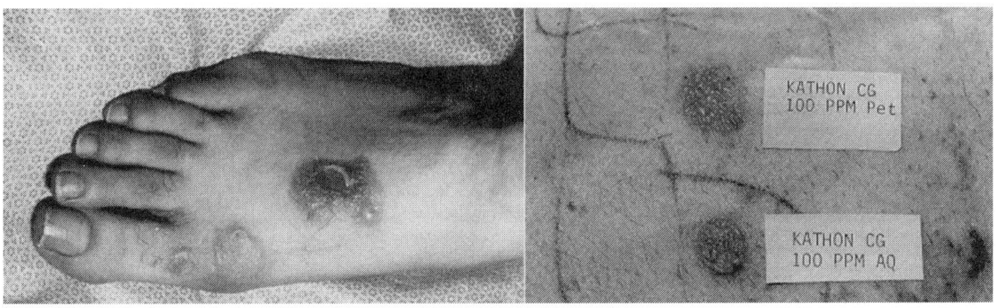

FIGURE 8.15 Irritant and allergic dermatitis associated with a mixture of 2-methyl-4-isothiazolin-3-one and 5-chloro-2-methyl-4-isothiazolin-3-one (Kathon CG). (Reprinted from Primka and Taylor, *American Journal of Contact Dermatitis*, 1997. With permission from W.B. Saunders.)

Iprodione (Rovral®, CAS#36734-19-7, MW 330.17, vp 9.75E-08, log P 3.1). Iprodione is both a contact and systemic fungicide used on a broad spectrum of crops. In the Draize test, the 50% formulation produced no recorded erythema or edema in 72 h of post-application observation and a 41.6% flowable formulation was a non-sensitizer in the Buehler assay. A cluster of acute dermal irritation reported to the California illness registry associated with the use of iprodione, *B. Thuringiensis*, and sulfur as a dipping solution for transplanted strawberries. Of 20 workers reporting dermatitis of the hands and forearms during the illness investigation, only 5 had sought treatment. Whether the reported cases were caused by one or the combination of the associated chemicals could not be determined.

Thiabendazole (Gustafson Rival Flowable Sipt. Fung. and other trade names, CAS# 148-79-8, MW 201.2, vp 3.99E-09 mmHg, log P NA). Thiabendazole is an agricultural fungicide with systemic activity against fusarium, penicillium, and other molds, formulated as a dust, wettable powder, and flowable dust. In the

Draize test, the 98.5% formulation caused slight erythema that cleared by 48 h. There were no cases associated with thiabendazole in the handler database. There were no sensitization studies done on thiabendazole per se, but the Buehler test was negative on a mixture of thiabendazole, captan, and pentachloronitrobenzene.

Myclobutanil (Ridomil®, Rally®, CAS# 88671-89-0, MW 288.78, vp 1.6E-06 mmHg, log P 2.94). Myclobutanil is a systemic fungicide used on grapes and tree fruits. The dermal sensitization data reviewed were not sufficient to determine whether myclobutanil is a skin sensitizer. It caused no erythema or edema in the Draize test, but an accidental direct exposure involving myclobutanil and an adjuvant occluded against the skin without decontamination did prove capable of causing dermatitis (92-1379).

Triadimefon (Bayleton®, CAS# 43121-43-3, MW 293.76, vp 1.5E-08 mmHg, log P 3.18). Triadimefon is a systemic fungicide for control of powdery mildew on cereals, deciduous fruit, and grapes. In the Draize test, a 50% formulation caused transient erythema that cleared by 48 h post application. In the Buehler assay, the 95% technical material is a sensitizer, but the same test on a more dilute formulation was negative. Case 82-1412, from the handler database, involved an applicator accidentally exposed to wind-blown spray suspected of an allergic reaction to triadimefon, but no provocation test was conducted.

Anilazine (Dyrene®, Nutro Turf Fungicide® CAS# 101-05-3, MW 275.54, vp 2.48E-08 mmHg, log P 3.02). Anilazine is a foliar and turf fungicide that has not been registered in California since 1993, but is still registered elsewhere in the U.S. Data on dermal irritation and sensitization from animal studies were not available for review, but the product has been reported as a human sensitizer in field workers by Schuman.[175, 176] The outbreak, reported in 1980, affected 14 of 26 migrant workers employed at a single farm growing tomatoes and strawberries in Tennessee. Although 11 pesticides were used on the farm, the only positive patch reaction was to anilazine (Dyrene), and 6 of the 7 workers patch tested demonstrated positive responses indicating the potency of anilazine as a senstizer. Seven additional workers also had dermatitis but were not patch tested. If all of the workers with dermatitis were sensitized, the total prevalence of sensitization would yave been 46%. A more recent case, affecting a midwestern lawn care applicator, was reported by Mathias.[177] A case in the California illness registry (92-732) involved a worker who developed dermatitis on exposed skin areas while dumping wettable powder into a mix tank prior to an aerial application. (The exposure was of a type commonly avoided by use of water soluble bags, which were apparently not available.) Although provocation testing was not done, examination demonstrated a dermatitis on the face, neck, and arms judged likely to be allergic in nature.

Bupirimate (Nimrod®, CAS# 58694-46-5, MW 316.4176, **vp NA, log P NA).** Bupirimate is a systemic fungicide especially effective against powdery mildews, but is not currently registered in the U.S. McFadden et al.[178] reported a case of allergic contact dermatitis from bupirimate in 1993. The patient noted upper respiratory irritation as well as a rash on the area of his face unprotected by goggles and and intense edema and erythema in the unprotected skin on his wrists. Patch testing to bupirimate (0.01, 0.1, and 1.0% pet.) gave positive reactions (?+, +, and ++, at day 4 post application) but was negative in one control.

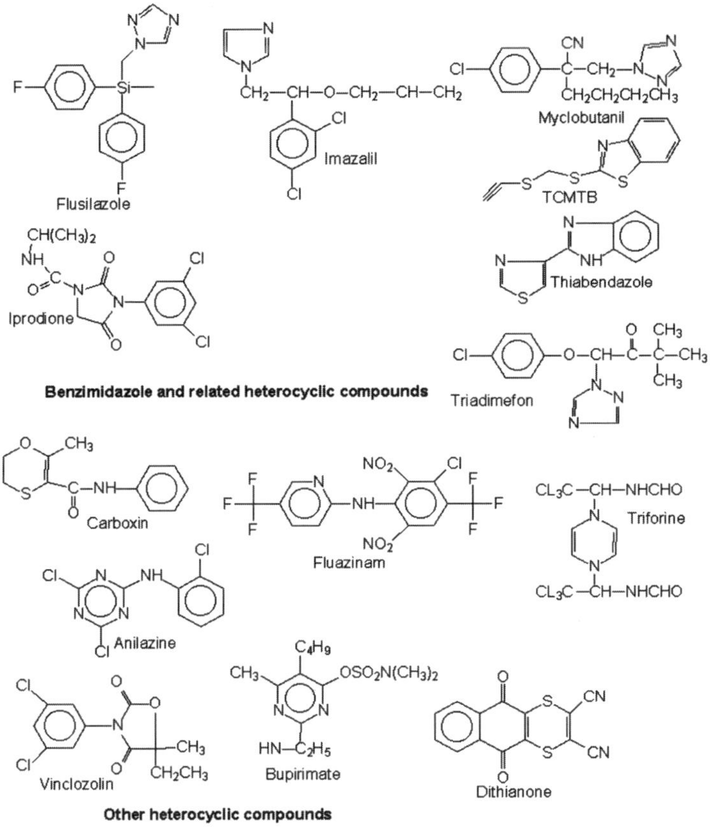

FIGURE 8.16 Heterocyclic fungicides.

Carboxin (Vitavax®, Kemikar®, and other brand names, CAS# 5234-68-4, MW 235.31, vp 1.9E-07 mmHg, log P 2.17). Carboxin is a systemic fungicide and seed protectant. It is a non-irritant in the Draize test.[179] A mixture containing carboxin, PCNB, and metalaxyl was a non-sensitizer in the Buehler assay.[180] There were no cases associated with its carboxin listed in the California handler database.

Chloroneb (NuFlo®, CAS# 2675-77-6, MW 207.1, vp 0.003 mmHg, log P NA). Chloroneb is a fungicide used for control of seedling diseases in beans, cotton, and soybeans. A formulation containing 30% chloroneb and 3.5% metalaxyl was a category III irritant in the Draize test. The same mixture was a non-sensitizer in the Buehler assay. There were no cases associated with chloroneb in the handler database.

In QSAR terms, the reactive element in chloroneb is a methoxy group (Figure 8.16), akin to the easily metabolized phenol element (IAROH) described in Table 1.4.

Triforine (Funginex®, CAS# 26644-46-2, MW 435, vp 1.95E-07 mmHg, log P 2.2). Triforine is a fungicide for *Ascochyta* blight. In a mass examination of chrysanthemum growers using triforine, 17% reacted to 0.02% aqueous triforine. Cross sensitization between triforine and dichlorvos (DDVP) was also shown.[181]

Fluazinam (Shirlan®, CAS# 79622-59-6, MW 465.0954, **vp NA, log P NA).** Fluazinam is a newly introduced broad-spectrum antifungal. Allergic contact dermatitis from fluazinam was documented in 1995 by van Ginkel and Sabapathy.[182] Positive patch tests to both the active ingredient and the commercial formulation, Shirlan, were demonstrated among 7 of 9 farmers who developed dermatitis about 1–4 weeks after initial contact. The irritant effect of fluazinam was ruled out by negative patch testings with fluazinam 2.5, 0.5, 0.1, 0.02, and 0.04% pet. in 10 consecutive control volunteers.

Methylene bis thiocyanate (MBTC); (Premium Woodlife Voc®, CAS# 6317-18-6, MW, vp NA log P NA). MBTC is a reactive compound, used as a bactericide both in water cooling systems and in pulp and paper mill operations. It is also used as a fungicide in wood preservation products. A 10% formulation was corrosive in the Draize test. The only sensitization study reviewed involved a mixture of methylene bis (thiocyanate)(0.2%), chlorpyrifos (0.1%), and TCMTB (0.2%) and did not show any evidence of an allergic response. Nevertheless, a 10% formulation used as an industrial preservative was labeled as a sensitizer.

TCMTB (Busan® and other brand names, CAS# 21564-17-0, MW, vp NA, log P NA). TCMTB (thiocyanomethylthiobenzothiazole) is a compound used as both a fungicide and anti-bacterial agent. It is contained in both wood preservation products and in industrial preservatives. A 30% formulation was corrosive in the Draize test. A 4-h exposure provoked intense erythema and moderate edema within 24 h that became gradually more intense throughout the 7 days of observation. A dilute wood preservative formulation (containing <1% TCMTB) was a non-sensitizer in the Buehler assay, but a 30% formulation of TCMTB was labeled as a sensitizer. There were no cases associated with its use included in the handler database.

Vinclozolin (Ronilan®, CAS# 50471-44-8, MW 286.1, vp 1.2E-07 mmHg, log P 3). Vinclozolin is a fungicide used for control of *Botrytis, Sclerotinia*, and *Monilia* on grapes, strawberries, and other crops. It is a minimal irritant in the Draize test. It is labeled as a sensitizer, but no animal test data were available for review. The handler database did not contain any cases associated with the use of vinclozolin.

8.3.8 Fungicides of Miscellaneous Structure

Dithianone (Thynon®, Delan®, and other trade names, CAS# 3347-22-6, MW, vp NA, log P NA). Dithianone is a broad-spectrum fungicide that affects fungal enzymes by reacting with thiol groups and inhibiting cellular respiration.[93] In QSAR terms, its principal reactive element is a conjugated olefin (ICONS) moiety (Figure 8.16). It is not registered in California, but a registration for a 95% technical formulation is listed as a registration by the U.S. EPA. Calnan[183] reported dermatitis in a female horticulturalist working in a fruit orchard, with a history of allergy to *Primula* and several other ornamental plants, who developed dermatitis on the hands, arm, face, and ears while pruning dithianone treated fruit trees, and cleaning spray equipment used to apply dithianone. Patch tests were negative to a patch series including organochlorines, organophosphates, and pyrethrum,[183] but positive to 1% dithianone in petrolatum.

Fosetyl-aluminium (Aliette®, CAS# 39148-24-8, MW 354.1, vp 9.75E-08 mmHg, log P 2.28). Fosetyl-aluminum is a systemic fungicide and bactericide, active against oomycetes, alternaria, and penicillium on avocado, strawberries, and other crops. It is formulated as a water dispersable granule, liquid wettable powder, and liquid injectable. In the Draize test, the 80% formulation caused slight irritation at 1 h and 24 h post application, with complete resolution by 48 h. It was a non-sensitizer in the Buehler assay and there were no cases associated with its use recorded in the California handler database.

Sulfur (Britz Brand Dusting Sulfur®, Cooke Sulfur Dust®, Yellow Jacket Wettable Dusting Sulfur II®, CAS# 7704-34-9, MW 32.064, vp 3.96 x 10^{-6} mmHg, log P NA). Elemental sulfur exists in various crystalline forms, with rings containing 6, 8, 10, or 12 sulfur atoms, or repeating concatenated chains.[184] There are 80 active formulations of elemental sulfur currently registered for use as a fungicide or acaricide on fruit, vegetable, and ornamental crops. The large number of cases associated with elemental sulfur in California agriculture is striking and would seem to imply that sulfur is a potent skin irritant. Scattered cases have also been reported in applicators and field workers in the state of Washington. However, a 90% formulation caused only minimal irritation in the Draize test. In range finding for the guinea pig maximization test, Matsushita[185] has shown that a 25% concentration of wettable powder produces a 2+ irritant reaction on intradermal injection; the same study identified 5% elemental sulfur as a moderately strong experimental allergen. However, registration studies show that elemental sulfur is not a sensitizer in the Buehler assay.[186]

Two case reports implicate elemental sulfur as a human contact allergen. Schneider[187] reported two cases of contact allergy in patients who used medications containing elemental sulfur to treat superficial fungal dermatoses. Both patients had positive patch test reactions to 5% elemental sulfur in various vehicles. A control series was not reported. Wilkinson[188] reported the case of a professional gardener with a previous history of atopic eczema who developed an eczematous eruption involving the elbow flexures and the right hand. He had a positive patch test reaction to 5% sulfur in petrolatum, but a control series was not reported. Gregorczyk and Swieboda[189] described 15 cases of desquamative dermatitis among 425 Polish sulfur miners in which irritant dermatitis due to elemental sulfur may have played a part. Reactions to sulfur were also the most common positive patch tests in a recent study of California nursery workers.[49] These disparate pieces of information suggest that active irritants (e.g., sulfuric acid by oxidation or hydrogen sulfide by reduction) or allergens (e.g., sulfites) may be formed from elemental sulfur by oxidation or reduction (hydrogen sulfide). Cases in the handler database alternately suggest irritant (82-1211, 83-1252) and allergic (87-174) mechanisms.

The possibility of irritant or allergic reactions to transformation products of elemental sulfur is also suggested by reports of contact dermatitis associated with bathing in green sulfur springs. Lesions characteristically occurred 24 h after bathing, suggesting a possible delayed contact reaction, but had positive sulfur provocation tests.[190] Vena et al.[191] reported frank allergic reactions to an oxidized form of sulfur known as sulfites.[191]

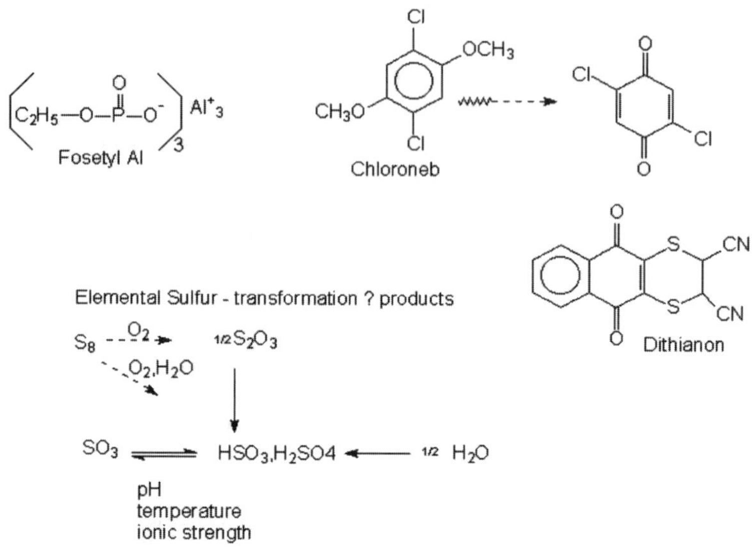

FIGURE 8.17 Miscellaneous fungicidal compounds.

8.4 FUMIGANTS

The classes of compounds used as fumigants (Figure 8.18) overlap somewhat with compounds used as fungicides and as anti-microbial agents. For example, the fumigant dazomet is an analog of the thiazoline group of antimicrobial compounds. Likewise, the phenol compounds include pentachlorophenol and numerous related phenol anti-microbial phenols.

The divergent compounds used as fumigants and biocides also have in common inherent chemical reactivity and a wide spectrum of biological effects, including severe dermal and systemic toxicity. The irritant effects of these compounds are particularly notorious when occluded against the skin.

8.4.1 Halogenated Fumigants

The halogenated fumigants contain the reactive IX structural element (Table 1.4). Some compounds are capable of causing sensitization, as well as acute irritation.

Dichloropropene (Telone ®, CAS# 542-75-6, MW 110, vp 27.75278 mmHg, log P 2.06). Dichloropropene is a fumigant used principally for soil sterilization in the production of root vegetables and other crops. Active product labels, some containing as much as 30% chloropicrin, identify it as a skin corrosive.[192] No animal sensitization studies were available for review, but the products registered in California are labeled as sensitizers. A case report by Nater and Gooskens[193] describes an allergic reaction to dichloropropene, and suggests that its allergenicity was based on its unsaturated carbon-carbon bond, and reactivity akin to that of epichlorohydrin. Gryzwa and Rudzki[194] describe dermatitis due to an adhesive containing the related compound dichloropropane. Cases in the California handler database appear most

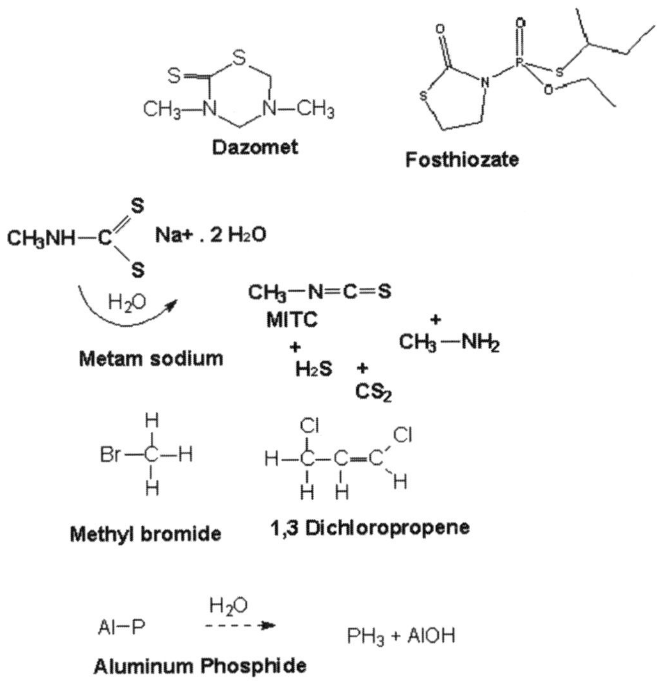

FIGURE 8.18 Fumigant compounds.

consistent with a simple irritant mechanism, as illustrated in the case described below and in a case investigated in 1979 by California Department of Pesticide Regulation enforcement staff (Figure 8.19).

88-2091 — An employee was unclogging one of the tubes through which the Telone (dichloropropene) is injected on his spray rig when material spilled on his foot. He wore leather rather than chemically-resistant boots. The next day he developed blisters and swelling of the foot.

Methyl bromide (Brom-o-gas®, Preplant Soil Fumigant® and other trade names, CAS# 74-83-9, MW 94.94,vp 1702.67 mmHg, log P NA). Although methyl bromide is scheduled to be phased out as an ozone depleting chemical by 2005,[195] current use is still extensive in agricultural and structural pest control. For example, 15,663,832 pounds of the material were reported used in California during 1997, the year for which the most recent data are available.[196]

An industrial accident in 1942 revealed the capacity of methyl bromide to cause skin burns and irritation.[197] The numerous cases in the handler database tended to involve prolonged occlusion of methyl bromide against the skin, as illustrated in the case described below:

> 82-32 — Methyl bromide leaked into an employee's boots when he changed cylinders on the tractor. Feet became inflamed, ulcerated, and infected over several days before employer noticed problem and sent the worker for medical treatment.

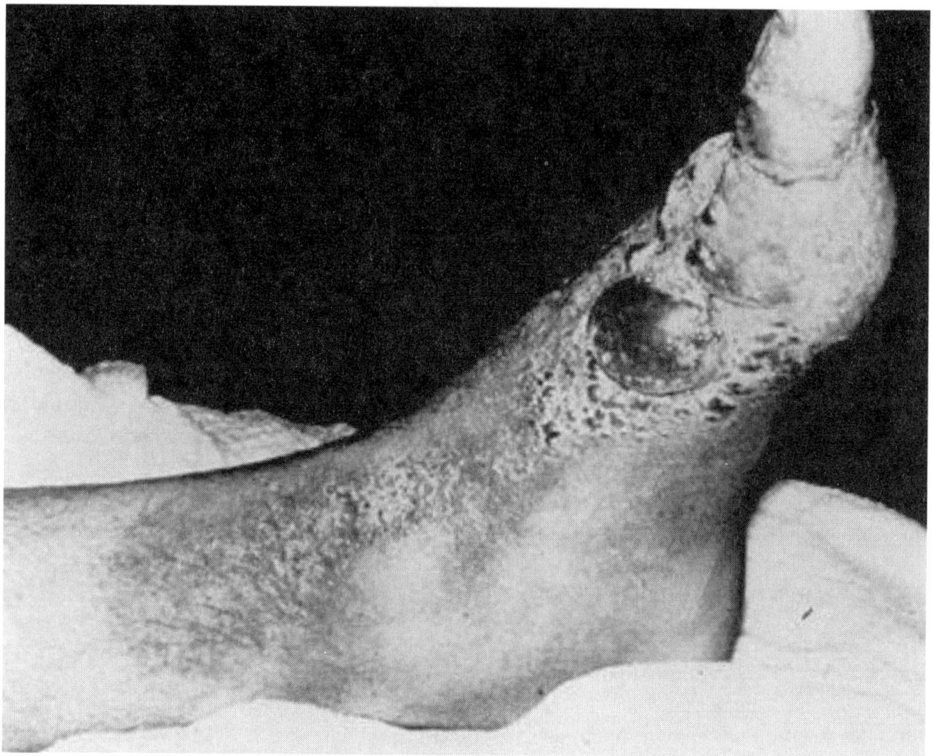

FIGURE 8.19 Severe burn from applying dichloropropene in Kern County worker wearing "huarache" sandals, 1979 CDPR archive, case history enforcement field staff. (From O'Malley, M.A., *State-of-the-Art Reviews in Occupational Medicine*, 1997. With permission from Hanley and Belfus.)

Skin reactions to airborne methyl bromide are not usually felt to be significant. However, in 1988 Hezeman-Boer[118] described dermal irritation from methyl bromide in six Dutch workers whose skin exposure occurred while wearing respiratory protection inside a fumigated building. Within a few hours all patients developed skin lesions, consisting of sharply demarcated erythema with multiple vesicles and large bullae (Figure 8.20). There was a striking predisposition for parts of the skin that were relatively moist or subject to mechanical pressure, such as axillae, groin, and abdomen. Plasma bromide levels after exposure strongly suggested percutaneous absorption of methyl bromide. Based upon monitoring performed on a separate group of workers wearing SCBA inside a treated structure in California, gas concentrations associated inside a fumigated structure can range up to 9500 ppm of methyl bromide.[198]

Methyl bromide can readily be absorbed through the skin[199] and has been responsible for a number of serious occupational poisonings in California following either inhalation or respiratory exposure.[200] Non-occupational poisonings are usually associated with accidental inhalation exposures.[201,202]

Ethylene Oxide (T-Gas®, CAS# 75-21-8, MW 44.053, vp[203] 1469 mmgHg, log P NA). Ethylene oxide (ETO) is a reactive, electrophilic (the IUNIQ element described in Table 1.4) gas, used as a fumigant in food processing and as a biocide in hospital sterilization equipment. Pesticidal uses are fairly minor compared to other fumigants: California pesticide use data for 1997 included reports of 1471 pounds in 29 separate applications.[196] No dermal irritation or sensitization study was available for review. However, irritant dermatitis has been associated with ETO production and with residual ETO in sterilized medical equipment. Case reports have also described allergic contact dermatitis and immediate hypersensitivity in hospital workers handling rubber products sterilized with ETO.[204,205]

A case described in the California handler database involved a chemical burn following accidental direct contact with ethylene oxide gas:

> 87-2720 — A hospital worker stuck her hand in an ethylene oxide sterilizer, before it had been aerated, and got a chemical burn on the exposed area of her hand. She was not wearing gloves.

8.4.2 Solid Pro-Fumigants

A number of compounds used as fumigants are formulated as solids or liquid concentrates and liberate fumigant gases upon decomposition. These include the phosphides, thiazoline fumigants, and the carbamate compound metam sodium.

8.4.2.1 Phosphide Fumigants

These include aluminum, zinc, and magnesium phosphide. All of these compounds are registered for use as commodity fumigants and as rodenticides and are formulated as solid tablets that release phosphine gas on contact with air and water. Aluminum phosphide, in practice, is used far more extensively than the other two compounds as a commodity fumigant. Zinc phosphide is used primarily as a rodenticide. Magnesium phosphide is used relatively little, being inherently more reactive and more difficult to handle. (Fumigant handlers in the Sacramento area have commented that magnesium phosphide is notoriously prone to "flash" when a container is opened. For this reason they prefer to handle aluminum phosphide for commodity fumigations.)

Aluminum phosphide[206] (Phostoxin®, CAS# 20859-73-8, MW 57.95, vp negligible as salt, log P NA). Animal sensitization and irritation data were not reviewed. Dermatitis cases have been reported in California mill workers with potential exposure to aluminum phosphide treated rice, but none were definitely linked to fumigant exposure or contact with partially spent residue.[207] However, dermatitis in the context of systemic phosphine toxicity has been reported by Akoboshi.[208]

8.4.2.2 Thiadizine Compounds

Dazomet (Basamid®, CAS# 533-74-4, MW 162.3, vp 2.77528E-06 mmHg, log P 0.15). Dazomet is a fumigant that releases irritant amines and mercaptans as it breaks down in soil. It is used principally as a pre-plant treatment for ornamental beds and nurseries. The technical material causes minimal irritation in the Draize test, but most of the dazomet products reviewed (20-24% active ingredients) were labeled as skin irritants. No dermal sensitization studies were available for review, but most of the dazomet products were labeled as skin sensitizers, presumably based upon human case reports. A dermatitis case reported to the California handler database was possibly related to dazomet accumulated under protective clothing.

90-2448 — A worker was applying a dusty granular form of dazomet, wearing cotton coveralls, respirator, goggles, hat, rubber gloves, and boots. He was sweating and developed a rash at the belt-line as well as front of legs, abdomen, and arms.

Cases of allergic and irritant dermatitis have also been reported from England, France, and Australia,[209-211] as described in Chapter 9.

Fosthiozate (IKI-1145® and other brand names, CAS# 98886-44-3, MW 283.33986**, vp NA, log P NA).** Fosthiozate is a compound with a heterocyclic 5 member nitrogen and sulfur ring resembling the isothiazolin compounds (Section 8.3.7). It has had limited use as a soil nematocide for root vegetable crops, but has not been registered in the U.S. since 1995.[6] California registration data show that fosthiozate caused only minimal irritation in the Draize test, but the technical material was a sensitizer in the guinea pig maximization test. There were no cases associated with its use in the California handler database.

8.4.2.3 Thiocarbamate Compounds

Metam sodium (Vapam®, CAS# 137-42-8, MW 129.18, vp negligible as salt, log P <1). The soil nematocide metam sodium is also effective against weeds and soil fungi. Upon reaction with water, metam produces a complex mixture of irritant compounds, including methylisothiocyanate (MITC), carbon disulfide, hydrogen sulfide, and methyl amine. In the Draize test submitted for registration of a 42% formulation, test animals demonstrated erythema, edema, and necrosis at 24 h, persisting until completion of the test at 14 days post application. The Buehler assay reported for the same formulation did not include irritation controls. There was so much irritation present in both the induction (10% dilution) and challenge phase of the test (1% dilution) that it was difficult to distinguish between irritant and allergic reaction. However, challenge response was slightly greater at 24 h than during the corresponding phase of the induction, despite the lower dilution tested, providing some evidence that sensitization occurred.

Cases of contact dermatitis associated with metam sodium have been reported in several jurisdictions around the world. In Germany, the cases stemmed from use of metam in the production of root vegetables.[212-214] These included cases of probable allergic contact dermatitis with positive reactions to patches with 0.1% metam sodium in aqueous solutions. The authors reported that patches containing 0.5% metam invariably resulted in a bullous irritant reaction.

Cases of dermatitis were also reported in workers wading into the Sacramento river to clean up metam sodium spilled into the river following a train derailment near Dunsmuir, California, in July 1991. There were 34 cases of dermatitis in the handlers reported to the California illness registry between 1982-1995. Between 1991-1995, one case of dermatitis associated with this compound was reported in a Washington applicator.[215] Typical cases from the California registry include

84-20 — An applicator sprayed Vapam® while wearing leather boots; the material soaked through the boots and caused irritation on the skin of his feet.
85-1697 — After making a pre-plant application with metam sodium to a cabbage field, a fumigator noticed pressure remaining in his hose; the material splashed on his face, body, and inside his shoes. He washed his face and body, immediately, but did not change his shoes. He developed burning irritant dermatitis of the feet.

8.5 HERBICIDES

Depending upon their chemical structure, herbicides can function by virtue of their general chemical reactivity or by such plant-specific mechanisms as inhibition of electron transport in photosynthesis. Many herbicides in the former category can cause systemic illnesses as well as eye and skin reactions in humans and other mammalian species. Compounds in the second group often have limited systemic toxicity. Selectivity in function as herbicides may depend upon details of plant metabolism — e.g., the presence of a glutathion pathway for detoxification — or the ability of certain herbicides to interfere with normal functioning of growth hormones, called auxins, for certain families of plants. Herbicide formulations are diverse, including liquids, soluble concentrates, granules and non-granular solids, and a variety of spreading agents or adjuvants to enhance the contact between the active herbicide and the target plant. In some instances, the spreading agents or adjuvants and preservatives present in a formulation may have a greater capacity to irritate or sensitize the skin than the active herbicidal component. [93,96,216,217]

8.5.1 BIPYRIDYLS

Paraquat (Gramoxone®, CAS# 1910-42-5, MW 257.2, vp negligible, log P 4.5) and diquat (Real-Kill Vegetation Killer®, CAS# 85-00-7, MW 344.06, vp negligible, log P 4.6). The bipyridyl compounds, paraquat and diquat (Figure 8.21), have reactive ammonium ions in their core structures. Superoxides generated by the bipyridyl compounds damage plant cell membranes and cytoplasm, and are especially active during photosynthesis. Both paraquat and diquat are contact herbicides and desiccants used on cotton, potatoes, and alfalfa, clover, and soybeans. They have no biological activity once absorbed onto clay soils or organic matter. However, some degradation on leaf surfaces results from photodecomposition.[93]

Exposures occur by direct contact of the skin with liquid concentrates during mixing and/or loading, and inhalation or skin contact with spray mist during application. As indicated above, both compounds adhere tightly to soil and plant material, and exposure upon contact with residue is less likely to cause either topical or systemic effects. Although cases of systemic poisoning following skin contact have been reported,[218-222] the double ammonium ions present in the compounds limit dermal absorption. The vast majority of systemic poisonings occur following ingestion.[223,224] Details of notable cases of systemic poisoning following dermal contact with paraquat indicate that the cases occurred under somewhat unusual circumstances. Garnier et al.[218] reported two cases of paraquat poisoning that resulted from

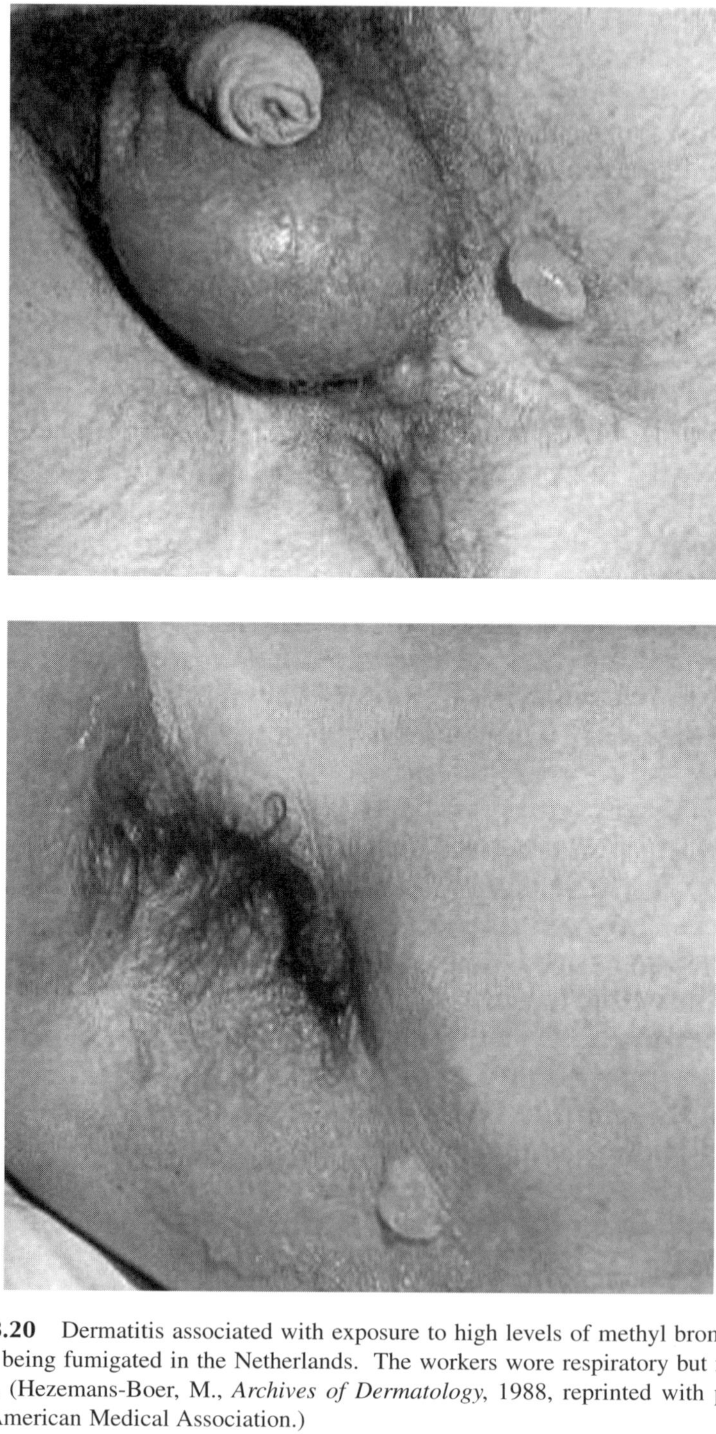

FIGURE 8.20 Dermatitis associated with exposure to high levels of methyl bromide inside a building being fumigated in the Netherlands. The workers wore respiratory but not dermal protection. (Hezemans-Boer, M., *Archives of Dermatology*, 1988, reprinted with permission from the American Medical Association.)

skin absorption. One patient died from respiratory failure 26 days after applying paraquat on his whole body as a treatment for scabies.[218] The other, with a previous history of psoriasis, developed extensive generalized pustular erythroderma and a moderate, transitory renal and respiratory impairment 13 days after exposure. Tungsanga et al.[221] reported a man who developed renal and respiratory failure and hepatic damage after mistakenly washing his perineum with a concentrated paraquat solution. The patient later recovered. In contrast, in the case described by Newhouse et al.,[222] fatal dermal absorption occurred through small dermal abrasions during routine spraying of paraquat in a fruit orchard.[22]

Acute irritant contact dermatitis and chemical burns are the most notable reported dermal effects of paraquat.[215,225] Discoloration of the fingernail, nail deformity, onycholysis, and loss of the nail have been reported in workers whose fingernails came into contact with paraquat sprays and concentrates.[226-229] A case with periungual eczematous dermatitis with striking nail lesions due to paraquat was reported by Botella et al.[230] (Figure 8.22). Dermatitis associated with exposure to high levels of methyl bromide inside a building being fumigated in the Netherlands. The workers wore respiratory but not dermal protection. (Hezemans-Boer, M., Archives of Dermatology, 1988, reprinted with permission from the American Medical Association.) Patch testing to paraquat 0.001 and 0.1% in water was negative.[230] A necrotic ulcer of the scrotum following direct contact with paraquat solution was reported by Sharvill.[231] Several chronic skin conditions have also been associated with exposure to paraquat. Workers in factories where paraquat is manufactured have been reported to have an increased risk of occupationally induced keratoses, Bowen's disease, and squamous cell carcinoma of the skin. Paraquat itself is probably not the etiologic agent for these premalignant and malignant skin lesions.[232,233] George[234] also reported a case from Nigeria in which exposure to paraquat coincided with a case of leukoderma and began to resolve with cessation of contact with the material.

8.5.2 Amino and Nitro Substituted Benzenes (Figure 8.23)

Nitralin (Planavin®, CAS# 4726-14-1, MW, vp NA, log P NA). Nitralin was registered as a selective, pre-emergent herbicide, but has not been actively registered for use in the U.S. since 1981.[235] Allergic contact dermatitis to nitralin and its precursor 4-chloro3,5-dinitrophenylmethyl sulfone was reported in a man working in a factory producing this herbicide. He also showed cross sensitivity to DNCB.[236]

Nitrofen (TOK®, CAS# 1836-75-5, MW, vp 284.1, vp 7.95E-06 mmHg, log P NA). Nitrofen was used as pre- or post-emergence herbicide on a wide variety of crops, but has not been actively registered since 1984.[237] Nitrofen was shown to produce liver toxicity in rats, dogs, and mice,[238] pancreatic tumors in female rats, and in teratology studies was associated with the occurrence of diaphragmatic hernias, heart anomalies, hydronephrosis, and ophthalmic anomalies.[239]

Propanil (WHAM! ®, CAS# 709-98-8, MW 218.08, vp 9.08E-07 mmHg, log P 2.29). The amide herbicide propanil (N- (3,4-dichlorophenyl) propionamide) is a post-emergence herbicide, still actively registered in the U.S., used against broadleaf weeds and grasses in rice. Morse et al.[240] in a survey of occupationally acquired disease in workers at a pesticide plant found that 17 (61%) of 28 production workers with exposure to dichloroaniline and propanil had chloracne. The chloracnegenic potential of propanil was demonstrated with rabbit ear tests.[240] Contaminants related

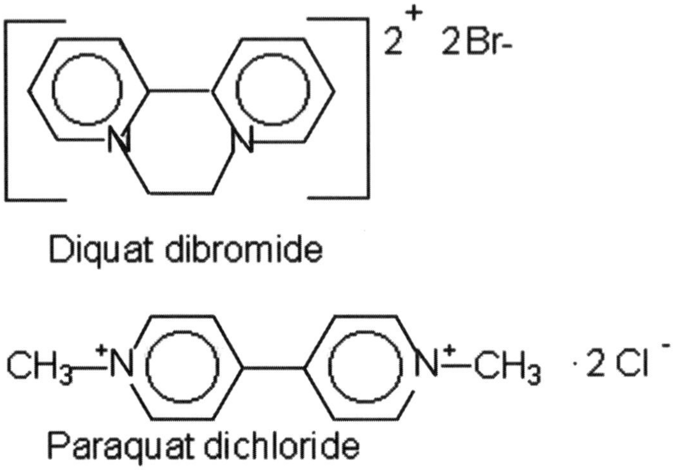

FIGURE 8.21 Bipyridyl compounds.

to 3,3',4,4'-tetrachloroazobenzene are responsible for the chloracnegenic characteristic of propanil. The planar structure produced by the nitrogen-nitrogen double bond and two attached chlorinated benzene rings is probably sterically similar to the ring structures in other chloracnegenic materials such as PCBs and TCDD.[241]

8.5.3 Phenoxy Compounds

Salts of the chlorophenoxyacetic acids (2,4-D, 2,4,5-T, MCPA) are readily absorbed by the roots of broadleaf plants and the numerous ester derivatives are readily absorbed by the foliage and are translocated to the plant growth zones and function as hormonal growth inhibitors (auxins).[93] The selective effect of plants on broadleaf plants has made them popular for use on grain crops and they are used extensively on turf as well. Structures for selected compounds are shown in Figure 8.22.

2,4-D (Weed and Feed products®, CAS# 94-75-7, MW 221.04, vp 1.43E-07 mmHg, log P NA). 2,4-dichlorophenoxyacetic acid (2,4-D) is manufactured by direct chlorination of phenol and 2,4,5-T from a different process — by hydroxylation of 1,2,4,5-tetrachlorobenzene, yielding 2,4,5-trichlorophenol.[242] This difference in the manufacturing process is critical because of the notorious tendency of the latter reaction to get out of control, leading to occasional major process accidents, contamination of the process stream with 2,3,7,8 tetrachlorodibenzodioxin (TCDD), and associated outbreaks of chloracne (see Table 8.4).[243-266] TCDD is a well-documented animal carcinogen, and is also associated with an increase in human cancer mortality.[267-269] Because of the ongoing problems with its production, 2,4,5-T was removed from the U.S. market between 1981 and 1985.[6]

In some U.S. plants manufacturing 2,4,5-trichlorophenol and 2,4,5-T, a high incidence of porphyria cutanea tarda (PCT), characterized by thinning and blistering

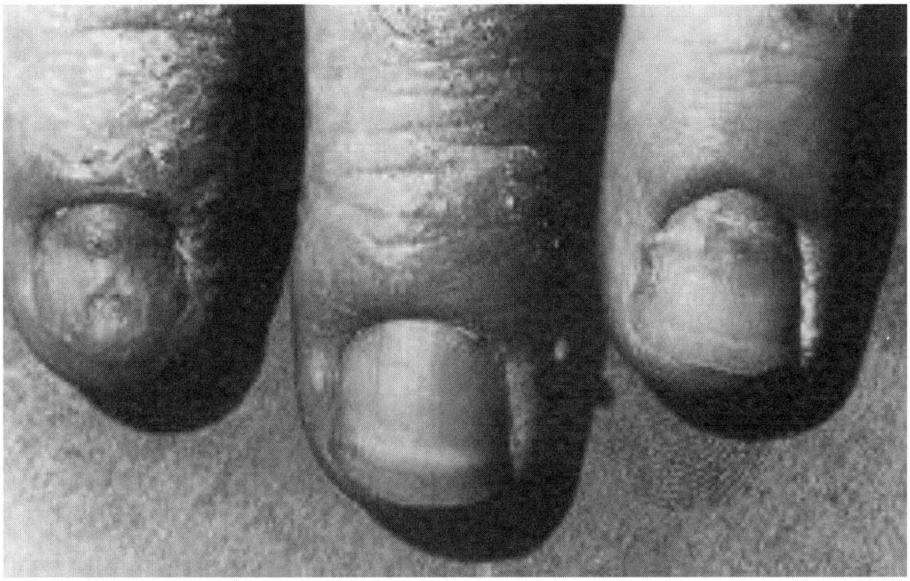

FIGURE 8.22 Nail injury from paraquat. (Case reported by Botella et al., Contact dermatitis to paraquat, *Contact Dermatitis*, 12, 123,1985. Reprinted with permission from *Contact Dermatitis*, Munskgaard.)

of the skin in areas exposed to the sun, was observed.[270] The condition occurred principally in a plant that also manufactured hexachlorobenzene, which may have been the actual cause of the condition.[271] Consultants employed by the company operating the plant also proposed the theory that the high incidence of porphyria resulted from the intake of alcohol or the use of prescription medication.[272] Porphyria associated with exposure to 2,4,5-T has also been described in a case report by McConnell,[273] and in a small cohort from a 2,4,5-T manufacturing plant from Czechoslovakia.[274,275]

In the Draize test, a 97% technical formulation of 2,4-D did not provoke any observed irritation. The 2,4-D products registered in California are labeled as sensitizers,[276] presumably based upon the historical data on allergenicity in humans. The Buehler sensitization tests conducted for several end-use formulations were completely negative. An animal sensitization for a 56.7% solution of amine salts of 2,4-D showed a borderline sensitization reaction, with mild erythema at 24 h (scored as 0.2 erythema, increasing to 0.4 erythema at 48 h).[277] The 2,4-D derivative MCA (2-methyl-4-chlorophenoxyacetic acid, CAS# 94-74-6) caused skin erythema in rabbits in a 3-week dermal toxicity study when applied at 0.5 g/kg/day, 1.0 g/kg/day, and 2 g/kg/day.[278]

In clinical reports, contact dermatitis, rather than chloracne or porphyria, has been the principal skin problem associated with 2,4-D and its related compounds. Among 30 Indian farmers with chronic contact dermatitis of 2.5 years average duration

TABLE 8.4
Process Accidents and Process Control Problems in the Manufacture of 2,4,5-trichlorphenol

1949 — West Virginia — nitro process explosion, contamination of reaction site, and outbreak of chloracne in the plant workforce involved in cleanup

1953 — BASF, Germany — process explosion, and outbreak of chloracne in plant work force

1963 — The Netherlands — process reaction

1964-1965 — Midland, Michigan — increased production of 2,4,5-T for Operation Ranch Hand associated with higher levels of TCDD in reaction process stream and outbreak of chloracne in production work force

1969-1975 — Austria — ongoing outbreak of chloracne associated with 2,4,5-T production

1960's — New Jersey — ongoing problems with chloracne in 2,4,5-trichlorophenol production process

1973 — Coalite works, England — process accident followed by outbreak of chloracne in work force

1975 — Chloracne following phenoxy herbicide exposure at the state AgroChemical Association Manufacturing Complex Khimprom in Ufa, Russia

1976 — Seveso, Italy — process explosion in 2,4,5-trichlorophenol production process
1988 — Germany — process explosion and chloracne outbreak

evaluated by Sharma, 3 proved to have contact sensitivity to 2,4-D (1% in petrolatum). No reactions were seen among 20 healthy controls without dermatitis.[36] Five cases of irritant bullous dermatitis also affected German forestry workers applying a mixture of 2,4,5-T and 2,4- D. Two cases had positive patch test reactions to 0.4% concentration of the product mixture in diesel oil. The first demonstrated +++ reaction at 48 and 72 h. A weakly positive reaction was observed in the second case.[279]

8.5.4 Acetanilides, Acetamides, and Amides

The acetanilide compounds, alachlor, allidochlor, metolachlor, and acetochlor (Figure 8.23), have a high degree of structural similarity and all act by inhibition of

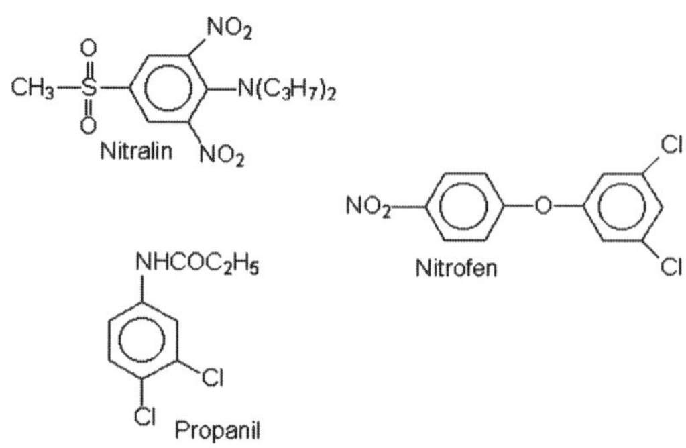

FIGURE 8.23 Amino and nitro substituted herbicides.

plant protein synthesis and root elongation. Corn plants possess relative tolerance to the compounds because of a glutathione detoxification pathway.[93] The compounds have a similar range of functional uses as selective pre-emergent herbicides for annual grasses on corn, peas, beans, and other field and nursery crops. California registration data indicate that an 85.1% formulation of metolachlor was a category III irritant in the Draize rabbit dermal irritation assay test and a 45% formulation of alachlor caused acute corrosion (category I).[280] No animal sensitization studies were available on alachlor, but all of the product labels reviewed indicate that the material is a sensitizer, because of human case reports in the public literature (see below). A Buehler sensitization upon a mixture of metolachlor and flumetsulam was interpreted as causing sensitization because of increasing erythema in challenge with 50% dilution of test formulation compared to induction phase tests conducted with 100% dilution.

Alachlor (Lasso®, CAS# 15972-60-8, MW 269.77, vp 3.1E-05 mmHg, log P 2.64-2.92). Alachlor is predominantly a maize herbicide. In the Draize test submitted for registration, a 45% solution provoked marked irritation that persisted until the conclusion of the study (72 h after the initiation of the study), and alachlor was graded as corrosive based on the available data. Registered products are also identified as sensitizers, based upon reports in the published scientific literature.

Clinical observations included a report by Iden et al.[281] who identified five patients with allergic contact dermatitis associated with exposure to alachlor and noted cross reactions to propachlor in three of the five. Won etal.[282] documented allergic contact dermatitis from alacholor in a 32-year-old woman farmer who accidentally contaminated her clothing while working with alachlor and paraquat. Patch tests showed positive reactions for both 0.1 and 0.2% alachlor. Of ten control patch tests with alachor performed, one had a late reaction thought to be active sensitization.

Metolachlor (Dual®, Pennant®, and other trade names, CAS# 51218-45-2, MW, vp NA, log P NA). Propachlor is used for selective pre-emergent control of grasses and broadleaf weeds in corn, sorghum, beans, lentils, peas, peanuts, peppers, tomatoes, nursery ornamentals, and other crops. In the Draize test, an 85.1% formulation produced grade 1-2 edema and grade 1 erythema at 24 h, gradually resolving until the skin was completely clear at 9 days post application. A Buehler sensitization study reported for registration of a formulation of metolachlor (79.9%) and flumetsulam (2.1%) did not include quantitative scoring information, but described an increase in erythema score from induction to challenge, suggesting sensitization occurred.

Propachlor (Bexton®, Ramrod®, CAS# 1918-16-7, MW 211.69, vp 0.000225 mmHg, log P 1.62-2.30). Propachlor is a pre-emergent herbicide used for control of grasses and some broadleaf weeds in corn and sorghum. In 1979, Schubert[283] reported the case of a 29-year-old farm worker who developed hand dermatitis after handling propachlor for 8 days. The reaction healed after he stopped using the material. Testing with 1% propachlor was positive after 48 h and no reactions were observed in 5 control subjects. Heise[284] reviewed the results of animal sensitization and irritation tests on propachlor, together with human case reports describing sensitization, and recommended that skin contact with the material be minimized.

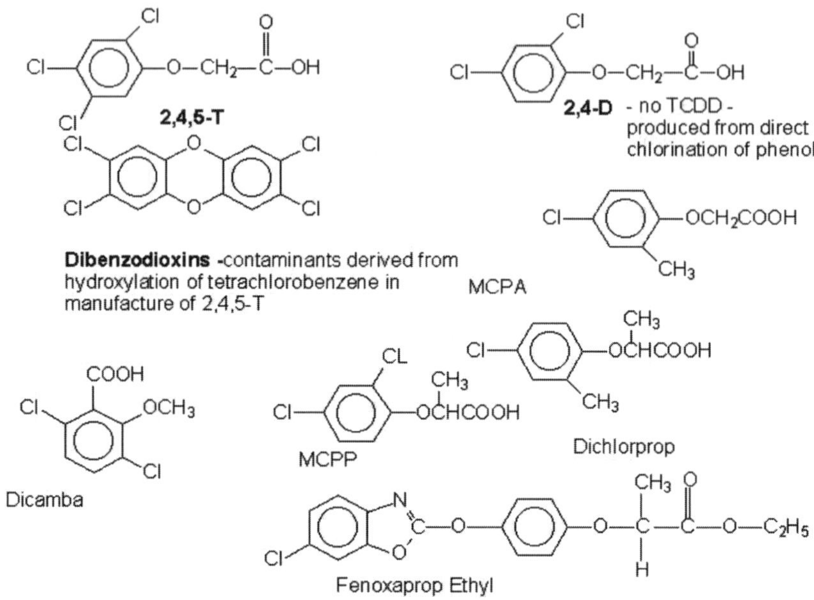

FIGURE 8.24 Phenoxy herbicide compounds.

Allidochlor (Randox® CAS# 93-71-0, MW, vp NA, log P NA). Allidochlor is an older acetamide compound, used as a selective pre-emergent, that has not been registered in the U.S. since 1985.[285] Spencer reported three cases of allergic contact dermatitis associated with handling allidochlor, with cross sensitization to propachlor demonstrable upon patch testing.[286]

8.5.5 Triazines

The triazine herbicides — atrazine, cyanazine, propazine, and simazine — structurally resemble the triazine fungicide compound, anilazine (Dyrene®), which is a potent sensitizer. Anilazine, nevertheless, has a second aromatic ring (see Figure 8.24) not present in the herbicidal triazines that has an undetermined effect on its ability to cause sensitization. The recently developed triazine biocide Irgarol-1051 (2-*tert*-Butylamino-4-cyclopropylamino-6-methylthio-1,3,5-triazine) is also a sensitizer and has reactive substituents, e.g., an amino-cyclopropane ring attached to the triazine nucleus.[287]

Atrazine (Aatrex®, CAS#1912-24-9, MW 215.68, vp 2.85E-07 mmHg, log P 2.34-2.8). Atrazine is a pre-emergent herbicide used for control of broadleaf and grassy weeds in field crops such as corn and sorghum, for forestry plantings, and as a non-selective pre-emergent for roadside right-of-way and other non-crop applications. Over 64 million acres of cropland were treated with atrazine in the U.S. in 1990. Severe contact dermatitis to atrazine has been reported in a farmer with a history of dermatitis caused by propachlor. This patient had a positive patch test reaction to 1:1000 dilution of a commercial atrazine formulation.[288]

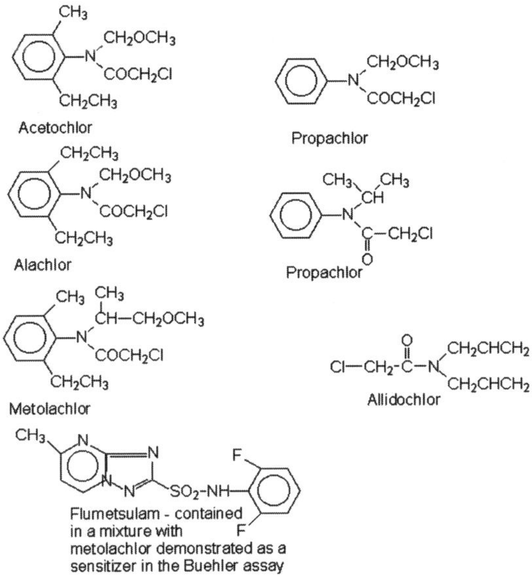

FIGURE 8.25 Acetanilide herbicide compounds.

In 1972, 124 cases of contact dermatitis were reported among workers manufacturing the triazine herbicides propazine and simazine. Cases varied in duration from 3-4 days to 7-10 days and from slight erythema to frankly bullous eruptions. It was unclear whether the eruptions were attributable primarily to herbicides manufactured or to process intermediates.[289] Other groups of formualtion workers studied by Catenacci et al.[290] did not suffer from similar dermatologic complaints.

8.5.6 Other Heterocyclic Ring Compounds (Figure 8.25)

Amitrole (aminotriazole, Cytrol®, Weedazol®, CAS# 61-82-5, MW 84.04, vp 4.43E-07 mmHg, log P NA). Amitrole is a systemic herbicide used for control of weeds on roadside rights of ways and uncultivated non-agriculture areas. It inhibits chlorophyll formation and regrowth from buds.[291] English et al.[292] reported allergic contact dermatitis caused by amitrole. There are no active registrations of the material in U.S. at this time.

Generalized cutaneous scleroderma accompanied by esophageal involvement and muscle symptoms in a sprayer handling the herbicides amitrole, bromacil, and diuron, has been reported. The authors believed there was an occupational association because of the extent of reported chemical exposure.[293] Suzuki and Ohkido[294] also described a case of contact dermatitis in a 36-year-old worker employed in a factory making aminotriazole as well as the poultry feed additive aminoguanidine-hydrochloride. Upon provocation testing, he had positive reactions to the chemical intermediates — hydrazine-sulfate and hydrated hydrazine — at concentrations from 0.0005 to 5.000%, with erythema and papules or vesicles in addition to infiltration and also an allergic reaction. The reaction to aminotriazole was considered irritant rather than allergic.[294] That is consistent with the occasional cases of apparent irritant dermatitis reported among California applicators.

87-258 — While spraying weeds a worker got spray on his clothes. He later developed a rash on his chest and abdomen. He wore a short sleeve shirt, long pants, shoes, and a pair of heavy-duty work gloves, as required by the label.

86-1528 — A man developed severe dermatitis, pityriasis rubra pilaris, after he was inadvertently sprayed by walking under a boom during an application of atrazine.

Triclopyr (Garlon®, CAS# 69633-04-1, MW 256.48, vp 1.26E-06 mmHg, log P NA). Triclopyr is a selective systemic herbicide absorbed by foliage and roots and translocated throughout the plant. Susceptible species of broadleaf weeds treat the compound as a growth hormone or auxin, in a fashion akin to the sterically similar phenoxy herbicide compounds. Grasses are unaffected at normal application rates.[291] In the Draize test, a 61.6% formulation of triclopyr caused transient erythema (grade 1-2) immediately after the 4 h application, gradually resolving over 7 days.[295] In the Buehler test for a 44% formulation of a triclopyr triethylamine salt, erythema or edema was observed at 24 and 48 h post application in the challenge phase. The formulation was graded as a non-sensitizer. Several formulations are "overlabeled" as sensitizers, despite the absence of positive animal sensitization studies or evidence of sensitization in humans. Triclopyr cases reported to the California illness registry demonstrate transient irritation, comparable to that observed in the Draize assay described above.

88-2832 — The worker was spraying weeds along the roadside when a big rig drove by causing a shift in the wind direction. The spray blew back in her face and produced mild facial erythema.

92- 445 — While using a backpack sprayer, the hose of the tank failed, spilling material onto the worker's buttocks and lower back. It soaked through his coveralls causing a rash on the exposed area. He returned to the maintenance yard and showered. He never the less developed a burning, itching sensation, rash on buttocks, and a small patch on his face.

8.5.7 Short Chain Carboxylic Acid Derivatives (Figure 8.26)

Glyphosate isopropylamine salt (Roundup®, CAS# 38641-94-0, MW 169.1, vp 7.5E-08 mmHg, log P 1.6**).** Glyphosate isopropylamine salt (Roundup®, CAS# 38641), commonly formulated with a surfactant or spreading agent to enhance leaf surface contact, inhibits 5-enolpyruvylshikimate-3-phosphate-synthase (EPSPS) and functions by preventing synthesis of the aromatic amino acids, tyrosine and phenylalanine, needed for manufacture of proteins.[296] Mammals obtain aromatic amino acids from dietary sources, so that glyphosate has very low systemic toxicity for humans. Glyphosate is largely non-selective in its effect on plant biochemistry so that it has a broad range of action similar to that of paraquat, although its onset-of-action is markedly slower. Ivy-like plants with thick, waxy leaf surfaces are relatively resistant to glyphosate because they do not permit translocation of salts or polar compounds. This limitation can be overcome by adding a detergent to the mixture

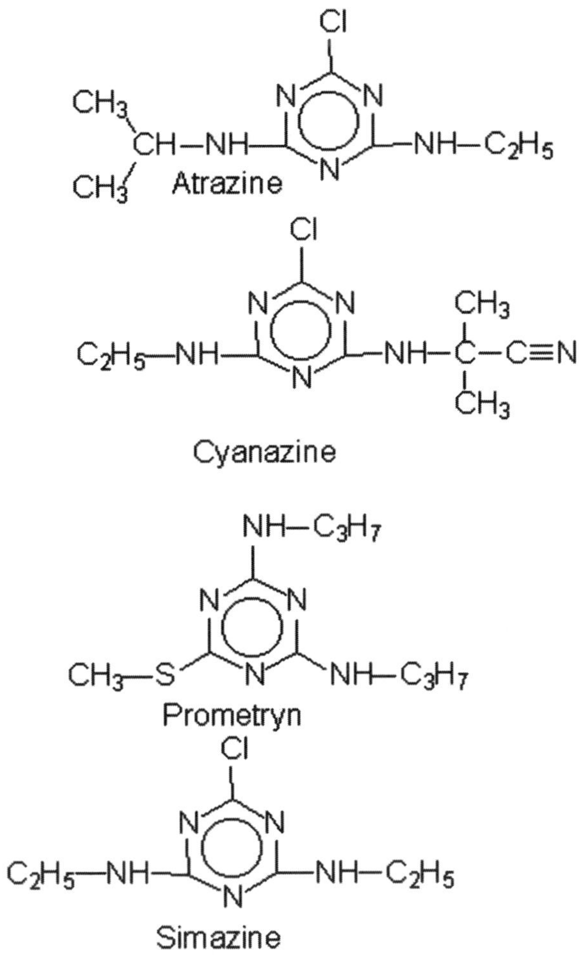

FIGURE 8.26 Triazine herbicides.

for spraying. Glyphosate is used on an extremely wide scale in the U.S. for both agricultural and non-agricultural weed control, but paraquat is still preferred in many areas of the world because of its lower cost.[297]

In 1986, contact dermatitis attributed to glyphosate was reported in 33 workers mixing or loading this herbicide in California.[298] Maibach[299] extensively investigated glyphosate and found it to be a non-sensitizer in the Draize Repeat Insult Patch Test and less irritating to the skin than household detergent and a commercial household cleaner in the population of volunteers tested. Hindson and Diffey[300] reported a phototoxic reaction to a standard formulation of glyphosate that proved upon evaluation to be due to the preservative 1,2-benzisothiazoline-3-one rather than to the active ingredient.[301] Smith and Oehme,[302] in investigating the overall toxicity of glyphosate, concluded there is little to no danger to man or the environment when used correctly. Penagos[303] reported 5 patients with occupational contact dermatitis who had positive patch tests to glyphosate (10% water) in a review of 847 workers at a banana plantation in Panama. The reactions were considered probably relevant to occupational contact with the compound.[303]

Dalapon (Dowpon®, CAS# 75-99-0, MW 164.95, vp 7.5E-06 mmHg, log P NA). Dalapon is an herbicide and growth regulator that selectively controls annual and perennial grasses by interfering with the production of pantothenic acid after being absorbed by leaves and roots of treated plants. It is not currently registered for use in the U.S. It is used in the Panamanian banana industry and Penagos[304] reported 2 cases of allergic contact dermatitis caused by dalapon in 1995.[304]

8.5.8 Carbanilates

Barban (Carbyne®, CAS# 101-27-9 4). Barban is a selective carbanilate herbicide that has not been registered in the U.S. since 1986.[305] Hogan and Lane[306] reported two cases of allergic contact dermatitis due to the herbicide barban confirmed with positive patch test. They also found that barban is a potent allergen in guinea pig sensitization studies. In an earlier case, described by Brancaccio,[307] a pesticide applicator with allergic contact dermatitis of the hands and wrists due to barban experienced residual loss of skin pigment in the affected areas. The depigmentation may not have been due to the herbicide per se. Treatment included application of triamcinolone prescribed by his physician and self-application of an over-the-counter cream that contained ortho-phenylphenol.

In QSAR terms, the sensitizing element in barban is likely a reactive chlorine (IX) attached to the compounds terminal methyl group (Figure 8.27).

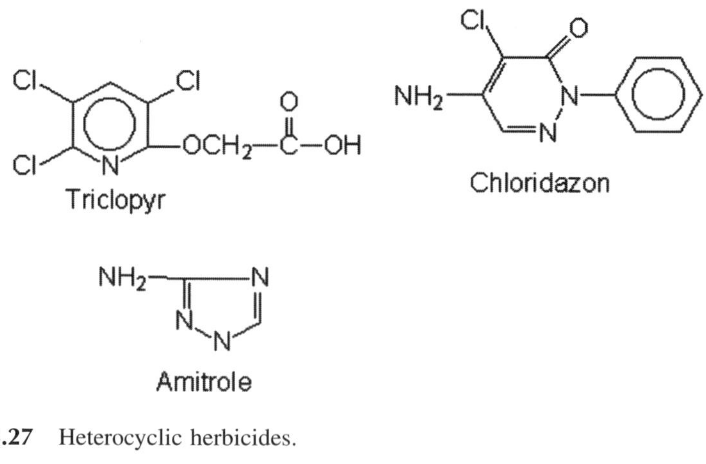

FIGURE 8.27 Heterocyclic herbicides.

Phenmedipham (SpinAid®, Betanal®, CAS# 13684-63-4). Phenmedipham is a post-emergent herbicide, which inhibits photosynthetic electron transport, and is used for control of mustard, green foxtail, and broadleaf weeds in red table beets, sugar beets, ornamental plants, and spinach. Registration data on dermal sensitization and dermal irritation were not available for review. However, the product label for an actively registered formulation containing 15.9% phenmedipham[6] did not identify it as either a sensitizer or an irritant, and there were no cases reported in the handler database.

Koch and Bahmer[308] described a case of photoallergic dermatitis and a fulminant hepatitis after the use of phenmedipham. Although photopatch testing revealed sensitization to phenmedipham,[308] the patient was also taking the antidepressant maprotiline (Ludiomil®) — reported elsewhere as a cause of hepatitis[309] — which may have been responsible for the reported liver injury. Nater and Grosfeld[310] reported allergic contact dermatitis in two separate applicators handling phenmedipham. The clinical syndrome resembled photosensitvity in the first case and the patient had a prolonged reaction to provocation testing. In the second case, a farmer with severe hand dermatitis, there was also a vigorous (3+) reaction to phenmedipham. A reaction to phenmedipham was also included in the case series reported by Jung et al.[311]

8.5.9 Aniline-Urea Herbicides

These compounds function by inhibiting photosynthetic electron transport and act as systemic herbicides absorbed by roots and foliage. At application rates below 5 kg/ha, they act as selective controls for broadleaf weeds and grasses, but have little selectivity at higher application rates (10-30 kg/ha) used on non-crop areas.[312]

Diuron (Ginstar®, CAS# 330-54-1). California registration data indicate that diuron is a moderate irritant (category III) in the Draize assay, causing initial erythema, followed by sloughing and scaling at 48 h with complete clearing occurring after 12 days.[313] A mixture of diuron (6%) and the related cotton herbicide and growth regulator thiuron (12%) demonstrated severe irritation in the Draize assay.[314] Published experimental data also indicate that multiple applications of diuron to mouse skin are associated with the development of benign skin tumors.[315] Contact dermatitis associated with diuron and related compounds have not, to the knowledge of the authors, been reported, but diuron was among the compounds, along with bromacil and aminotriazole, circumstantially associated with a case of scleroderma, as described above.[293]

Oxydiazol or methazole (Probe®, CAS# 20354-26-1). Methazole is a selective broadleaf herbicide for use on cotton, registered nationally in the U.S., but not in California. Chloracne has been reported in workers manufacturing oxydiazol (methazole). This resulted from the presence of 3,4,3',4' tetrachloroazoxybenzene (TCAB), an extraneous byproduct of reactions used to synthesize the herbicidal precursor 3,4-dichloroaniline.[316] Cases of chloracne related to mixing, loading, or spraying methazole have not been reported, but sporadic cases might be difficult to recognize given the high intrinsic prevalence of acne vulgaris (see Appendix IV).

8.5.10 Miscellaneous Herbicides

Chloridazon (Pyrazon, Pyramin, CAS# 1698-60-8). Chloridazon is an herbicide used to protect beets and can persist in the soil for several months. It functions by inhibiting photosynthetic electron transport.[317] Allergic contact dermatitis to chloridazon spray has been reported in a farmer who developed facial dermatitis after handling the material and had a recurrent reaction while thinning treated plants two months later. Patch tests to both the formulated product and active ingredient at 1% gave vesiculobullous reactions. No reactions were seen in 20 control subjects.[318]

Dichlobenil (Casoron®, CAS# 1194-65-6). Dichlobenil is an herbicide used for aquatic weed control and for landscape, forestry, and nursery weed control. Six men engaged in mixing or bagging dichlobenil developed dermatitis within one week to five months after first exposure. Although the condition involved comedones and was spoken of as chloracne, no cysts were observed, and judging from the description and one photograph, the dermatitis was not severe. The possibility that this condition may have been associated with a process contaminant may not have been explored.[319] From the structure of dichlobenil it is possible that the contaminant is a biphenyl derivative, analogous in its structure to TCDD.*

Trichlorobenzyl chloride (Actox Granular Selective Weed Killer For Corn®, CAS# 1344-32-7). Trichlorobenzyl chloride was formerly registered for use for pre-emergent application in combination with allidochlor (Randox®, see above). Spencer observed skin irritation in controls tested with 0.1 and 1% trichlorobenzyl chloride in the course of investigating the irritant and allergic properties of allidochlor.[286]

8.6 CONCLUSIONS

As outlined in Section 1.4, the skin effects of pesticides depend upon their chemical structures. Important categories, in terms of dermal effects, include phthalimido, thiocarbamates, triazines, phenoxies, pyrethroids, aniline-urea compounds, pyrethrins, copper compounds, nitro and nitrile chlorinated benzene derivatives, heterocyclic and multicyclic fungicides, thiadizines, bipyridyls, acetanilides, and short chain carboxylic acid derivatives. Observed skin effects range from transient irritation to corrosion of the skin, direct stimulation of the cutaneous nerves, contact and systemic allergy, porphyria, acne, and pigmentary disorders. Predictions of human skin effects from the results of tests in animals appear most straightforward for strong irritants and allergens. Information regarding the skin effects of many compounds is still incomplete. Likewise, the analysis of the reaction structural elements present in individual compounds is based upon simple structural analysis, based upon the principles described in Chapter 1. More complete structure-acitivity analysis awaits synthesis of new analog compounds, related recognized allergens, and additional experimental investigations.

REFERENCES

1. **U.S.D.A.,** The ARS Pesticide Properties Database, Available online at: http://www.arsusda.gov/rsml/ppdb.html, 2000

* Although the two ring structures are linked by a single covalent carbon-carbon bond, the resonance between the rings tends to keep the rings in a coplanar confirmation.

2. **Tomlin, C.,** Chlorpyrifos in *The Pesticide Manual: a World Compendium*, British Crop Protection Council, Farnham, Surrey, U.K., 1997, 235.
3. **Rycroft, R. J.,** Contact dermatitis from organophosphorus pesticides, *Br. J. Dermatol.*, 97, 693, 1977.
4. **California Department of Pesticide Regulation,** Registration data for corresponding compounds, 1998.
5. **Matsushita, T., Aoyama, K., Yoshimi, K., Fujita, Y., and Ueda, A.,** Allergic contact dermatitis from organophosphorus insecticides, *Ind. Health*, 23, 145, 1985.
6. **California Department of Pesticide Regulation,** Product label database, 1998.
7. **California Department of Pesticide Regulation,** Registration data for technical (98%) chlorpyrifos, 1998.
8. **California Department of Pesticide Regulation,** Registration data for 62% chlorpyrifos, 1998.
9. **Maddy, K., Kahn, C., and Fredrickson, A.,** A study of the decay of Naled (Dibrom) residues on grape foliage and soil in Kern County, California, August 1975, Sacramento, CA, 1975, California Department of Food and Agriculture, 345.
10. **Mick, D. L., Gartin, T. D., and Long, K. R.,** A case report: occupational exposure to the insecticide naled, *J. Iowa. Med. Soc.*, 60, 395, 1970.
11. **Edmundson, W. F. and Davies, J. E.,** Occupational dermatitis from naled. A clinical report, *Arch. Environ. Health*, 15, 89, 1967.
12. **Milby, T. and Epstein, W.,** Allergic contact sensitivity to malathion, *Arch. Environ. Health*, 9, 434, 1964.
13. **Kligman, A. M.,** The identification of contact allergens by human assay. 3. The maximization test: A procedure for screening and rating contact sensitizers, *J. Invest. Dermatol.*, 47, 393, 1966.
14. **Lahti, A. and Maibach, H. I.,** Contact urticaria from diethyl fumarate, *Contact Dermatitis*, 12, 139, 1985.
15. **Schanker, H., Rachelefsky, G., Siegel, S., Katz, R., Spector, S., Rohr, A., and Rodriquiz, C.,** Immediate and delayed type hypersensitivity to malathion, *Ann. Allergy*, 69, 526, 1992.
16. **California Department of Pesticide Regulation,** Registration data for technical malathion, 1998.
17. **California Department of Pesticide Regulation,** Registration data for 87.5% granules, 47.5% emulsifiable concentrate, 50% wettable powder, 5% granular, agricultural use 48% concentrate, lawn and garden 5% diazinon, 1998.
18. **California Department of Pesticide Regulation,** Registration data for 2% diazinon dust, 1998.
19. **Collins, A., Nichol, A., and Elsbury, S.,** Porphyria cutanea tarda and agricultural pesticides, *Aust. J. Derm.*, 23, 70, 1982.
20. **Nichol, A., Elsbury, S., Angel, L., and Elder, G.,** The site of inhibition of porphyrin biosynthesis by an isomer of diazinon in rats, *Biochem. Pharmacol.*, 32, 2653, 1983.
21. **O'Malley, M.,** Investigation of a death by ingestion of parathion in the Lost Hills area of Kern County, California, January 20, 1990, Sacramento, CA, 1993, California Department of Pesticide Regulation, Worker Health and Safety Branch, HS-1625.
22. **O'Malley, M.,** Systemic illnesses associated with exposure to parathion in California, 1982-1989, Sacramento, CA, 1992, Department of Pesticide Regulation, Worker Health and Safety Branch, HS-1626.
23. **Dikshith, T. and Datta, K.,** Effect of parathion on the skin of guinea-pigs, *Experientia*, 28, 169, 1972.

24. **Palmintieri, G.,** Experimental eczematous sensitization with diethyl-p-nitrophenyl thiophosphate (parathion), Presented at XXVth convention of Occupational Medicine, Taormina October 15-18, 1962.
25. **Pevny, I.,** Pesticide allergy — allergic contact eczema of a vintner (author's transl), *Derm. Beruf. Umwelt.*, 28, 186, 1980.
26. **Svindland, H. B.,** Subacute parathion poisoning with erysipeloid-like lesion, *Contact Dermatitis*, 7, 177, 1981.
27. **Bhargava, R. K., Singh, V., and Soni, V.,** Erythema multiforme resulting from insecticide spray [letter], *Arch. Dermatol.*, 113, 686, 1977.
28. **Farber, G. and Burks, J.,** Flea-collar dermatitis, *Cutis*, 9, 809, 1972.
29. **Cronce, P. C. and Alden, H.,** Flea-collar dermatitis, *JAMA*, 206, 1563, 1968.
30. **Mathias, C. G.,** Persistent contact dermatitis from the insecticide dichlorvos, *Contact Dermatitis*, 9, 217, 1983.
31. **Meister, R.,** Thiometon, in *Farm Chemicals Handbook '92*, Meister, R., Ed., Meister Publishing Company, Willoughby, OH, 1992, c 335.
32. **Maric, M.,** Toxic exanthema caused by ekatin, *Med. Pregl.*, 23, 51, 1970.
33. **California Department of Pesticide Regulation,** Registration data for technical dimethoate, 1998.
34. **Haenen, C., de Moor, A., and Dooms-Goossens, A.,** Contact dermatitis caused by the insecticides omethoate and dimethoate, *Contact Dermatitis*, 35, 54, 1996.
35. **Vandekar, M.,** Observations on the toxicity of carbaryl, folithion and 3-isopropylphenyl N-methylcarbamate in a village-scale trial in Southern Nigeria., *Bull. WHO*, 33, 107, 1965.
36. **Sharma, V. K. and Kaur, S.,** Contact sensitization by pesticides in farmers, *Contact Dermatitis*, 23, 77, 1990.
37. **Willems, P. W., Geursen-Reitsma, A. M., and van Joost, T.,** Allergic contact dermatitis due to methiocarb (Mesurol), *Contact Dermatitis*, 36, 270, 1997.
38. **Bruynzeel, D. P.,** Contact sensitivity to lannate, *Contact Dermatitis*, 25, 60, 1991.
39. **Paulsen, E., Andersen, K. E., and Skoven, I. G.,** Concomitant sensitization to lannate and Gerbera, *Contact Dermatitis*, 37, 128, 1997.
40. **Bowen, S. and Moursund, M.,** Chloracne in the manufacture of DDT, *AMA Arch. Dermatol.*, 75, 743, 1956.
41. **Tsankov, N., Kazandjieva, J., and Gantcheva, M.,** Contact pemphigus induced by dihydrodiphenyltrichlorethane, *Eur. J. Dermatol.*, 8, 442, 1998.
42. **Kwiatkowska, E. J. and Plonka, T. J.,** A case of alopecia due to DDT, *Przegl. Dermatol.*, 58, 187, 1971.
43. **Karimov, A. M. and Mirakhmedov, U. M.,** The effect of pesticides on the skin of agricultural workers, *Vestn. Dermatol. Venereol.*, 46, 50, 1972.
44. **Behrbohm,** Uber allergische Krahnkheiten durch DDT und HCH, *Allerg. Asthma*, 8, 237, 1962.
45. **Abrams, K., Hogan, D. J., and Maibach, H. I.,** Pesticide-related dermatoses in agricultural workers, *Occup. Med.*, 6, 463, 1991.
46. **Cronin, E.,** Pesticides, in *Contact Dermatitis*, Churchill Livingstone, Edinburgh, 1980, 393.
47. **Cronin, E.,** *Contact Dermatitis*, Churchill Livingstone, Edinburgh, 1980.
48. **O'Malley, M., Rodriguez, P., and Maibach, H. I.,** Pesticide patch testing: California nursery workers and controls, *Contact Dermatitis*, 32, 61, 1995.
49. **O'Malley, M. and Rodriguez, P.,** Contact dermatitis in California nursery workers: Part II pilot field study, Sacramento, CA, 1998, California Department of Pesticide Regulation, Worker Health and Safety Branch, HS-1767.

50. **Hayes, W. and Laws, E.,** *Handbook of Pesticide Toxicology,* Academic Press, San Diego, CA, 1990, 791.
51. **Farkas, J.,** Irritative contact dermatitis to scabicides as a sort of postscabies dermatitis, *Derm. Beruf. Umwelt.,* 31, 189, 1983.
52. **Hayes, W. and Laws, E.,** *Handbook of Pesticide Toxicology,* Academic Press, San Diego, CA, 1990, 786.
53. **California Department of Pesticide Regulation,** Registration data for technical (57%) pyrethrin, 1998.
54. **California Department of Pesticide Regulation,** Registration data for 1% ready-to-use pyrethrin spray, 1.4% of pyrethrins (510), and a mixture of 12% pyrethrins and 48% piperonyl butoxide, and a mixture of k salts of fatty acids (1%) and pyrethrins 0.01%, 1998.
55. **Magnusson, B., Blohm, S. G., Fregert, S., Hjorth, N., Hovding, G., Pirila, V., and Skog, E.,** Routine patch testing. IV. Supplementary series of test substances for Scandinavian countries, *Acta Derm. Venereol.,* 48, 110, 1968.
56. **Feinberg, S.,** Pyrethrum sensitization: Its importance and relation to pollen allergy, *JAMA,* 102, 1557, 1934.
57. **Mitchell, J.,** Allergic contact dermatitis from compositae, *Trans. St. John's Hosp. Derm. Soc.,* 55, 174, 1969.
58. **Epstein, E.,** Urticaria due to an insecticide (pyrethrum), *Urol. Rev.,* 48, 829., 1938.
59. **Mitchell, J. C., Dupuis, G., and Towers, G. H.,** Allergic contact dermatitis from pyrethrum (Chrysanthemum spp.). The roles of pyrethrosin, a sesquiterpene lactone, and of pyrethrin II, *Br. J. Dermatol.,* 86, 568, 1972.
60. **Flannigan, S. A. and Tucker, S. B.,** Influence of the vehicle on irritant contact dermatitis, *Contact Dermatitis,* 12, 177, 1985.
61. **Hayes, W. and Laws, E.,** *Handbook of Pesticide Toxicology,* Academic Press, San Diego, CA, 1990, 585.
62. **Gammon, D. W.,** Correlations between *in vitro* and *in vivo* mechanisms of pyrethroid insecticide action, *Fundam. Appl. Toxicol.,* 5, 9, 1985.
63. **Cagen, S. Z., Malley, L. A., Parker, C. M., Gardiner, T. H., Van Gelder, G. A., and Jud, V. A.,** Pyrethroid-mediated skin sensory stimulation characterized by a new behavioral paradigm, *Toxicol. Appl. Pharmacol.,* 76, 270, 1984.
64. **Flannigan, S. A. and Tucker, S. B.,** Variation in cutaneous perfusion due to synthetic pyrethroid exposure, *Br. J. Ind. Med.,* 42, 773, 1985.
65. **Flannigan, S. A., Tucker, S. B., Key, M. M., Ross, C. E., Fairchild, I. E., Grimes, B. A., and Harrist, R. B.,** Synthetic pyrethroid insecticides: a dermatological evaluation, *Br. J. Ind. Med.,* 42, 363, 1985.
66. **California Department of Pesticide Regulation,** Registration data for permethrin, technical, 1998.
67. **California Department of Pesticide Regulation,** 1995 Annual Pesticide Use Report Indexed by Chemical, December, 1996, Sacramento, CA, 1998.
68. **Vero, F. and Genovesse, S.,** Occupational dermatitis in cigar makers due to contact with tobacco leaves, *Arch. Dermatol.,* 43, 257, 1941.
69. **Smith, E., Smith, K., Maibach, H., Anderson, P., Cleary, G., and Wilson, D.,** The local side effects of transdermally absorbed nicotine, *Skin Pharmacol,* 5, 69, 1992.
70. Green tobacco sickness in tobacco harvesters — Kentucky, 1992, *MMWR,* 42, 237, 1993.
71. **McKnight, R., Koetke, C., and Donnelly, C.,** Familial clusters of green tobacco sickness., *J. Agromed.,* 3, 51, 1996.

72. **Ballard, T., Brandt, V., Ehlers, J., Freund, E., Halperin , W., and Auslander, M.,** Green tobacco sickness: Occupational nicotine poisoning in tobacco workers, *Arch. Environ. Health*, 50, 384, 1995.
73. **Ghosh, S., Gokani, V., Parikh, J., Doctor, P., Kashyap, S., and Chatterjee, B.,** Protection against "green symptoms" from tobacco in Indian harvesters: A preliminary intervention study, *Arch. Environ. Health*, 42, 121, 1987.
74. **Bernstein, I. L., Bernstein, J. A., Miller, M., Tierzieva, S., Bernstein, D. I., Lummus, Z., Selgrade, M. K., Doerfler, D. L., and Seligy, V. L.,** Immune responses in farm workers after exposure to *Bacillus thuringiensis* pesticides, *Environ. Health Perspect.*, 107, 575, 1999.
75. **Bender, C. and Peck, S.,** Health symptoms reported during *BTK* spraying spring 1994 in the capital regional district, *Environ. Health Rev.*, Summer, 42, 1996.
76. **EXTOXNET,** Pesticide Information Profiles, Abamectin, Available online at: http://ace.ace.orst.edu/info/extoxnet/, 2000.
77. **EXTOXNET,** Pesticide Information Profiles, Azadirachtin, Available online at: http://ace.ace.orst.edu/info/extoxnet/, 2000.
78. **Benezra, C., Ducombs, G., Sell, Y., and Foussereau, J.,** *Plant Contact Dermatitis*, B.C. Decker, Toronto, 1985.
79. **California Department of Pesticide Regulation,** Product Label Database — ready to use deer and rabbit repellent containing 0.216% of allyl isothiocyanate and 0.625% of capsicum oleoresin, 1998.
80. **Eming, S. A., Piontek, J. O., Hunzelmann, N., Rasokat, H., and Scharffetter-Kachanek, K.,** Severe toxic contact dermatitis caused by garlic [letter], *Br. J. Dermatol.*, 141, 391, 1999.
81. **Perez-Pimiento, A. J., Moneo, I., Santaolalla, M., de Paz, S., Fernandez-Parra, B., and Dominguez- Lazaro, A. R.,** Anaphylactic reaction to young garlic, *Allergy*, 54, 626, 1999.
82. **Kanerva, L., Estlander, T., and Jolanki, R.,** Occupational allergic contact dermatitis from spices, *Contact Dermatitis*, 35, 157, 1996.
83. **Delaney, T. A. and Donnelly, A. M.,** Garlic dermatitis, *Austr. J. Dermatol.*, 37, 109, 1996.
84. **Verder-Carlos, M. and O'Malley, M.**, A report on the illnesses related to the application of rotenone to Lake Davis, Sacramento, CA, 1998, California Environmental Protection Agency, Department of Pesticide Regulations, Worker Health and Safety Branch., HS–1772.
85. **Hayes, W.,** *Pesticides Studied in Man*, Williams & Wilkins, Baltimore, 1982, 85.
86. **Grouse, G., Sparks, T., DeAmicis, C., Kirst, H., Martynow, J., Creemer, L., Worden, T., and Anzeveno, P.,** Chemistry and insecticidal activity of the spinosyns, in *Pesticide Chemistry and Bioscience*, Brooks, G., and Roberts, T., Eds., Royal Society of Chemistry, London, 1999, 160.
87. **USEPA,** Spinosad; Pesticide Tolerance. Available online at: http://www.epa.gov/fedrgstr/EPA- PEST/1999/September/Day-23/p24696.htm, 2000.
88. **Maibach, H. I. and Johnson, H. L.,** Contact urticaria syndrome. Contact urticaria to diethyltoluamide (immediate-type hypersensitivity), *Arch. Dermatol.*, 111, 726, 1975.
89. **California Department of Pesticide Regulation,** Registration data for 40% formulation of oxythioquinox ornamental miticide, 1998.
90. **Maddy, K.**, Human exposure incidents relating to use of pesticides containing propargite (Omite® and Comite®), Sacramento, CA, 1976, California Department of Food and Agriculture, HS-222.
91. **Hayes, W.,** *Pesticides Studied in Man*, Williams & Wilkins, Baltimore, 1982, 41.

92. **Emmett, E.,** Occupational skin cancer: A review, *J. Occup. Med.*, 17, 44, 1975.
93. **Tomlin, C. and British Crop Protection Council,** *The Pesticide Manual: A World Compendium*, 11th ed., British Crop Protection Council, Farnham, Surrey, U.K., 1997.
94. **Miller, E. C. and Miller, J. A.,** Mechanisms of chemical carcinogenesis, *Cancer*, 47, 1055, 1981.
95. **Miller, E. C. and Miller, J. A.,** Searches for ultimate chemical carcinogens and their reactions with cellular macromolecules, *Cancer*, 47, 2327, 1981.
96. **Rosenberg, J. and O'Malley, M.,** Pesticides, in *Occupational Medicine*, LaDou, J., Ed., Lange-Appleton, 1997, chap. 32.
97. **Simeray, A.,** Action of an agricultural fungicide (orthocide) in 250 cases of pityriasis versicolor, *Bull. Soc. Fr. Dermatol. Syphiligr.*, 73, 337, 1966.
98. **California Department of Pesticide Regulation,** Registration data for captan, technical, 1998.
99. **Marzulli, F. and Maibach, H.,** Antimicrobials: experimental contact sensitization in man, *J. Soc. Cosmet. Chem.*, 20, 67, 1973.
100 **Jordan, W. P., Jr. and King, S. E.,** Delayed hypersensitivity in females. The development of allergic contact dermatitis in females during the comparison of two predictive patch tests, *Contact Dermatitis*, 3, 19, 1977.
101. **Croy, I.,** Etiology of urticaria due to captan-based antifungal agents, case report, *Z. Gesamte. Hyg.*, 19, 710, 1973.
102. **Fregert, S.,** Allergic contact dermatitis from the pesticides captan and phaltan, *Contact Dermatitis Newsletter*, 2, 28, 1967.
103. **O'Malley, M. A., Mathias, C. G. T., and Coye, M. J.,** Epidemiology of pesticide-related skin disease in california agriculture: 1978-1983, *Principles of Health and Safety in Agriculture,* Dosman, J. A. and Cockcroft, D. W., Eds., 1989.
104. **Hirano, S. and Yoshikawa, K.,** Patch testing with European and American standard allergens in Japanese patients, *Contact Dermatitis*, 8, 48, 1982.
105. **Rudner, E.,** North American group results, *Contact Dermatitis*, 3, 208, 1977.
106. **Groundwater, J. R.,** Difolatan dermatitis in a welder; non-agricultural exposure, *Contact Dermatitis*, 3, 104, 1977.
107. **Cottel, W.,** Difolatan, *Contact Dermatitis Newsletter*, 11, 252, 1972.
108. **Camarasa, G.,** Difolatan dermatitis, *Contact Dermatitis*, 1, 127, 1975.
109. **Takamatsu, M., Futatsuka, M., Arimatsu, Y., Maeda, H., and Inuzuka, T.,** Epidemiologic survey on dermatitis from a new fungicide used in tangerine orchards in Kumamoto prefecture, *Kumamoto Igakkai Zasshi*, 42, 854, 1968.
110. **Arimatsu, M., Misumi, N., Ueda, A., and Nomura, S.,** Skin injuries among fungicide applicators in citrus orchards, *J. Jpn. Ass. Rural Med.*, 25, 342, 1976.
111. **Kasai, T. and Sugimoto, T.,** Studies on dermatitis caused by difolatan, *Nippon Kogyo Eisei Gakkai Nenkai Hokoku*, 46, 122, 1973.
112. **Matsushita, T., Nomura, S., and Wakatsuki, T.,** Epidemiology of contact dermatitis from pesticides in Japan, *Contact Dermatitis*, 6, 255, 1980.
113. **Lee, S., Chin, Y., Chang, W., and Kim, J.,** A Study on hypersensitivity of Korean farmers to various agrochemicals. Determination of concentration for patch test of fruit-tree agrochemicals and hypersensitivity of orange orchard farmers in Che-ju Do. Korea, *Seoul Med. J.*, 22, 137, 1981.
114. **Stoke, J. C.,** Captafol dermatitis in the timber industry, *Contact Dermatitis*, 5, 284, 1979.
115. **Verhagen, A.,** Contact dermatitis in Kenya, *Trans. St. John's Hosp. Dermatol. Soc.*, 60, 86, 1974.
116. **California Department of Pesticide Regulation,** Product Label Database — Difolatan technical, flowable and flowable seed, spells, 1998.

117. **Lisi, P., Caraffini, S., and Assalve, D.,** Irritation and sensitization potential of pesticides, *Contact Dermatitis*, 17, 212, 1987.
118. **California Department of Pesticide Regulation,** Registration data for folpet, technical (88% active ingredient), 1998.
119. **van Ketel, W. G.,** Allergic dermatitis from a new pesticide, *Contact Dermatitis*, 1, 297, 1975.
120. **van Ketel, W. G.,** Active sensitization by o-o, diethyl-phthalimido-phosphothioate (Plondrel), *Contact Dermatitis*, 3, 51, 1977.
121. **Matsushita, T. and Aoyama, K.,** Examination on cross sensitivity between benomyl and other major pesticides in hypersensitive contact dermatitis, *Nippon Noson Igakkai Zasshi*, 28, 464, 1979.
122. **Savitt, L.,** Contact dermatitis due to benomyl insecticide, *Arch. Dermatol.*, 105, 1972.
123. **Fregert, S.,** Allergic contact dermatitis from two pesticides, *Contact Dermatitis Newsletter*, 13, 1973.
124. **van Ketel, W. G.,** Sensitivity to the pesticide benomyl, *Contact Dermatitis*, 2, 290, 1976.
125. **van Joost, T., Naafs, B., and van Ketel, W. G.,** Sensitization to benomyl and related pesticides, *Contact Dermatitis*, 9, 153, 1983.
126. **Zweig, G., Gao, R. Y., and Popendorf, W.,** Simultaneous dermal exposure to captan and benomyl by strawberry harvesters, *J. Agric. Food Chem.*, 31, 1109, 1109.
127. **Everhart, L. P. and Holt, R. F.,** Potential benlate fungicide exposure during mixer/loader operations, crop harvest, and home use, *J. Agric. Food Chem.*, 30, 222, 1982.
128. **Hargreave, P.,** Benomyl residues on lichtis after post-harvest dipping, *Aust. J. Exp. Agric. Anim. Husb.*, 23, 95, 1983.
129. **California Department of Pesticide Regulation,** Registration data for 85% formulation of thiophanate methyl, a mixture of mancozeb (63%) and thiophanate methyl (15%); and thiophanate methyl (2.5%), 1998.
130. **Adams, R. and Fischer, T.,** Diagnostic patch testing, in *Occupational Skin Disease, 2nd Edition*, Adams, R., Ed., W.B. Saunders, Philadelphia,, 1990, 223.
131. **Fisher, A.,** *Contact Dermatitis*, third ed., Lea & Febiger, Philadelphia, 1986.
132. **Schultz, K. and Hermann, W.,** Tetramethylthiuramdisulphide, ein Thioharnstoffderivat also Ekzemnoxe bei Hafenarbeiten., *Berufsdermatosen*, 6, 130, 1958.
133. **Shelley, W.,** Golf course dermatitis due to thiram fungicide, *JAMA*, 188, 415, 1964.
134. **Rudzki, E. and Napiorkowska, T.,** Dermatitis caused by the Polish fungicide Sadoplon 75, *Contact Dermatitis*, 6, 300, 1980.
135. **Riordan, A. T. and Nahass, G. T.,** Occupational vitiligo following allergic contact dermatitis, *Contact Dermatitis*, 34, 371, 1996.
136. **Matsushita, T., Arimatsu, Y., and Nomura, S.,** Experimental study on contact dermatitis caused by dithiocarbamates maneb, mancozeb, zineb, and their related compounds, *Int. Arch. Occup. Environ. Health*, 37, 169, 1976.
137. **Scarpa, C. and Ippolito, F.,** La dermatite da ditiocarbammati, nuova entiti clinica, *Dermatologia, Napoli*, 10, 257, 1959.
138. **Laborie, F., Laborie, R., and Dedieu, E.,** Allergie aux fongicides de la gamme du menebe et du zinebe, *Arch. Mal. Prof. Med. Tavail. Securite Soc.*, 25, 419, 1964.
139. **Nater, J. P., Terpstra, H., and Bleumink, E.,** Allergic contact sensitization to the fungicide maneb, *Contact Dermatitis*, 5, 24, 1979.
140. **Adams, R. M. and Manchester, R. D.,** Allergic contact dermatitis to maneb in a housewife, *Contact Dermatitis*, 8, 271, 1982.
141. **Peluso, A. M., Tardio, M., Adamo, F., and Venturo, N.,** Multiple sensitization due to bis-dithiocarbamate and thiophthalimide pesticides, *Contact Dermatitis*, 25, 327, 1991.

142. **Piraccini, B. M., Cameli, N., Peluso, A. M., and Tardio, M.,** A case of allergic contact dermatitis due to the pesticide maneb, *Contact Dermatitis*, 24, 381, 1991.
143. **Tsyrkunov, L. P.,** Toxico-allergic dermatitis induced by ethylenediamine in the manufacture of the herbicide zineb, *Gig. Tr. Prof. Zabol.*, 45, 1987.
144. **Burry, J. N.,** Contact dermatitis from agricultural fungicide in South Australia, *Contact Dermatitis*, 2, 289, 1976.
145. **Lisi, P. and Caraffini, S.,** Pellagroid dermatitis from mancozeb with vitiligo, *Contact Dermatitis*, 13, 124, 1985.
146. **Iliev, D. and Elsner, P.,** Allergic contact dermatitis from the fungicide Rondo-M and the insecticide Alfacron, *Contact Dermatitis*, 36, 51, 1997.
147. **Higo, A., Ohtake, N., Saruwatari, K., and Kanzaki, T.,** Photoallergic contact dermatitis from mancozeb, an agricultural fungicide, *Contact Dermatitis*, 35, 183, 1996.
148. **California Department of Pesticide Regulation,** Dermal sensitization studies in the registration database for elemental copper, copper ammonium carbonate, copper hydroxide, copper naphthenate, copper oxide, copper oxychloride, copper sulfate, copper triethanolamine complex, and cupric oxide, 1998.
149. **Fisher, A.,** Copper Dermatitis, in *Contact Dermatitis*, Lea & Febiger, Philadelphia, 1986, 727.
150. **Rademaker, M.,** Occupational contact dermatitis among New Zealand farmers, *Austr. J. Dermatol.*, 39, 164, 1998.
151. **Kambe, Y., Ide, T., and Takeshige, M.,** Investigation on allergy due to pesticides by patch test II, *Nippon Noson Igakkai Zasshi*, 25, 628, 1976.
152. **California Department of Pesticide Regulation,** Registration data for 84.8% formulation of elemental copper, 1998.
153. **Bruynzeel, D. P. and van Ketel, W. G.,** Contact dermatitis due to chlorothalonil in floriculture, *Contact Dermatitis*, 14, 67, 1986.
154. **Penagos, H., Jimenez, V., Fallas, V., O'Malley, M., and Maibach, H. I.,** Chlorothalonil, a possible cause of erythema dyschromicum perstans (ashy dermatitis), *Contact Dermatitis*, 35, 214, 1996.
155. **Johnsson, M., Buhagen, M., Leira, H. L., and Solvang, S.,** Fungicide-induced contact dermatitis, *Contact Dermatitis*, 9, 285, 1983.
156. **Spindeldreier, A. and Deichmann, B.,** Contact dermatitis against a wood preservative with a new fungicidal agent (author's transl), *Derm. Beruf. Umwelt.*, 28, 88, 1980.
157. **Lord, J., Moats, R., and Jones, J.,** Too much golf, *J. Forensic. Sci. Soc.*, 24, 359, 1984.
158. **Dannaker, C. J., Maibach, H. I., and O'Malley, M.,** Contact urticaria and anaphylaxis to the fungicide chlorothalonil, *Cutis*, 52, 312, 1993.
159. **Malten, K.,** DNCB in cooling water, *Contact Dermatitis Newsletter*, 15, 1974.
160. **Zimmerman, M.,** Dinitrochlorobenzene in water systems, *Contact Dermatitis Newsletter*, 7, 165, 1970.
161. **Basketter, D. A., Scholes, E. W., Fielding, I., Dearman, R. J., Hilton, J., and Kimber, I.,** Dichloronitrobenzene: a reappraisal of its skin sensitization potential, *Contact Dermatitis*, 34, 55, 1996.
162. **EXTOXNET,** Pesticide Information Profiles, Quintozene, **Available online at**: http://ace.ace.orst.edu/info/extoxnet/, 2000
163. **O'Malley, M. and Rodriguez, P.,** Contact dermatitis in California nursery workers: Part I surveillance of skin disease, Sacramento, CA, 1998, California Department of Pesticide Regulation, Worker Health and Safety Branch, HS-1767.
164. **Cripps, D., Gocmen, A., and Peters, H.,** Porphyria turcica. Twenty years after hexachlorobenzene intoxication, 116, 46, 1980.
165. **Wahlberg, J.,** Yellow staining of hair and nails and contact sensitivity to dinobuton, *Contact Dermatitis Newsletter*, 16, 481, 1974.

166. **Dahlquist, I.,** Contact allergy to 3-ethylamino-1,2-benzisothiazol-hydrochloride, a veterinary fungicide, *Contact Dermatitis*, 3, 277, 1977.
167. **Basketter, D. A., Rodford, R., Kimber, I., Smith, I., and Wahlberg, J. E.,** Skin sensitization risk assessment: A comparative evaluation of 3 isothiazolinone biocides, *Contact Dermatitis*, 40, 150, 1999.
168. **Chew, A. and Maibach, H.,** 1,2-Benzisothiazolin-3-one (Proxel): irritant or allergen? A clinical study and literature review, *Contact Dermatitis*, 36, 131, 1997.
169. **Sanz-Gallen, P., Planas, J., Martinez, P., and Gimenez-Arnau, J. M.,** Allergic contact dermatitis due to 1,2 benzisothiazolin-3-one in paint manufacture, *Contact Dermatitis*, 27, 271, 1992.
170. **Foussereau, J., Brandle, I., and Boujnah-Khouadja, A.,** Allergic contact eczema caused by isothiazolin-3-one derivatives, *Derm. Beruf. Umwelt.*, 32, 208, 1984.
171. **Toren, K., Brisman, J., and Meding, B.,** Sensitization and exposure to methylisothiazolinones (Kathon) in the pulp and paper industry — a report of two cases, *Am. J. Ind. Med.*, 31, 551, 1997.
172. **California Department of Pesticide Regulation,** Pesticide label database for a 25% formulation of octhilinone, 1998.
173. **Mathias, C. G., Andersen, K. E., and Hamann, K.,** Allergic contact dermatitis from 2-n-octyl-4-isothiazolin-3-one, a paint mildewcide, *Contact Dermatitis*, 9, 507, 1983.
174. **Thormann, J.,** Contact dermatitis to a new fungicide, 2-n-octyl-4-isothiazolin-3-one, *Contact Dermatitis*, 8, 204, 1982.
175. **Schuman, S. H., Dobson, R. L., and Fingar, J. R.,** Dyrene dermatitis [letter], *Lancet*, 2, 1252, 1980.
176. **Schuman, S. H. and Dobson, R. L.,** An outbreak of contact dermatitis in farm workers, *J. Am. Acad. Dermatol.*, 13, 220, 1985.
177. **Mathias, C. G.,** Allergic contact dermatitis from a lawn care fungicide containing dyrene, *Am. J. Contact. Dermatitis*, 8, 47, 1997.
178. **McFadden, J. P., Kinoulty, M., and Rycroft, R. J.,** Allergic contact dermatitis from the fungicide bupirimate, *Contact Dermatitis*, 28, 47, 1993.
179. **California Department of Pesticide Regulation,** Registration data for technical carboxin, 1998.
180. **California Department of Pesticide Regulation,** Registration data for mixture of carboxin (15%), pcnb (15%), metalaxyl (3.12%) tested undiluted in the Buehler assay, 1998.
181. **Ueda, A., Aoyama, K., Manda, F., Ueda, T., and Kawahara, Y.,** Delayed-type allergenicity of triforine (Saprol), *Contact Dermatitis*, 31, 140, 1994.
182. **van Ginkel, C. J. and Sabapathy, N. N.,** Allergic contact dermatitis from the newly introduced fungicide fluazinam, *Contact Dermatitis*, 32, 160, 1995.
183. **Calnan, C.,** Dithianone sensitivity, *Contact Dermatitis Newsletter*, 6, 119, 1969.
184. **Cotton, F. and Wilkinson, G.,** *Advanced Inorganic Chemistry*, Third ed., Interscience Publishers, New York, 1972.
185. **Matsushita, T., Yoshioka, M., Aoyama, K., Arimatsu, Y., and Nomura, S.,** Experimental study on contact dermatitis caused by fungicides benomyl and thiophanate-methyl, *Ind. Health*, 18, 141, 1977.
186. **California Department of Pesticide Regulation,** Registration data for a 12% flowable concentration of sulfur wettable powder, 1998.
187. **Schneider, H.,** Schwefelallergie [Sulfur allergy], *Hautartz*, 29, 340, 1978.
188. **Wilkinson, D.,** Sulphur sensitivity, *Contact Dermatitis*, 1, 58, 1975.
189. **Gregorczyk, L. and Swieboda, K.,** Uber den EinfluB von Schwefelverbindungen auf die Haut und auf die Schleimhaute [On the effect of sulfur compounds on the skin and mucuous membranes], *Pols. Tyg. Lek.*, 23, 463, 1968.

190. **Sun, C. and Ming-Shen, S.,** Sulfur spring dermatitis, *Contact Dermatitis*, 32, 31, 1995.
191. **Vena, G., Foti, C., and Angelini, G.,** Sulfite contact allergy, *Contact Dermatitis*, 31, 172, 1994.
192. **California Department of Pesticide Regulation,** Product label database: Formulation containing 4.2% of inert ingredients, 30% of chloropicrin, 65.8% of 1,3-dichloropropene, 1998.
193. **Nater, J. P. and Gooskens, V. H. J.,** Occupational dermatosis due to a soil fumigant, *Contact Dermatitis*, 2, 227, 1976.
194. **Gryzwa, Z. and Rudzki, E.,** Dermatitis from dichloropropane, *Contact Dermatitis*, 7, 151, 1981.
195. **U.S. EPA, O. o. P. P.,** Support for Development of Methyl Bromide Alternatives, **Available online at**: http://www.epa.gov/oppfead1/international/#B8, 1998.
196. **California Department of Pesticide Regulation**, 1997 Annual Pesticide Use Report Indexed by Chemical, Sacramento, CA, 1998, California Department of Pesticide Regulation.
197. **Watrous, R.,** Methyl bromide — local and mild systemic toxic effects, *Ind. Med.*, 11, 575, 1942.
198. **Hezemans-Boer, M., Toonstra, J., Meulenbelt, J., Zwaveling, J. H., Sangster, B., and van Vloten, W. A.,** Skin lesions due to exposure to methyl bromide, *Arch. Dermatol.*, 124, 917, 1988.
199. **Longley, E. and Jones, A.,** Methyl bromide poisoning in man, *Ind. Med. Surg.*, 34, 499, 1965.
200. **Hine, C.,** Methyl bromide poisoning, *J. Occupational Medicine*, 11, 1, 1969.
201. **Horowitz, B., Albertson, T., O'Malley, M., and Swenson, E.,** An unusual exposure to methyl bromide leading to fatality, *J. Toxicol. Clin. Toxicol.*, 36, 353, 1998.
202. **O'Malley, M.,** Clinical evaluation of pesticide exposure and poisonings., *Lancet*, 349, 1161, 1997.
203. **Schlesinger, G.,** Vapor pressures, critical temperatures and critical pressures of organic compounds, in *Handbook of Chemistry and Physics*, The Chemical Rubber Company, Cleveland, OH, 1971, D151.
204. **Fisher, A.,** Ethylene oxide, in *Contact Dermatitis*, Lea & Febiger, Philadelphia, 1986, 631.
205. **Fisher, A.,** Ethylene oxide dermatitis, in *Contact Dermatitis*, Lea & Febiger, Philadelphia, 1986, 473.
206. **EXTOXNET,** Pesticide Information Profiles, Aluminum Phosphide, Available online at: http://ace.ace.orst.edu/info/extoxnet/, 2000.
207. **O'Malley, M.**, Illnesses and injuries associated with exposures to phosphine and phosphine degradation products, Sacramento, CA, 1998, Department of Pesticide Regulation, Worker Health and Safety Branch, HS-1756.
208. **Akoboshi, S., Takahashi, M., Seki, K., Kameda, S., Tashiro, M., and Harada, A.,** A case of acute phosphine toxicosis with chronic course, *Proc. Annu. Meet. Jpn. Assoc. Ind. Health*, 50, 285, 1977.
209. **Black, H.,** Dazomet and chloropicrin, *Contact Dermatitis Newsletter*, 14, 410, 1973.
210. **Garnier, R., Prince, C., Reygagne, A., Azoyan, P., Dally, S., and Efthymiou, M. L.,** Contact dermatitis from dazomet: Seven cases, *Arch. Mal. Prof. Med. Trav.*, 54, 649, 1993.
211. **Warin, A. P.,** Allergic contact dermatitis from Dazomet, *Contact Dermatitis*, 26, 135, 1992.
212. **Jung, H. D.,** Professional dermatoses in agriculture of the agricultural and industrial district of Neubrandenburg, *Dtsch. Gesundheitswes.*, 30, 1540, 1975.
213. **Wolff, F. and Jung, H. D.,** Acute contact dermatitis after contact with nematin, *Z. Gesamte Hyg.*, 16, 423, 1970.

214. **Schubert, H.,** Contact dermatitis to sodium N-methyl dithiocarbamate, *Contact Dermatitis*, 6, 370, 1978.
215. **O'Malley, M. A.,** Skin reactions to pesticides, *Occup. Med.*, 12, 327, 1997.
216. **Castillo, L., Chaverri, F., Ruepert, C., and Wesseling, C.,** *Manual de Plaguicidas Guia para America Central*, EUNA (Editorial Universidad Nacional), Heredia, Costa Rica, 1995.
217. **Meister, R.,** *Farm Chemicals Handbook '95*, Meister Publishing Company, Willoughby, OH, 1995.
218. **Garnier, R., Chataigner, D., Efthymiou, M. L., Moraillon, I., and Bramary, F.,** Paraquat poisoning by skin absorption report of two cases, *Vet. Hum. Toxicol.*, 36, 313, 1994.
219. **Smith, J. G.,** Paraquat poisoning by skin absorption: A review, *Hum. Toxicol.*, 7, 15, 1988.
220. **Tungsanga, K., Sitprija, V., Suvanpha, R., Pochanugool, C., Chusilp, S., Benyajati, C., Kritalugsana, S., and Paoboon, N.,** Paraquat poisoning: experience in fourteen patients, *J. Med. Assoc. Thailand*, 64, 215, 1981.
221. **Tungsanga, K., Chusilp, S., Israsena, S., and Sitprija, V.,** Paraquat poisoning: Evidence of systemic toxicity after dermal exposure, *Postgrad. Med. J.*, 59, 338, 1983.
222. **Newhouse, M., McEvoy, D., and Rosenthal, D.,** Percutaneous paraquat absorption. An association with cutaneous lesions and respiratory failure, *Arch. Dermatol.*, 114, 1516, 1978.
223. **Wohlfahrt, D.,** Paraquat poisoning in Papua New Guinea, *Papua New Guinea Med. J.*, 24, 164, 1982.
224. **Stevens, J. and Sumner, D.,** Herbicides, in *Handbook of Pesticide Toxicology*, Vol. 3, Hayes, W. and Laws, E., Eds., Academic Press, San Diego, CA, 1990, 1317.
225. **Penagos, H.,** Dermatitis de contacto toxico agudo por biperidilo (Gramoxone), *Dermatitis Contacto (Boletin del GCIDC)*, 1, 16, 1993.
226. **Hearn, C. and Keir, W.,** Nail damage in spray operators exposed to paraquat, *Br. J. Ind. Med.*, 28, 1971.
227. **Samman, P. and Johnston, E.,** Nail damage associated with handling of paraquat and diquat, *Br. Med. J.*, 1, 818, 1969.
228. **Dobbelaere, F. and Bouffioux, J.,** Leukonychia in bands due to paraquat, *Arch. Belg. Dermatol.*, 30, 283, 1974.
229. **Baran, R.,** Letter: Nail damage caused by weed killers and insecticides, *Arch. Dermatol.*, 110, 467, 1974.
230. **Botella, R., Sastre, A., and Castells, A.,** Contact dermatitis to paraquat, *Contact Dermatitis*, 13, 123, 1985.
231. **Sharvill, D.,** Reaction to paraquat, *Contact Dermatitis Newletter*, 9, 210, 1971.
232. **Bowra, G., Duffield, D., Osborn, A., and Purchase, I. F. H.,** Premalignant and neoplastic skin lesions associated with occupational exposure to "tarry" byproduct during manufacture of 4, 4'-bipyridyl, *Br. J. In. Med.*, 39, 76, 1982.
233. **Wang, J., Li, W., Hu, F., and Hu, K.,** Occupational risk and the development of premalignant skin lesions among paraquat manufacturers, *Br. J. Ind. Med.*, 44, 1987.
234. **George, A. O.,** Contact leucoderma from paraquat dichloride?, *Contact Dermatitis*, 20, 225, 1989.
235. **California Department of Pesticide Regulation,** Product label database for a dispersible liquid herbicide and wettable herbicide powder, 1998.
236. **Nishioka, K., Asagami, C., Kurata, M., and Fujita, H.,** Sensitivity to the weed killer DNA-nitralin and cross-sensitivity to dinitrochlorobenzene, *Arch. Dermatol.*, 119, 304, 1983.

237. **California Department of Pesticide Regulation,** Pesticide Label Database; data on TOK® 25 and TOK®, 1998.
238. **Hurt, S., Smith, J., and Hayes, A.,** Nitrofen: a review and perspective, *Toxicology,* 29, 1, 1983.
239. **Kavlock, R. and Gray, L. J.,** Postnatal evaluation of morphological and functional effects of prenatal exposure to nitrofen in the Long-Evans rat, *J. Toxicol. Environ. Health,* 11, 679, 1983.
240. **Morse, D. L., Baker, E. L., Jr., Kimbrough, R. D., and Wisseman, C. L. D.,** Propanil-chloracne and methomyl toxicity in workers of a pesticide manufacturing plant, *Clin. Toxicol.,* 15, 13, 1979.
241. **Hill, R. H., Rollen, Z. J., Kimbrough, R. D., Groce, D. F., and Needham, L. L.,** Tetrachloroazobenzene in 3,4-dichloroaniline and its herbicidal derivates: propanil, diuron, linuron, and neburon, *Arch. Environ. Health,* 36, 11, 1981.
242. **Shreve, R. and Brink, J.,** Agrichemical industries, in *Chemical Process Industries,* McGraw-Hill, New York, 1977, 419.
243. **Poland, A., Smith, D., Metter, G., and Possick, P.,** A health survey of workers in a 2,4-D and 2,4,5-T plant, with special attention to chloracne, porphyria cutanea tarda and psychologic parameters, *Arch. Environ. Health,* 22, 316, 1971.
244. **Holmstedt, B.,** Prolegomena To Seveso, *Arch. Toxicol.,* 44, 211, 1980.
245. **Jensen, N.,** Chloracne: three cases, *Proc. R. Soc. Med.,* 65, 687, 1972.
246. **Bertazzi, P. A., Bernucci, I., Brambilla, G., Consonni, D., and Pesatori, A. C.,** The Seveso studies on early and long-term effects of dioxin exposure: A review, *Environ. Health Perspect.,* 106 Suppl. 2, 625, 1998.
247. **Neuberger, M., Kundi, M., and Jager, R.,** Chloracne and morbidity after dioxin exposure (preliminary results), *Toxicol. Lett.,* 96,97, 347, 1998.
248. **Bond, G., McLaren, E., Brenner, F., and Cook, R.,** Incidence of chloracne among chemical workers potentially exposed to chlorinated dioxins, *J. Occup. Med.,* 31, 771, 1989.
249. **Moses, M., Lilis, R., Crow, K. D., Thornton, J., Fischbein, A., Anderson, H. A., and Selikoff, I. J.,** Health status of workers with past exposure to 2,3,7,8-tetrachlorodibenzo-*p*-dioxin in the manufacture of 2,4,5-trichlorophenoxyacetic acid: Comparison of findings with and without chloracne, *Am. J. Ind. Med.,* 5, 161, 1984.
250. **Thiess, A. M. and Link, R.,** Mortality study of persons exposed to dioxin in a trichlorophenol-process accident that occurred in the BASF AG on November 17, 1953, *Am. J. Ind. Med.,* 3, 179, 1982.
251. **Coenraads, P. J., Olie, K., and Tang, N. J.,** Blood lipid concentrations of dioxins and dibenzofurans causing chloracne, *Br. J. Dermatol.,* 141, 694, 1999.
252. **Thomke, F., Jung, D., Besser, R., Roder, R., Konietzko, J., and Hopf, H. C.,** Increased risk of sensory neuropathy in workers with chloracne after exposure to 2,3,7,8-polychlorinated dioxins and furans, *Acta Neurol. Scand.,* 100, 1, 1999.
253. **Needham, L. L., Gerthoux, P. M., Patterson, D. G., Jr., Brambilla, P., Turner, W. E., Beretta, C., Pirkle, J. L., Colombo, L., Sampson, E. J., Tramacere, P. L., Signorini, S., Meazza, L., Carreri, V., Jackson, R. J., and Mocarelli, P.,** Serum dioxin levels in Seveso, Italy, population in 1976, *Teratog. Carcinog. Mutagen.,* 17, 225, 1997.
254. **Rosas Vazquez, E., Campos Macias, P., Ochoa Tirado, J. G., Garcia Solana, C., Casanova, A., and Palomino Moncada, J. F.,** Chloracne in the 1990s, *Int. J. Dermatol.,* 35, 643, 1996.
255. **Coenraads, P. J., Brouwer, A., Olie, K., and Tang, N.,** Chloracne. Some recent issues, *Dermatol. Clin.,* 12, 569, 1994.

256. **Jung, D., Konietzko, J., Reill-Konietzko, G., Muttray, A., Zimmermann-Holz, H. J., Doss, M., Beck, H., Edler, L., and Kopp-Schneider, A.,** Porphyrin studies in TCDD-exposed workers, *Arch. Toxicol.*, 68, 595, 1994.
257. **McDonagh, A. J., Gawkrodger, D. J., and Walker, A. E.,** Chloracne — study of an outbreak with new clinical observations, *Clin. Exp. Dermatol.*, 18, 523, 1993.
258. **Collins, J. J., Strauss, M. E., Levinskas, G. J., and Conner, P. R.,** The mortality experience of workers exposed to 2,3,7,8-tetrachlorodibenzo-*p*-dioxin in a trichlorophenol process accident [see comments], *Epidemiology*, 4, 7, 1993.
259. **Manz, A., Berger, J., Dwyer, J. H., Flesch-Janys, D., Nagel, S., and Waltsgott, H.,** Cancer mortality among workers in chemical plant contaminated with dioxin [see comments], *Lancet*, 338, 959, 1991.
260. **Mocarelli, P., Needham, L. L., Marocchi, A., Patterson, D. G., Jr., Brambilla, P., Gerthoux, P. M., Meazza, L., and Carreri, V.,** Serum concentrations of 2,3,7,8-tetrachlorodibenzo-p-dioxin and test results from selected residents of Seveso, Italy, *J. Toxicol. Environ. Health*, 32, 357, 1991.
261. **Neuberger, M., Landvoigt, W., and Derntl, F.,** Blood levels of 2,3,7,8-tetrachlorodibenzo-p-dioxin in chemical workers after chloracne and in comparison groups, *Int. Arch. Occup. Environ. Health*, 63, 325, 1991.
262. **Zober, A., Messerer, P., and Huber, P.,** Thirty-four-year mortality follow-up of BASF employees exposed to 2,3,7,8-TCDD after the 1953 accident, *Int. Arch. Occup. Environ. Health*, 62, 139, 1990.
263. **Assennato, G., Cervino, D., Emmett, E. A., Longo, G., and Merlo, F.,** Follow-up of subjects who developed chloracne following TCDD exposure at Seveso, *Am. J. Ind. Med.*, 16, 119, 1989.
264. **Caramaschi, F., del Corno, G., Favaretti, C., Giambelluca, S. E., Montesarchio, E., and Fara, G. M.,** Chloracne following environmental contamination by TCDD in Seveso, Italy, *Int. J. Epidemiol.*, 10, 135, 1981.
265. **May, G.,** Chloracne from the accidental production of tetrachlorodibenzodioxin, *Br. J. Ind. Med.*, 30, 276, 1973.
266. **Goldmann, P. J.,** Most severe acute chloracne, a mass intoxication caused by 2,3,6,7-tetrachlordibenzodioxine, *Hautarzt*, 24, 149, 1973.
267. **Zahm, S., Ward, M., and Blair, A.,** Pesticides and cancer, *Occup. Med., State Art Rev.*, 12, 269, 1997.
268. **Fingerhut, M. A., Halperin, W. E., Marlow, D. A., Piacitelli, L. A., Honchar, P. A., Sweeney, M. H., Greife, A. L., Dill, P. A., Steenland, K., and Suruda, A. J.,** Cancer mortality in workers exposed to 2,3,7,8-tetrachlorodibenzo-*p*-dioxin [see comments], *N. Engl. J. Med.*, 324, 212, 1991.
269. **Steenland, K., Piacitelli, L., Deddens, J., Fingerhut, M., and Chang, L. I.,** Cancer, heart disease, and diabetes in workers exposed to 2,3,7,8-tetrachlorodibenzo-*p*-dioxin [see comments], *J. Natl. Cancer Inst.*, 91, 779, 1999.
270. **Bleiberg, J., Wallen, M., Brodkin, R., and Applebaum, I.,** Industrially acquired porphyria, *Arch. Dermatol.*, 89, 793, 1964.
271. **Calvert, G. M., Sweeney, M. H., Fingerhut, M. A., Hornung, R. W., and Halperin, W. E.,** Evaluation of porphyria cutanea tarda in U.S. workers exposed to 2,3,7,8-tetrachlorodibenzo-*p*-dioxin, *Am. J. Ind. Med.*, 25, 559, 1994.
272. **Jones, R. and Chelsky,** Further discussion concerning porphyria cutanea tarda and TCDD exposure, *Arch. Environ. Health*, 41, 100, 1986.

273. **McConnell, R., Anderson, K., Russell, W., Anderson, K., Clapp, R., Silbergeld, E., and Landrigan, P.,** Angiosarcoma, porphyria cutanea tarda, and probable chloracne in a worker exposed to waste oil contaminated with 2,3,7,8-tetrachlorodibenzo-*p*-dioxin., *Br. J. Ind. Med.*, 50, 699, 1993.
274. **Pazderova-Vejlupkova, J., Nemcova, M., Pickova, J., Jirasek, L., and Lukas, E.,** The development and prognosis of chronic intoxication by tetrachlorodibenzo-*p*-dioxin in men, *Arch. Environ. Health*, 36, 5, 1981.
275. **Jirasek, L., Kalensky, J., Kubec, K., Pazderova, J., and Lukas, E.,** Acne chlorina, porphyria cutanea tarda and other manifestations of general poisoning during the manufacture of herbicides. II, *Cesk Dermatol.*, 49, 145, 1974.
276. **California Department of Pesticide Regulation,** Registration data for 96.7% technical formulation of 2,4- D, 1998.
277. **California Department of Pesticide Regulation,** Registration data for ready-to-use formulation of 2,4-D amine salts, 1998.
278. **Verschuuren, H., Kroes, R., and Den Tonkelaar, E.,** Short-term oral and dermal toxicity of MCPA and MCPP, *Toxicology*, 3, 349, 1975.
279. **Jung, V. H. and Wolf, F.,** Kontaktezekzeme durch das Herbizid Selest 100 in der Forstwirstschaft, *Dtsch. Gesundheitwes.*, 30, 1540, 1975.
280. **California Department of Pesticide Regulation,** Registration data for 45% formulation of alachlor and 85% formulation of metolachlor, 1998.
281. **Iden, D. L. and Schroeter, A. L.,** Allergic contach dermatitis to herbicides [letter], *Arch. Dermatol.*, 113, 983, 1977.
282. **Won, J. H., Ahn, S. K., and Kim, S. C.,** Allergic contact dermatitis from the herbicide alachlor, *Contact Dermatitis*, 28, 38, 1993.
283. **Schubert, H.,** Allergic contact dermatitis due to propachlor (author's transl), *Dermatol. Monatsschr.*, 165, 495, 1979.
284. **Heise, H., Mattheus, A., and Puskeiler, T.,** Irritant and sensitizing effect of the herbicide propachlor, *Z. Gesamte Hyg.*, 29, 675, 1983.
285. **California Department of Pesticide Regulation,** Product label database information for Randox herbicide, 1998.
286. **Spencer, M. C.,** Herbicide dermatitis, *JAMA*, 198, 1307, 1966.
287. **Andersson, T.,** Occupational allergic contact dermatitis from 2-*tert*-butylamino-4-cyclopropylamino-6-methylthio-1,3,5-triazine, *Contact Dermatitis*, 36, 272, 1997.
288. **EXTOXNET,** Pesticide Information Profiles, Atrazine, Available online at: http://ace.ace.orst.edu/info/extoxnet/, 1998.
289. **Elizarov, G. P.,** Occupational skin diseases caused by simazine and propazine, *Vestn. Dermatol. Venerol.*, 46, 27, 1972.
290. **Catenacci, C., Tringali, S., and Terzi, R.,** Retrospective study of morbidity in a group professionally exposed to chlorotriazine herbicides, *G. Ital. Lav. Ergon.*, 19, 23, 1997.
291. **Tomlin, C.,** Amitrole, in *The Pesticide Manual: A World Compendium*, British Crop Protection Council, Farnham, Surrey, U.K., 1997, 40.
292. **English, J. S., Rycroft, R. J., and Calnan, C. D.,** Allergic contact dermatitis from aminotriazole, *Contact Dermatitis*, 14, 255, 1986.
293. **Dunnill, M. and Black, M.,** Sclerodermatous syndrome after occupational exposure to herbicides — response to systemic steroids, *Clin. Exp. Dermatol.*, 19, 1994.
294. **Suzuki, Y. and Ohkido, M.,** Contact dermatitis from hydrazine derivatives, *Contact Dermatitis*, 5, 113, 1979.
295. **California Department of Pesticide Regulation,** Registration data for 61.6% formulation of triclopyr, 1998.

296. **Tomlin, C.,** Glyphosate, in *The Pesticide Manual: A World Compendium*, British Crop Protection Council, Farnham, Surrey, U.K., 1997, 346.
297. **Jeyaratnam, J.,** *International Conference on Pesticide Use in Developing Countries: Impact on Health and Environment*, San Jose, Costa Rica, 1983.
298. **Edmiston, S. and Maddy, K. T.,** Summary of illnesses and injuries reported in California by physicians in 1986 as potentially related to pesticides, *Vet. Hum. Toxicol.*, 29, 391, 1987.
299. **Maibach, H. I.,** Irritation, sensitization, photoirritation and photosensitization assays with a glyphosate herbicide, *Contact Dermatitis*, 15, 152, 1986.
300. **Hindson, C. and Diffey, B.,** Phototoxicity of glyphosate in a weedkiller, *Contact Dermatitis*, 10, 51, 1984.
301. **Hindson, T. C. and Diffey, B. L.,** Phototoxicity of a weedkiller: A correction, *Contact Dermatitis*, 11, 260, 1984.
302. **Smith, E. A. and Oehme, F. W.,** The biological activity of glyphosate to plants and animals: a literature review, *Vet. Hum. Toxicol.*, 34, 531, 1992.
303. **Penagos, H.,** Occupational dermatoses in the banana industry: A review of 847 patients, unpublished data, 1998.
304. **Penagos, H.**, Contact dermatitis among banana plantation workers in Panama, 1995.
305. **U.S. EPA/OPP Pesticide Products Database,** data for Carbyne 2EC Wild Oat Herbicide, Carbyne Emulsifiable Wild Oat Herbicide, Available online at: http://www.cdpr.ca.gov/docs/epa/m2.htm, 1998.
306. **Hogan, D. J. and Lane, P. R.,** Allergic contact dermatitis due to a herbicide (barban), *Can. Med. Assoc. J.*, 132, 387, 1985.
307. **Brancaccio, R. R. and Chamales, M. H.,** Contact dermatitis and depigmentation produced by the herbicide Carbyne, *Contact Dermatitis*, 3, 108, 1977.
308. **Koch, P. and Bahmer, F. A.,** Photoallergic dermatitis caused by the herbicide phenmedipham, *Derm. Beruf. Umwelt.*, 37, 203, 1989.
309. **Braun, J., Geiger, R., Wehner, H., Schaffer, S., and Berger, M.,** Hepatitis caused by antidepressive therapy with maprotiline and opipramol., *Pharmacopsychiatry*, 31, 152, 1998.
310. **Nater, J. P. and Grosfeld, J. C.,** Allergic contact dermatitis from Betanal (phenmedipham), *Contact Dermatitis*, 5, 59, 1979.
311. **Jung, H. D., Rothe, A., and Heise, H.,** Epicutaneous testing with plant protective and pest control agents (pesticides), *Derm. Beruf. Umwelt.*, 35, 43, 1987.
312. **Tomlin, C.,** Diuron, in *The Pesticide Manual: A World Compendium*, British Crop Protection Council, Farnham, Surrey, U.K., 1997, 443.
313. **California Department of Pesticide Regulation,** Registration data for 81% formulation of diuron, 1998.
314. **California Department of Pesticide Regulation,** Registration data for Ginstar EC formulation-12%, 1998.
315. **Antony, M., Shukla, Y., and Mehrotra, N. K.,** Tumor initiatory activity of a herbicide diuron on mouse skin, *Cancer Lett.*, 48, 125, 1989.
316. **Taylor, J. S., Wuthrich, R. C., Lloyd, K. M., and Poland, A.,** Chloracne from manufacture of a new herbicide, *Arch. Dermatol.*, 113, 616, 1977.
317. **Tomlin, C., Chloridazone,** in *The Pesticide Manual: A World Compendium*, British Crop Protection Council, Farnham, Surrey, U.K., 1997, 215.
318. **Bruze, M. and Fregert, S.,** Allergic contact dermatitis to chloridazon, *Contact Dermatitis*, 8, 427, 1982.
319. **Deeken, J. H.,** Chloracne induced by 2,6-dichlorobenzonitrile, *Arch. Dermatol.*, 109, 245, 1974.

9 International Experience with the Skin Effects of Pesticides

Michael O'Malley, Homero Penagos, Luis Conde Salazar, Alicia Cannavo, Luis Mirande, and Thomas J. Ferguson

CONTENTS

1-56670-293-3/00/$0.00+$.50

9.1 INTRODUCTION

Many obvious gaps exist in the knowledge about the international occurrence of skin effects of pesticides. These parallel, but also to some extent exceed, the gaps in pesticide illness surveillance information. Despite the gaps in coverage by surveillance programs and lack of published information on skin effects from some programs, a partial idea of the international distribution of skin conditions related to pesticides can be obtained from review of existing data. These include published case reports, case series, outbreak information, and formal epidemiologic studies of incidence and prevalence. The potential likelihood of pesticide skin problems can also be inferred from crop production patterns and pesticide use data, including historical import and export data. Although international use data are available from the United Nations Food and Agricultural Organization (FAO), the level of available detail varies depending upon the data submitted by national pesticide authorities.

To the extent that the information overlaps with that discussed in the portion of this work reviewing the literature by use and chemical structural category, the reader is referred to the relevant subsection of Chapter 8. Additional details are discussed below where these are pertinent to the epidemiologic perspective. Each section of the review considers in turn case reports and cases series, surveillance, illness outbreak investigations, incidence and prevalence data. Particular emphasis is given to the interplay between surveillance and outbreak investigations, and considered together because the proper conduct of surveillance includes field epidemiology as well as the tabulation of illness statistics. Outbreak investigations frequently present the opportunity for public health interventions not possible under less compelling circumstances.[1]

9.2 THE U.S.

As described in epidemiology, the distribution of dermatitis is categorized according to type of study design or type of information, beginning with background on pesticide use and crop production and concluding with case reports, surveillance information, cohort, case-control, and cross-sectional studies.

The sections below will discuss U.S. pesticide use data and the influence of crop type on pesticide use. Case reports, surveillance information, and incidence/prevalence and cohort studies will be reviewed by geographic section of the country, beginning with California, which has the most complete information available. Other states with at least limited surveillance information include Washington, Oregon, Arizona, Texas, Iowa, Wisconsin, South Carolina, and New York. Descriptive data on the distribution of cases within the subgroups of agricultural and non-agricultural

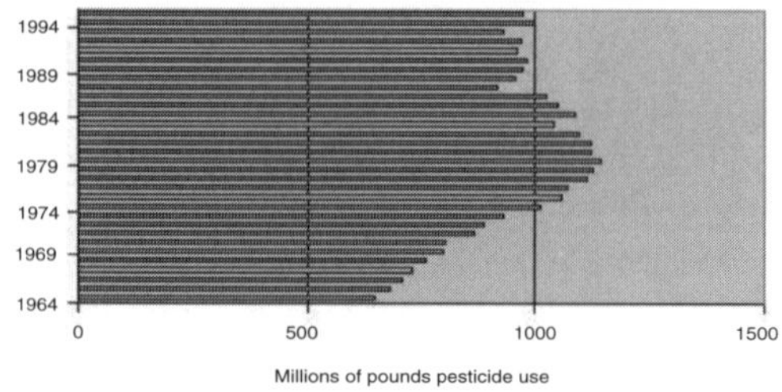

FIGURE 9.1 USEPA estimate of pesticide use in the U.S., 1964–1995.

Standard Industrial Classification (SIC) is confined to California, because published data from the other state systems did not include SIC coding. Although national pesticide illness surveillance data from the National Institute of Occupational and Safety and Health (NIOSH) is not currently available, data from California on rates of illness by standard industrial classification could, in theory, be applied to similar businesses nationally.

9.2.1 Use Data for U.S.

The 270 million residents of the U.S. account for 4.5% of the 6 billion population in the world.[2] In 1995, the U.S. accounted for a total $11,316,000,000 in pesticide sales, approximately 30% of the $37,696,000,000 world market for that year. This equated to about 1 billion pounds of pesticides applied, including both agricultural and non-agricultural pesticides, but not including antimicrobial compounds.[3]

Time trend data from the EPA (Figure 9.1) show a marked increase in estimated use from 1964 to 1974 and relatively stable, or slightly decreasing total use since then. Estimate of user expenditures on pesticides for the period between 1979 and 1995 showed an increase from $5,300,000,000 to $11,300,000,000 (Figures 9.2a and 9.2b).

Total U.S. pesticide production estimates for 1994-1995 was about 1.5 billion pounds, including 0.5 billion pounds produced for export. In the same time period, 0.2 billion pounds of pesticides were imported for use in the U.S. In 1993, the pesticide production industry (SIC 2879, agricultural chemicals) in the U.S. employed the equivalent of 27,158 full-time workers.[4] Ten states, principally in the eastern half of the U.S. (Figure 9.3), accounted for approximately 62% of the total. Trade data from FAO indicate that U.S. pesticide imports were relatively stable (less than $400,000,000 annually) between 1981 and 1997. During the same time period, pesticide exports from the U.S. increased from about $400,000 to approximately $1,600,000,000 (Figure 9.4).*

* Although these figures do not directly indicate the volume of material imported and exported, the data in Figures 9.1 and 9.2a and b suggest that the dollar cost of pesticides approximately doubled over this time period. Expenditures increased from $5,000,000,000 in 1979 to $10,000,000,000 in 1996. Estimates of pounds used were relatively stable during the same period, approximately 1 billion pounds annually. Together these numbers suggest that the four-fold increase in the dollar value of exports approximately corresponded to a doubling of the volume of material exported.

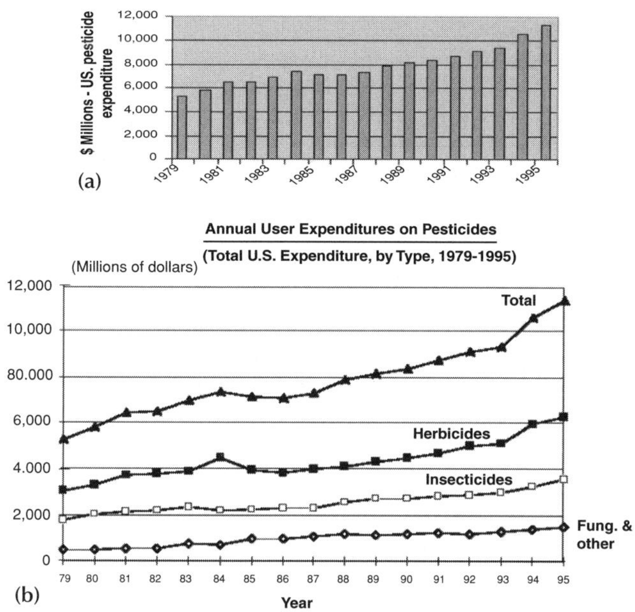

FIGURE 9.2 (a) USEPA estimate of dollars spent on pesticides in the U.S., 1979-1998. (b) Time trends in pesticide expenditures by use category (USEPA information).

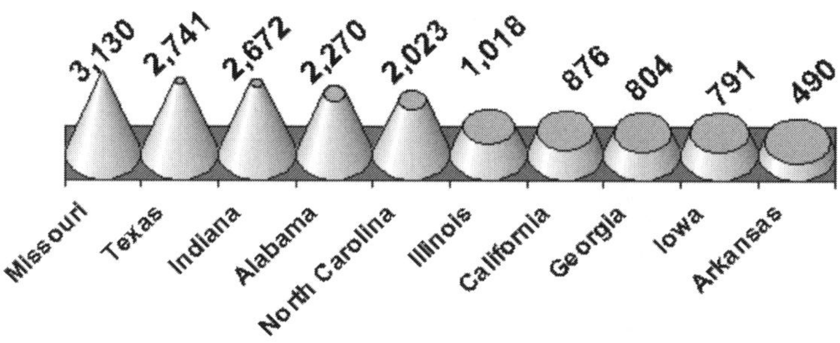

FIGURE 9.3 Leading states in production of agricultural chemicals in the U.S., based on annual average employment. (U.S. Bureau of Labor Statistics annual average employment data for 1993.)

9.2.2 Crop Production and Influence of Crop Type on Pesticide Use

Although cereal dominates the U.S. production figures, FAO data demonstrate a wide variety of crops grown, including fruit, pulses (legumes), oil crops, roots and tubers, tree nuts, and extensive production of beef, eggs, poultry, and milk (Figure 9.5). The EPA data clearly demonstrate the influence of crop type on use of pesticides, demonstrating an 11% increase in use associated with an increase in acreages of corn and soybeans, both high pesticide-use crops. Floods and unusual weather patterns in 1993-1994 in the midwest also led to an increase in weed

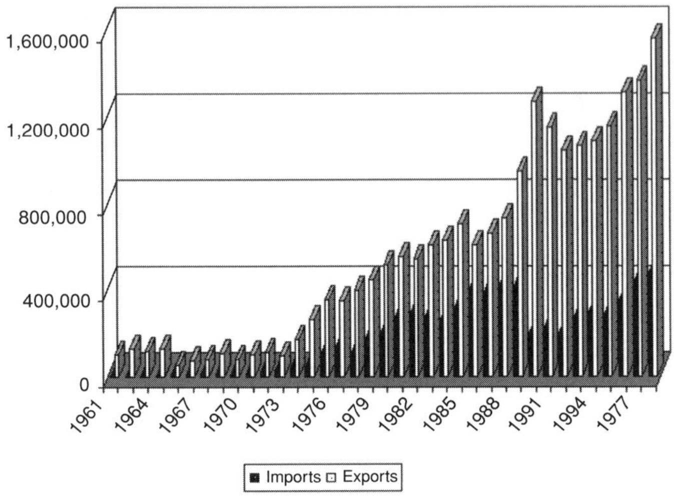

FIGURE 9.4 Time trends in U.S. import and export of pesticides, 1961–1997. (UN FAO data).

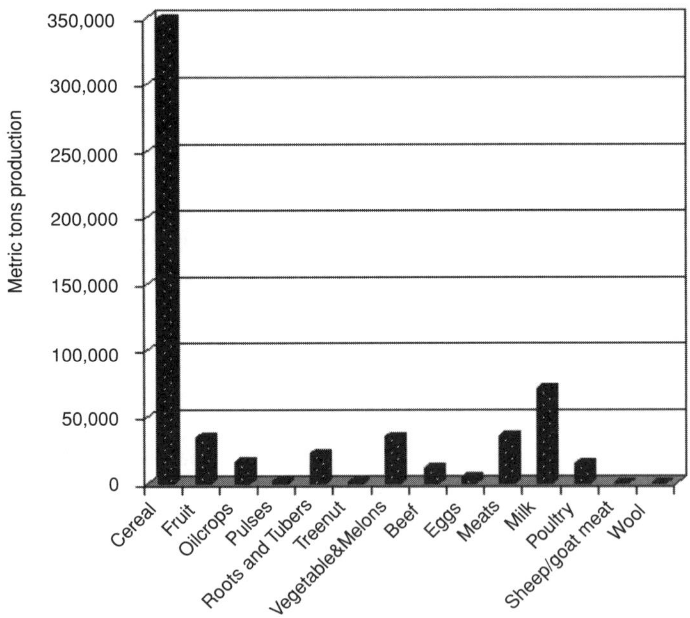

FIGURE 9.5 U.S. crop production data, 1998. (From the UN FAO.)

infestations and a subsequent increase in herbicide use. Details of regional crop production data are discussed below in conjunction with data from the state surveillance programs (Figure 9.6).

FIGURE 9.6 Distribution of pesticide surveillance programs in the U.S. (*States doing ongoing surveillance: California, Washington, Oregon, Arizona, Texas, and New York. +States with pilot or periodic surveillance programs: Wisconsin (1979-1980), Iowa (1989-1990), South Carolina (1983-1987), and Florida (1987-1989).)

9.2.3 California

In 1997, the 27,698,779 acres of California farm land included 10,803,804 acres of cropland acres, approximately 87% under irrigation.[5] Wheat, rice, barley, alfalfa, and grain crops accounted for 22% of the cropland, and cotton, vegetables, and orchard and fruit crops for the remainder. Cash sales in 1997 amounted to approximately $23 billion, with fruit, nuts, and berries accounting for 34%; dairy, cattle, poultry, and other livestock products for 27.3%; vegetables for 18%; nursery and greenhouse crops for 9.65%; grain (barley, wheat, oats, sorghum, and corn), hay and silage for 7.0%; cotton for 4.5%, and other crops for 1.31%.

9.2.3.1 Case Reports

Case reports from California reflect the wide variety of activities associated with pesticide use in the state. These include bullous dermatitis (see Figure 9.7)[6] and contact urticaria[7] associated with the insect repellent diethyl toluamide, contact urticaria associated with chlorothalonil residue exposures in a conifer seedling nursery,[8] irritant dermatitis in a truck driver accidentally exposed to DDVP (dichlorvos),[9] contact allergy to the fungicides norflurazon[10] and difolatan (accompanied by allergically mediated asthma),[11] and the thiazolin (Kathon®) antibacterial compound ochtilinone.[12] Chloracne, associated with handling pentachlorophenol treated lumber has also been reported.[13] The published cases represent a tiny portion of cases reported to the state illness registry (see, for example, the illustrations in Figures 8.19, 9.8, and 9.9), and the discussion below.

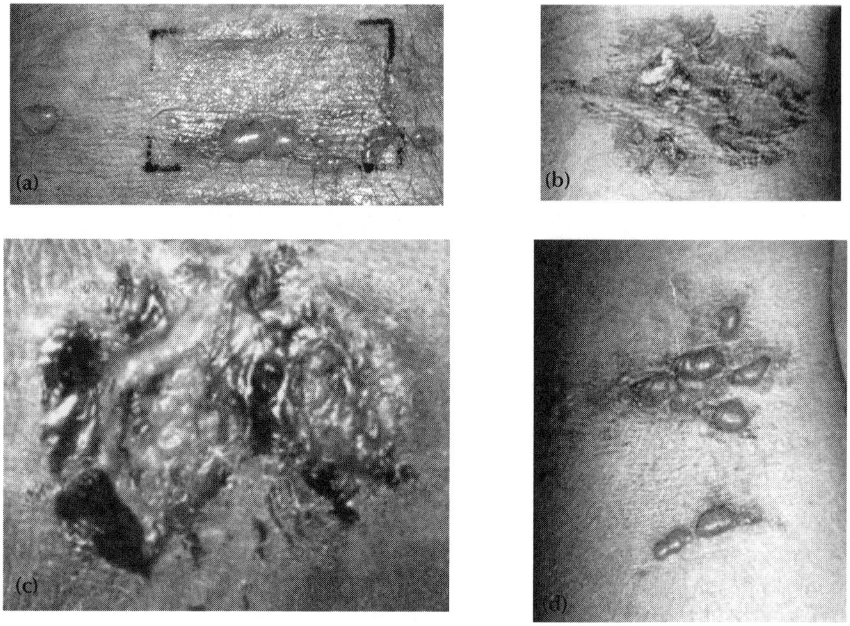

FIGURE 9.7 Effects of 75% DEET. (a) Positive reaction to DEET patch on upper arm. (b) Erosive reaction to patch test. (c) Necrotic reaction to DEET in antecubital fossa — described as a typical clinical case. (d) Bulbous reaction in antecubital fossa. (Reprinted from Lamberg, *Archives of Dermatology*, 1969. With permission from the American Medical Association.)

9.2.3.2 Descriptive Information from the Illness Registry and Propargite Outbreak Episodes

The diversity of agriculture in California allows the study of pesticide-related skin problems associated with many different types of agricultural production (Table 9.1), ranging from principally mechanized production of cereals (SIC 011) and livestock (SIC 02), to the labor-intensive production of fruit (SICs 0171, 0172, and 0175), vegetables (SIC 016), nut crops (SIC 0173), and ornamental nursery crops (SIC 018). There are also a variety of agricultural services including businesses that supply hand labor (SIC 0723) and others that provide agricultural pest control (SIC 0721). Unsurprisingly, because agricultural pest control workers handle pesticides on a daily basis, the latter group had the highest reported rate of pesticide-related skin disease (22.3 reported cases/10,000 employed),. Nevertheless, some of the employees in this group were exposed during hand labor activities rather than from pesticide handling work. Pesticide handlers accounted for a large percentage of cases for most agricultural production SICs. The highest rates of pesticide-related skin conditions among agricultural production SICs were found for grapes (15.8 reported cases per 10,000 employed), due to a large number of field residue cases. Notably low rates, below the average for all agricultural SICs (4.49/10,000) combined, were associated with grain production (4.12/10,000) and with the production of livestock (1.37/10,000).

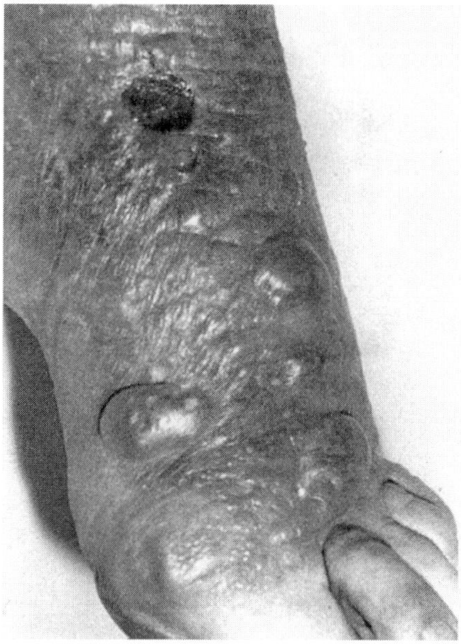

FIGURE 9.8 Dermatitis of the foot associated with accidental exposure to methyl bromide. (Courtesy of Pesticide Use Enforcement Field Staff, California Department of Pesticide Regulation.)

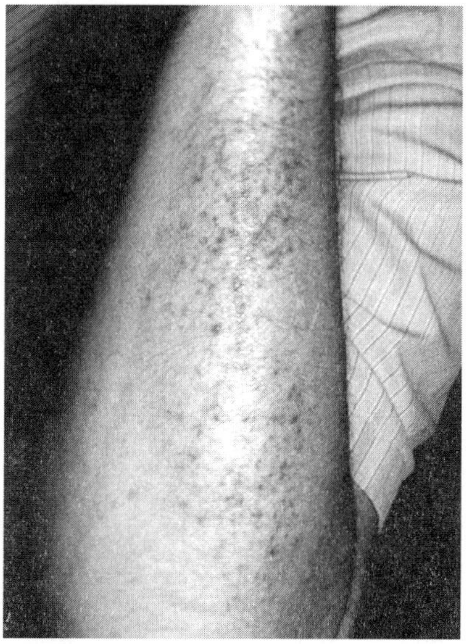

FIGURE 9.9 Reaction to sulfur/malathion/zineb mixture in pesticide applicator. (Reprinted from O'Malley, M.A., State-of-the Art Reviews *in Occupational Medicine*, 1997, with permission from Hanley and Belfus.)

TABLE 9.1
Possible, Probable, and Definite Dermatitis Cases by SIC 1988–1995

SIC description	SIC#	Handlers	Others	Cases	Population	Rate/10,000 employed
All grains	011	4	6	10	24254	4.12
All 013 crops	013	25	31	56	82738	6.77
Cotton	0131	20	30	50	52731	9.29
Sugar crops	0133	3	0	3	1761	17.04
Irish potatoes	0134	1	0	1	4545	2.20
Field crops, exc. grains	0139	1	2	3	36270	0.83
Vegetables and melons	0161	25	35	60	252938	2.37
All 017 crops	017	154	507	661	676753	2.54
Berry crops	0171	6	39	45	126558	3.56
Grapes	0172	76	337	413	262188	15.75
Tree nuts	0173	29	10	39	58301	6.69
Citrus	0174	14	33	47	44493	10.56
Deciduous tree fruit	0175	26	86	112	124810	8.97
Fruits and tree nuts, etc.	0179	2	2	4	91082	0.44
Ornamental nursery, etc.	018	75	66	141	261434	5.39
General farms primarily crop production	0191	48	23	71	216122	3.29
All crop production SICs	01	331	668	999	1514239	6.60
All livestock	02	22	10	32	232953	1.37
Soil preparation serv	0711	11	1	12	8224	14.59
Crop planting and protection	0721	78	13	91	40838	22.28
Crop harvesting	0722	0	14	14	60516	2.31
Crop preparation	0723	19	74	93	240283	3.87
Cotton ginning	0724	2	0	2	9134	2.19
Vet services	0742	10	0	10	94862	1.05
Livestock services, exc. spec	0751	0	0	0	4188	0.00
Animal specialty services	0752	8	0	8	31485	2.54
Farm lab contract	0761	16	178	194	625203	3.10
Farm management services	0762	4	2	6	74557	0.80
Horticultural/landscaping service	078	24	2	26	307506	0.85
All agricultural services	07	172	285	457	1496796	3.05
Forestry	08	1	1	2	9181	2.18
Fishing, hunting, trapping	09	0	1	1	10643	0.94
All agricultural SICs	01-09	527	967	591	3321904	1.78
Mining SICS	1000-1499	0	0	0	229694	0.00
Construction SICS	1500-1999	5	9	14	3387824	0.04
Manufacturing	2000-3999	69	50	119	12121047	0.10
Agricultural chemical manufacturing	2879	1	6	7	13461	5.20
Transportation/utilties	4000-4999	18	29	47	3557581	0.13
Wholesale/retail trade	5000-5999	161	65	226	17488047	0.13
Finance/real estate	6000-6999	13	6	19	4911636	0.04
Services	7000-8999	378	77	455	19112928	0.24
Disinfect/exterminating services	7342	30	2	32	89220	3.59
Government	9000-9999	39	38	77	13635604	0.06
Total employed (sum of major divisions)		1209	1239	2448	120645400	0.20

Table 9.2 shows the compounds most frequently identified as primary causes of possible, probable, or definite dermatitis among handlers. These included pesticide mixtures, antimicrobial compounds (sodium hypochlorite, quaternary ammonium, phenolic compounds, glutaraldehyde, pine oil, cyanuric acid, isothiazolin compounds), insecticides (propargite, chlorpyrifos), herbicides (glyphosate, diquat, fluazifop), biocides and fumigants (metam sodium and methyl bromide), and elemental sulfur. Among non-handlers (Table 9.3), slightly more than a third of the possible-definite cases were associated with pesticide mixtures. The single compounds most frequently identified as primary cause of skin reactions among non-handlers included sodium hypochlorite/chlorine releasers, propargite, sulfur, and metam sodium.

TABLE 9.2
Definite, Probable, Possible Dermatitis Cases Pesticide Handlers Reported to California Illness Registry 1988–1995

Pesticide	Cases
Adjuvants	
Pine oil	12
Petroleum oil	6
Adjuvants	1
Antimicrobials	
Chlorine/chlorine releasers	
Sodium hypochlorite	184
Calcium hypochlorite	10
Cyanuric acid	23
Chlorine	3
Other antimicrobials	
Quaternary ammonias	139
Phenolics	38
Gluteraldehyde	31
Isothiazolin compounds (Kathons®)	14
Trichloromelamine	9
Hydrogen chloride	4
Other	11
Agricultural bactericides	
Streptomycins	2
Fumigants	
Metam sodium	34
Methyl bromide	27
Sulfur dioxide	8
1,3-Dichloropropene	8
Sulfuryl fluoride	4
Fungicides	
Sulfur	32
Chlorothalonil	6
Benomyl	6
Creosote	5
Copper sulfate	4
Copper hyodroxide	4
Iprodione	3
Fenarimol	3
Anilazine	3
Other	27
Herbicides and plant growth regulators	
Glyphosate	89
Diquat	11
Fluazifop-butyl	11
Trifluralin	9
Paraquat	7
Indole-3-acetic acid	4
Oryzalin	3
2,4-D	3
Oxadiazon	3
Other	21
Insecticides and miticides	
Propargite	31
Abamectin	5
Pyrethrin/pyrethroids	
Permethrins	4
Pyrethrins	9
Cyfluthrin	5
Organophosphates	
Chlorpyrifos	10
Diazinon	6
Naled	3
Other	30
Pesticide mixtures	270
Other and not specified	30
Total	**1225**

TABLE 9.3
California Non-Handler Cases from 1988–1995

Primary pesticide	Cases
Aldehydes	4
Chlorine/chlorine releasers	104
Isothiaziline (Kathon®)	8
Phenolic disinfectants	7
Quaternary ammonia	13
Acid/inorganic/peroxides	4
Antimicrobials	140
Metam sodium	20
Sulfur dioxide	3
Other	5
Fumigants	28
Sulfur	166
Copper compounds	15
Benomyl	7
Captan	3
Imidazole	4
Chlorothalonil	4
Creosote	21
Thiocarbamate	9
Other fungicides	
Fungicides	254
Pyrethroids	29
Organochlorines	7
Organophosphates	49
Carbamates	27
Propargite	158
Insecticides total	302
Pesticide mixtures	476
Plant growth regulators	5
Glyphosate	11
EPTC	3
Trifluralin	3
Herbicides	30
Other, unspecified, unknown	56
Total	**1291**

Because the pesticide contact for non-handlers was invariably less direct than for handlers, the proportion of non-handler cases classified as definite or probable, rather than possible (Table 9.4), in relation to the reported pesticide exposure, was significantly lower (30.7%) than for handlers (57.4%). Full investigation of non-handler cases was often difficult, requiring extensive residue sampling and evaluation of complex work-history information. Examples include a 1988 episode (Table 9.5, priority number 31-Tul-88) involving workers who developed cumulative irritant dermatitis while harvesting nectarines with persistent residue levels ranging up to 1.66 mg/cm^2. Because the propargite had been applied several weeks earlier, the episode prompted a change in the post-application re-entry interval on nectarines and other tree fruit.[14]

TABLE 9.4
Case Classification by Work Activity

Relation	Handler	Non-Handler	Total
Probable/ definite	704	396	1100
Possible	521	895	1416
Total	**1225**	**1291**	**2516**

TABLE 9.5
Cluster Illnesses, 1988–1995

Priority #	Ref. ID	# of cases	Priority illness
1995			
	36-Fre-95 95-1322	62	Crews of workers turning cane (a high contact work activity) in a table grape vineyard encountered propargite residues ranging from 0.37 to 0.66 mg/cm^2 developed an acute dermatitis
	36-Tul-95 95-1144	6	A crew of 29 workers were picking nectarines when six of the workers developed rashes. They reported their rashes two days later, but waited 7 more days before seeing a doctor. Materials applied 53 or more days previously included sulfur, methyl parathion, and adjuvant
	24-Fre-95 95-915	10	Ten workers in a crew of 25 to 30 developed symptoms after working in a garbanzo bean field treated with chlorothalonil 18 days before; both chlorothalonil and garbanzo beans are known to cause skin irritation
1994			
	44-Mer-94 94-1152	6	Seven harvesters on 4 different crews developed skin rashes while harvesting cantaloupes; no pesticides were used on fields suspected of causing the problem, no residue samples were taken because field had been disced.
	3-Imp-94 94-50	5	A few crew members with irritant symptoms after exposure to mevinphos because of a re-entry violation; others had symptoms of systemic organophosphate exposure
1992			
	52-Tul-92 92-973	20	A group of 20 orange harvesters reported burning, itchy rashes on body working in a field previously treated with chlorpyrifos, formetanate hydrochloride, and methomyl
	23-SJ-92 92-612	5	While unloading a van, five army depot employees were exposed to a box leaking carbaryl dust; three suffered itching skin, one had erythema, and one was asymptomatic
	58-Tul-92 92-1824	8	An 18-member harvesting crew entered an unposted plum orchard 17 days after treatment, 4 days before expiration of the 21-day re-entry interval for propargite on plums; 8 of the 18 workers developed rashes
	92-580	8	8 irrigators developed skin irritation while working in a garbanzo bean field; a herbicide was applied to the soil 6 months prior to this incident (before planting); the beans are known to produce an acid which can be a skin irritant
	92-465	5	Five workers developed rashes following harvesting of parsley that had not been treated with any pesticides; parsley, especially older and diseased plants, secrete juices which can cause irritation of skin
1991			
	6-SD-91 91-2130	14	15 inmates were clearing brush from roadsides around significant amounts of poison oak (dormant); guard apparently exposed during pat-downs; roadside sprayed with herbicide16 days before, samples showed low levels of residue; cases were classified as unrelated to the herbicide exposure
	33-Sis-91 91-1679	10	Cases of dermatitis were reported and associated with the metam sodium spill into the Sacramento River near Dunsmuir; these included community residents with principally airborne exposure and workers with direct contact with contaminated river water

TABLE 9.5
Cluster Illnesses, 1988–1995 (continued)

Priority #	Ref. ID	# of cases	Priority illness
	91-1197	7	7 workers harvesting onions in a field infested with Mayweed (*Anthemis cotula*) developed blisters, burning and pain on both feet and ankles; there were no pesticides reported to have been used on the field
1990			
	90-1850	6	6 workers sought treatment after pulling leaves in vineyard previously treated with benomyl, sulfur, and triadimefon
1988			
	66-Ven-88 88-2213	22	Pickers developed rashes in a lemon orchard previously treated with rotenone and petroleum oil and worked in fields with similar spray histories 3 days prior to the incident; the attending physician reported that dermatitis cases ranged from minor to moderate
	41-Ker-88 88-1665	7	8 harvesters from a crew of 59 developed rashes after working for 2 days in a block of grapes previously treated with endosulfan, methomyl, and sulfur; dislodgeable foliar residue samples showed trace levels (0.037mg/cm^2) sulfur present
	47-Fre-88 88-1594	10	While chopping weeds, identified as cocklebur, in a cotton field, a work crew of 13 developed rash; no propargite was found on dislodgeable residue samples, but 53 ppm was identified in a whole leaf residue sample, possibly the result of drift from an application half a mile away
	42-Kin-88 88-1438	10	10 workers developed itching and dermatitis while picking peaches; foliage samples ranged from .08 - 0.74mg/cm^2 propargite, with a mean level of 0.28 mg/cm^2 and median level of 0.23mg/cm^2.
	34-SBN-88 88-1216	7	Workers harvesting cabbage in a field previously treated with maneb, adjuvant, endosulfan, and methomyl developed skin irritation of undetermined cause
	31-Tul-88 88-1313	47	During a 14-day period, nectarine harvesters encountered propargite residues ranging up to 1.66mg/cm^2, with median values from 0.4-0.6 mg/cm^2, and developed dermatitis due to cumulative irritation; no dermatitis was seen in a crew exposed to residues below 0.2mg/cm^2
	41-Gle-88 88-754	18	A crew of workers pulling weeds in a field of sugar beets, last treated with pesticides months earlier, developed bullous dermatitis from contact with *Anthemis cotula* (Mayweed).

At the same time the re-entry interval was extended on tree fruit, the re-entry interval for grapes was extended from 14 to 30 days. Subsequently, there was a marked decrease in the overall number of reported dermatitis cases in grape workers. This occurred not only for cases to propargite, but also for cases attributed to sulfur and other pesticides (Figures 9.7a,b,c, and d and 9.8). Understanding of the situation was further complicated by the occurrence of a large propargite-associated outbreak in 1995 that accounted for most of the grape dermatitis cases in that year. The outbreak was possibly due to a re-entry violation, an over-application, or possibly due to slow residue dissipation (Table 9.5, priority number 36-Fre-95).

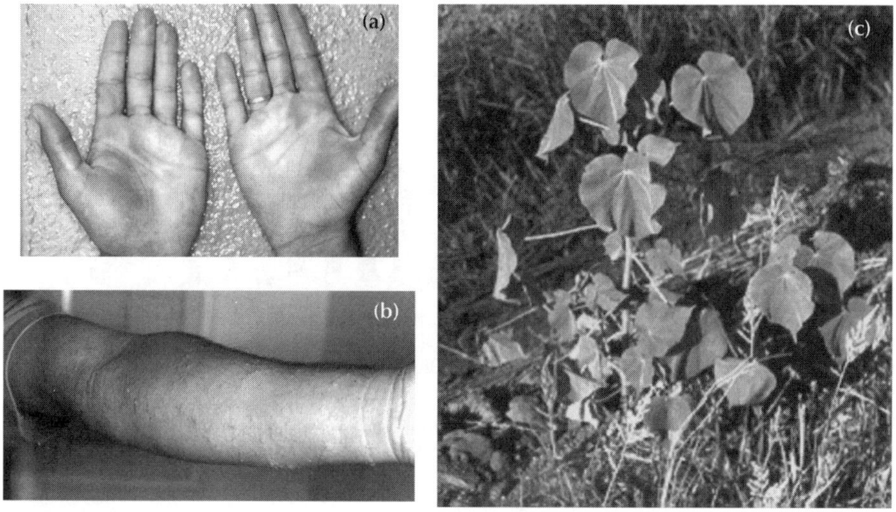

FIGURE 9.10 Dermatitis associated with pulling weeds, predominantly velvetleaf. (a) Hand dermatitis. (b) Follicular irritant dermatitis. (c) *Abutilon theophrasti* — velvetleaf. (Photos, Thomas Ferguson)

The 1995 episode occurred among workers turning cane in a vineyard, and involved a more acute dermatitis than in 1988, reflecting the high contact with foliage necessary for hand labor activities in grapes. Other factors underlying the outbreak included slow residue dissipation (half-lives of 9-11 days), wet foliage in a block where residue had failed to dissipate prior to the end of the 30-day applicaton re-entry interval, and two applications of propargite within some vineyard blocks within a one-week period. The double application was legal, but nevertheless effectively amounted to an over-application because the initially applied material would not be expected to dissipate prior to the second application.[15]

Propargite has a long history of provoking skin problems in California agriculture, beginning with dermatitis and conjunctivitis in mixer/loaders shortly after its initial introduction in the state.[16, 17] Other propargite related outbreaks included workers harvesting peaches (41-Kin-88) and plums (58-Tul-92, involving a violation of the 21 day re-entry interval set in 1988). An extended release formulation of propargite was also associated with a large scale outbreak in California orange harvesters, shortly after its introduction in 1985.[18]

9.2.3.3 Other Cluster Illnesses

Table 9.5 shows illness clusters involving 5 or more individual cases of dermatitis reported in California between 1988 and 1995. In addition to the two propargite episodes described above, there were several episodes that proved related to plant, rather than pesticide contact (Figures 9.10a, b, and c, 9.11, and 9.12). These included a roadside work crew exposed to poison oak (reference SD 91-2130),

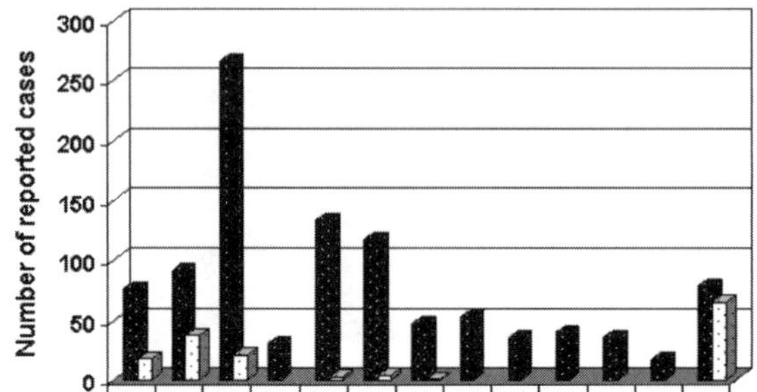

	83	84	85	86	87	88	89	90	91	92	93	94	95
■ Grape worker total	76	91	266	31	133	117	47	53	35	40	36	17	79
□ Identified propargite cases	18	37	21		3	4	1						65

FIGURE 9.11 Trends in reported cases of dermatitis in California grape workers, 1983-1995. Extension of propargite re-entry interval from 14 to 30 days occurred in 1998.

episodes of dermatitis (reference Fre-95-915 and 92-580) associated with harvesting garbanzo beans (known to be irritant[19]), harvesting parsley (reference 92-465) — also a possible irritant[19] — and pulling weeds such as cocklebur (reference Gle-88-754) and Mayweed (reference 91-1197 and 88-754). Prior work on the epidemiology of skin disease in California agricultural workers has demonstrated that contact with plants is as frequent a source of dermatitis as contact with pesticides, particularly in workers involved in the production of ornamental plants.[20]

Other noteworthy illness clusters included cases of dermatitis associated with the metam sodium spill in Sacramento in 1991 (33-Sis-91) and a small cluster of cases following accidental exposure to carbaryl (23-SJ-92). There were also several cases involving mixed pesticide exposure for which a clear cut cause could not be identified (66-Ven-88, 41-Ker-88, 44-Mer-94).

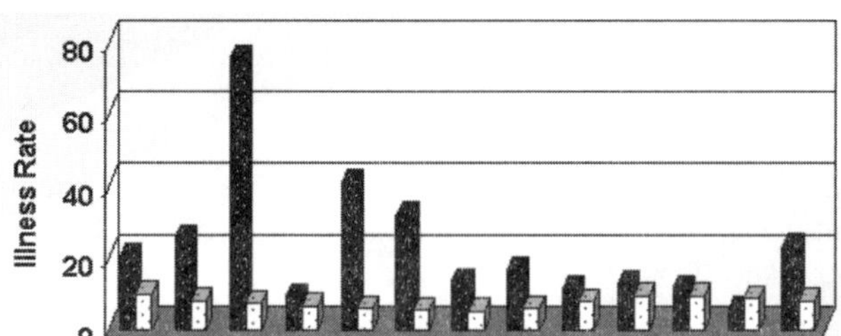

	83	84	85	87	88	89	90	91	92	94	95
■ Cases/10,000 employed	20.28	25.69	75.56	41.06	32.15	13.33	16.38	11.31	12.81	5.574	23.17
□ State unemployment rate %	9.7	7.8	7.2	5.8	5.3	5.1	5.8	7.7	9.3	8.6	7.8

FIGURE 9.12 California unemployment trends vs. the number of reported grape worker dermatitis cases/10,000 employed, based on annual average employment in SIC 0172.

9.2.3.4 Evaluation of Incidence and Prevalence

Variations in the number of reported cases, such as that discussed above for propargite dermatitis, and completeness of illness of the type portrayed in reporting provide another possible explanation for the variable rate of dermatitis in field workers portrayed in Figures 9.7a,b,c, and d. Although this decrease in cases was taken to result from the 1988 increase in the propargite re-entry interval, an alternative hypothesis entertained by surveillance program staff related the decrease to the effect of high statewide unemployment during the early 1990s on illness reporting. This hypothesis was not well supported by a comparison of illness rates and California unemployment data for 1983-1995 (Figure 9.8), but could not be actually discounted based upon the simple descriptive data available. Studies explicitly evaluating the question of illness reporting, incidence, and prevalence of pesticide illness are discussed below.

9.2.3.4.1 Tulare, Monterey, and Stanislaus County studies, 1970 and 1975

The initial studies, reported from Tulare County in 1970[21] and in Monterey/Stanislaus Counties in 1975,[22] relied entirely on questionnaire data and relatively general definitions of poisoning. The Tulare county study identified an annual incidence of work-related symptoms, presumed to be attributable to pesticide exposure, of 87%, compared to an annual incidence of the same symptoms in 34% of a control population. These symptoms required medical treatment at the rate of 249/1000 employed, compared to the official reported rate of 2 pesticide illnesses per 1000 employed. Using the same questionnaire, but a different sampling strategy and numerical method, the 1975 study identified 9.37 h lost from pesticide-related symptoms per 1000 h worked, compared to an officially reported rate of 0.056/1,000 worked. These studies, summarized in a review article published by Kahn,[23] are the source of a frequently cited repeated assertion that the proportion of cases reported is only 1% of those actually occurring in California. In 1981, this contention was discussed by Peoples and Maddy, epidemiologists associated with the state illness registry, who felt that most significant illness clusters were likely to be reported, despite possible under-reporting of individual cases. The argument was based principally on the absence of cholinesterase inhibition in randomly sampled groups of California field workers and did not address the issue of pesticide skin effects.[24] Kurtz and Esser,[25] by contrast, indicated that the principal problem with the illness registry data, rather than under-reporting, involved the hysterical over-reporting of symptoms. They illustrated the point by discussing a drift exposure episode involving organophosphates, in which the affected workers had normal cholinesterase tests.[25] This, in turn, was contested by Moses and Maizlish,[26] who pointed out that many of the workers in one of the episodes described by Kurtz actually proved to have rising cholinesterase values on serial follow-up examinations. Since the mid-1980s, this has widely been regarded as definitive evidence of significant exposure to organophosphates in workers without baseline values.[27,28]

9.2.3.4.2 Sensitivity to malathion in mosquito abatement district workers, 1963

To follow up reports of skin rash received at the California Department of Public Health in 1963, Milby and Epstein conducted a survey of 157 workers employed by mosquito abatement districts (spray operators, mechanics, and supervisors), and 43 poultry ranchers who had a history of using malathion within the previous 3 years. Four of the 157 mosquito abatement workers had positive tests, compared to 2 of the 43 poultry workers. Although all of the workers had a prior history of unexplained episodes of dermatitis, none of these were explicitly related to malathion exposure. In addition to the cross-sectional study described above, Milby and Epstein conducted an experimental investigation with 87 male volunteer subjects divided into 4 groups. Induction of sensitization was done by applying concentrations of malathion varying from 0.1 to 10% to members of each of the groups. Subjects were retested with 1% malathion at a separate site 30 days later. 46% of the subjects induced with 10% malathion had positive patch tests on re-test, compared to none of the group induced with 1% malathion, and 1% of the group induced with 1% malathion. The 46% prevalence of sensitization observed in the group induced with 10% malathion is especially noteworthy in comparison with the 2.5% prevalence of positive patch tests among the mosquito control workers and 4.7% among the poultry workers.[29]

9.2.3.4.3 *Tulare County farm workers*

A study of Tulare County farm workers by Mines and Kearney[30] included specific information about the occurrence of skin symptoms in a general health survey. They identified symptoms of rash or itching in 194 (45.5%) of 426 workers reporting on symptoms related to chemical exposures in the agricultural work place. Among a subset of workers identifying chemical exposures in a specific crop or work activity, rashes were reported most frequently among grape workers compared to workers pruning or harvesting deciduous tree fruit. The authors also emphasize the variation in access to medical care among the subgroups of Tulare farmworkers, finding that lone male migrant workers averaged 1.8 medical visits per year, compared to 3.8-5.3 visits per year for workers who resided in the county with their families. The more migratory workers also had the poorest work conditions and were especially likely to be working without valid immigration papers, live in housing supplied by labor contractors, and also depending upon labor contractors for food and transportation.*

9.2.3.4.4 *Incidence of dermatitis in grape workers*

Winter and Kurtz[31] reported on a study of dermatitis in California grape workers based on incident cases reported to the industrial nurse of the company medical service for a single large grower, employing about 1000 workers during a season. The overall rate of cases reported to the company nurse was 25/1000 workers (approximately 10 times the rate of cases reported for grape workers in Table 9.1). By work activity, the highest rates were observed for thinning vineyard foliage (9.48 rashes

---

* Members of a crew, made up of solo male workers, evaluated by the two of the authors (M.O. and T.J.F.) in a Yolo County labor camp had a dermatitis of the hands and forearms (Figure 9.12) associated with pulling weeds (Velvet leaf or *Abutilon theophrasti*) in a bean field near the town of Knights Landing. Although none of the cases were reported through the workers' compensation system, the cases were brought to the authors' attention by a health educator working for a Knight's Landing migrant clinic. They investigated through the county agricultural commissioner's office to evaluate the possible role of pesticides. Residue samples showed minimal levels of the thiocarbamate fungicide zineb, but no other compounds. Weed samples were identified by the state botany laboratory operated by California. The labor camp conditions exactly correspond to those described by Mines and Kearney in 1981 where the employer provided transportation, food, and housing.

TABLE 9.6
Examination and Questionnaire Survey of California Tomato and Grape Harvesters

Outcome	Grape	Tomato	Total
History of rash within last 3 months	95	8	103
Pustular eruptions	55	12	67
Eczematous eruptions except CD	14	3	17
Arthropod bites	17	1	18
Sun damaged skin	5	0	5
CD	4	1	5
Total subjects	**183**	**43**	**226**

Source: From Milby, T. And Epstein, W., *Arch. Environ.* Health, 9, 434, 1964.

per 100 person-weeks worked) and for cane-turning (4.92 rashes per 100 person-weeks worked). The cause of incident cases was evaluated by examining company application records for the 2 weeks prior to the study, pre-supposing that all applied chemicals took no longer than 2 weeks to dissipate. (The findings of several of the outbreak investigations discussed above contradicted the presumption of relatively rapid pesticide dissipation.) Temperature, rather than pesticide exposure, showed the closest correlation with the occurrence of rashes. This finding prompted speculation by the authors that the reported cases of rash were actually miliaria rubra, but none of the reported cases had skin examinations to confirm miliaria as the diagnosis.

A study published in 1989 by McCurdy et al.[32] evaluated the prevalence of occupational skin disease in grape workers compared to other agricultural workers. The comparison was made by examination and questionnaire administered to a group of tomato harvesters and a separate group of grape harvesters. The tomato harvesters principally worked on sorting machines and had substantially less contact with potentially contaminated crop foliage than the grape workers whose work required almost complete immersion in foliage.[32] Then 183 grape and 43 tomato workers completed an interviewer-administered questionnaire and waist-up physical examination. Acne or folliculitis was present in 30% of subjects. Eczema was noted in 10% and contact dermatitis was diagnosed in 2%. The prevalence of conditions on skin examination did not differ between the two groups of workers, but 52% of the grape workers vs. 19% of the tomato workers reported a rash occurring in the previous 3 months. From the exam and questionnaire data, and the formula relating prevalence and incidence (see Chapter 7, Section 7.2.1), the duration of the reported rashes was estimated to be approximately 2 weeks. Several limitations of the study methodology were apparent. Some skin conditions present on examination were not identified on the workers' questionnaire and the responses on the questionnaire were not specific. It was not possible to evaluate whether the history of dermatitis reported on the questionnaires represented work-related contact dermatitis (or more specifically, contact dermatitis due to pesticides).

9.2.3.4.5 Nursery Workers

O'Malley and Rodriguez[33] evaluated the prevalence of patch tests reactions to pesticides and plants in nursery dermatitis cases reported to the California pesticide illness registry. A battery of pesticides, plants, and the standard European patch antigens were tested on a sample of nursery workers. Of 349 potential participants with valid addresses, 29 (8.3%) agreed to participate. Because of the low participation rate, 11 additional subjects were also tested, including an agricultural extension nursery specialist and 10 co-workers of the subjects identified from compensation reports. Reactions to one or more test substances occurred in 18 (46.2%). This included 8 subjects (20.5%) reacting to materials from the European standard series (nickel, chromium, fragrance mix, wool, or thiuram, an ingredient in many rubber formulations that is also used as a fungicide), 5 (12.2%) reacting to either carnations or primrose in the plant series, and 10 (25.6%) reacting to either sulfur, PCNB, captafol, difolatan, or chlorpyrifos in the pesticide series. Reactions to pesticides and to sulfur, in particular, appeared to be correlated with a history of mixing, loading, or applying pesticides. The study findings suggest that allergic reactions to pesticides may account for as much as 25% of reported cases of dermatitis in nursery workers. However, because the study was hampered by a low participation rate, the findings cannot appropriately be generalized to the entire population of interest. As with the McCurdy study, questionnaire data frequently identified a history of unreported dermatitis.

REFERENCES

1. **McConnell, R.**, Pesticide research for change in Latin America, San Jose, Costa Rica, 1998.
2. **United Nations Population Fund,** State of the World Population, 1999, 6 billion. A Time for Choices. Available online at: www.unfpa.org, 1999.
3. **U.S. EPA**, Pesticides Industry Sales and Usage 1994 and 1995 Market Estimates, 1997, U.S. EPA, Office of Prevention, Pesticides Environmental Protection and Toxic Substances.
4. **U.S. Bureau of Labor Statistics**, Employment and wages annual averages, 1993, Washington, D.C., 1994, U.S. Bureau of Labor Statistics.
5. **United States Department of Agriculture,** 1997 Census of agriculture. Available online at: www.usda.gov, 1999.
6. **Lamberg, S. I. and Mulrennan, J. A., Jr.,** Bullous reaction to diethyl toluamide (DEET). Resembling a blistering insect eruption, *Arch. Dermatol.,* 100, 582, 1969.
7. **Maibach, H. I. and Johnson, H. L.,** Contact urticaria syndrome. Contact urticaria to diethyltoluamide (immediate-type hypersensitivity), *Arch. Dermatol.*, 111, 726, 1975.
8. **Dannaker, C. J., Maibach, H. I., and O'Malley, M.,** Contact urticaria and anaphylaxis to the fungicide chlorothalonil, *Cutis*, 52, 312, 1993.
9. **Mathias, C. G.,** Persistent contact dermatitis from the insecticide dichlorvos, *Contact Dermatitis*, 9, 217, 1983.
10. **Leow, Y. H. and Maibach, H. L.,** Allergic contact dermatitis from norflurazon (Predict), *Contact Dermatitis*, 35, 369, 1996.
11. **Groundwater, J. R.,** Difolatan dermatitis in a welder; non-agricultural exposure, *Contact Dermatitis*, 3, 104, 1977.
12. **Mathias, C. G., Andersen, K. E., and Hamann, K.,** Allergic contact dermatitis from 2-n-octyl-4-isothiazolin-3-one, a paint mildewcide, *Contact Dermatitis*, 9, 507, 1983.
13. **Cole, G. W., Stone, O., Gates, D., and Culver, D.,** Chloracne from pentachlorophenol-preserved wood, *Contact Dermatitis*, 15, 164, 1986.

14. **O'Malley, M., Smith, C., Krieger, R., and Margetich, S.,** Dermatitis among stone fruit harvesters, Tulare County, 1988, *Am. J. Contact Dermatitis*, 1, 100, 1990.
15. **O'Malley, M.**, Irritant Chemical Dermatitis among Grape Workers in Fresno County, August 1995, Sacramento, CA, 1998, California Department of Pesticide Regulation.
16. **Maddy, K.**, Summary of Human Occupationally Acquired Ililnesses Reported in 1974 by Physicians in California as Due to Exposure of Omite or Comite, 1974, California Department of Food and Agriculture.
17. **Maddy, K.**, Human Exposure Incidents Relating to Use of Pesticides Containing Propargite (Omite® and Comite®), Sacramento, CA, 1976, California Department of Food and Agriculture.
18. **Saunders, L. D., Ames, R. G., Knaak, J. B., and Jackson, R. J.,** Outbreak of Omite-CR-induced dermatitis among orange pickers in Tulare County, California, *J. Occup. Med.*, 29, 409, 1987.
19. **Mitchell, J. and Rook, A.,** *Botanical Dermatology: Plants and Plant Products Injurious to the Skin*, Greengrass, Vancouver, 1979.
20. **O'Malley, M. A., Mathias, C. G. T., and Coye, M. J.,** Epidemiology of pesticide-related skin disease in California agriculture: 1978-1983, *Principles of Health and Safety in Agriculture*, Dosman, J. A. and Cockcroft, D. W. Eds., 1989.
21. **Office of Pesticides, B. o. S. S. E.,** California Community Studies on Pesticides: Morbidity and Mortality of Poisonings, 1970, USPHS, January 15, 1970.
22. **Howitt, R. E.,** , Davis, Calif., 1975, 243 *.
23. **Kahn, E.,** Pesticide related illness in California farm workers, *J. Occup. Med.*, 18, 693, 1976.
24. **Peoples, S., Maddy, K., and Smith, K.**, An Overview of Occupational Exposures to Pesticides in California in 1981, Sacramento, CA, 1982, California Department of Food and Agriculture.
25. **Kurtz, P. and Esser, T.,** A variant of mass (epidemic) psychogenic illness in the agricultural work setting, *J. Occup. Med.*, 31, 331, 1989.
26. **Maizlish, N. and Moses, M.,** Fieldworker exposure to pesticides, *J. Occup. Med.*, 32, 87, 1990.
27. **Midtling, J., Barnett, P., Coye, M., Velasco, A., Romero, P., Clements, C., O'Malley, M., Tobin, M., Rose, R., and Monosson, I.,** Clinical management of fieldworker organophosphate poisoning, *West. J. Med.*, 142, 514, 1985.
28. **Coye, M., Barnett, P., Midtling, J., Velasco, A., Romero, P., Clements, C., O'Malley, M., Tobin, M., and Lowry, L.,** Clinical confirmation of organophosphate poisoning of agricultural workers, *Am. J. Ind. Med.*, 10, 399, 1986.
29. **Milby, T. and Epstein, W.,** Allergic contact sensitivity to malathion, *Arch. Environ. Health*, 9, 434, 1964.
30. **Mines, R. and Kearney, M.**, The health of Tulare county farmworkers: A report of 1981 survey and ethnographic research for the Tulare County Department of Health., Berkeley, CA, 1982, California Department of Health Services.
31. **Winter, C. K. and Kurtz, P. H.,** Factors influencing grape worker susceptibility to skin rashes, *Bull. Environ. Contam. Toxicol.*, 35, 418, 1985.
32. **McCurdy, S. A., Wiggins, P., Schenker, M. B., Munn, S., Shaieb, A. M., Weinbaum, Z., Goldsmith, D., McGillis, S. T., Berman, B., and Samuel, S.,** Assessing dermatitis in epidemiologic studies: occupational skin disease among California grape and tomato harvesters, *Am. J. Ind. Med.*, 16, 147, 1989.
33. **O'Malley, M. and Rodriguez, P.**, Contact dermatitis in California nursery workers: Part II pilot field study, Sacramento, CA, 1998, California Department of Pesticide Regulation, Worker Health and Safety Branch.

TABLE 9.7
1991–1995 Pesticide Dermatitis Cases in Eastern Washington by Activity

Activity description	Cases	%
Applicators	88	26.5
Mixer/loaders	15	4.5
Application equipment cleaning & repair	7	2.1
Drift	36	10.8
Field fumigation	2	0.6
Packing/processing	5	1.5
Field residues	159	47.9
Other residue	1	0.3
Accidental exposure to pesticide concentrate	13	3.9
Other	3	0.9
Unknown	3	0.9
Total	**332**	**100.0**

9.2.4 Washington

In the 1997 agricultural census, the state of Washington had 7,913,709 acres of cropland, approximately 17% under irrigation.[1] Corn, wheat, barley, and alfalfa accounted for approximately 50% of the acreage, but substantial acres were also planted in vegetables, apples, and other orchard crops, and potatoes. Total sales amounted to 4.8 billion dollars, with fruits, nuts, and berries accounting for 26%, cattle and dairy products for 26%, wheat, barley, and corn for 12.9%, nursery and greenhouse crops for 5.7%, vegetables for 5.7%, hay and silage for 5.3%, poultry and hogs for 3.8%, and other crops and livestock for 14%. Washington was the leading state for production of hops, with 55,000,000 pounds produced in 1997.

9.2.4.1 Surveillance Information

The Washington Department of Health database* contains reports on 332 cases of dermatitis associated with pesticides in eastern Washington, the principal agricultural area of the state, between 1991 and 1995—with the distribution of cases by work activity as shown in Table 9.7 and the principal compounds involved as shown in Table 9.8 and Table 9.9[2] As in California, the reported cases occurred principally in pesticide handlers and field workers exposed to pesticide residue. Insecticides accounted for half of the cases, with the most frequently reported compounds including azinphos-methyl, propargite, carbaryl, and chlorpyrifos (Table 9.10).

Individual cases of note investigated by the Washington Department of Health[3] included:

DOH, case 0061, 1995—A 24-year-old male, employed for the summer at a grain elevator, was spraying cyfluthrin in an empty grain tank for two days. Protective equipment included half-face respirator, goggles, and gloves. After the first day, he experienced burning on the face and neck but did not have a rash. The following day was windy and the stinging and burning was worse and he began to develop a rash. On evaluation by his healthcare provider, erythema with patchy hives on the face were noted. After he stopped spraying, the symptoms resolved in 1-2 days.

* Data provided courtesy of Lynden Baum, Washington Department of Health.

TABLE 9.8
Eastern Washington Cases

Use category	Cases
Antibacterial	5
Disinfectant	1
Fumigant	8
Fungicide	52
Growth regulator	27
Herbicide	27
Insecticide/acaricide	141
Not specified	71
Total	**332**

DOH, case 0078, 1997—A 55-year-old male was spraying Gramoxone® (paraquat) with a tractor in an apple nursery for 9 hours a day for 3 consecutive days. He felt spray on his face, perhaps because of the windy conditions. He developed an itchy rash on face and experienced irritation of the eyes and throat. The symptoms worsened despite a 2-day break from

TABLE 9.9
Applicator Dermatitis Cases in Eastern Washington 1991–1995

Chemical class	Cases
Anti-bacterial	2
Fumigant	
Thiocarbamate fumigant	1
Herbicides	
Triazole	7
Plant growth regulators	3
Phosphonates (glyphosate)	4
Carboximide	2
Dipyridyls	7
Fungicides	
Dithiocarbamates	
Mancozeb	1
Ziram	2
Other	
Captan	2
Myclobutanil	4
Sulfur	2
Triadimefon	3
Insecticides	
Propargite	5
Endosulfan	2
Growth regulator	3
Petroleum derivatives	4
N-methyl carbamate insecticides	7
Carbaryl	5
Formetanate hydrochloride	2
Organophosphates	23
Azinphos-methyl	10
Chlorpyrifos	6
Ethyl parathion	1
Malathion	3
Methyl parathion	1
Phosmet	2
Pyrethins/pyrethroids	3
Pyrethrins	1
Cyfluthrin	1
Esfenvalerate	1
Total (cases include mixtures)	**92**

TABLE 9.10
Eastern Washington Field Residue Dermatitis Cases

Chemical/class	Cases
Herbicides	
Carboximide	1
Chlorophenoxy herbicides	3
Dipyridyls	2
Triazoles	3
Pyridazinone	2
Fungicides	
Captan	4
Nitrile compounds	2
Chlorothalonil	1
Copper hydroxide	3
Fenarimol	4
Myclobutanil	3
Inorganic sulfur	5
Pyrimidine (bupirimate etc.)	4
Organic fungicides	3
Ziram	6
Insecticides	
Abamectin	4
Bacillus thuringiensis K.	1
Endosulfan	3
Imidacloprid	5
Petroleum oil	3
Propargite	1
Organophosphates/carbamates	
Azinphos-methyl	18
Chlorpyrifos	2
Diazinon	4
Malathion	2
Methyl parathion	1
N-Methyl carbamate insecticides	
Formetanate hydrochloride	2
Carbaryl	11
Pyrethroids	
Esfenvalerate	1
Biological	5
Miscellaneous	
Adjuvant	6
Growth regulator	10
Organic compound	3
Total cases (includes mixtures)	**159**

spraying. On seeking medical treatment, he was noted to have marked swelling of the face, dermatitis over the "V" of the neck, and mild infection of the conjunctivae. The rash was worse on re-examination the following day, with oozing noted on the left cheek; the patient received an injection of 80 mg IM Kenalog® (triamcinolone). On re-examination a week later, the rash was largely resolved.

9.2.4.2 Outbreak Investigations

9.2.4.2.1 Fungicides in fruit packing operations, 1952

In 1952, the environmental research laboratory at the University of Washington reported on a study[4] involving 1500 apple pickers and sorters exposed to fungicides in order to investigate a seasonal problem affecting packing house workers known as "left arm dermatitis" identified from claims filed with the state Department of Labor and Industries in 1948 and 1949. In 1950 and 1951, a survey evaluated 548 sorters and 837 packers in 28 packing operations in Washington and Oregon employing the fungicide, sodium chloro-2 phenyl phenate. A history of skin complaints was obtained from 7.5%, and about half of these had received medical treatment. Most continued working without problems, but some were so sensitized that they had to find other work. Of 30 patients patch tested with 0.3% sodium chloro-2 phenyl phenate, 3 had a positive reaction.

In addition to skin complaints, many also had upper respiratory or eye symptoms, usually at plants with poor ventilation over the apple washing areas or failure to separate already washed fruit from the packing/sorting area of the plants. Provision of adequate hygiene facilities was reported to minimize both the dermatitis and upper respiratory complaints. A non-chlorinated fungicide (sodium ortho phenyl phenate) was also found to reduce the frequency of irritation symptoms. In experimental studies, however, both chlorinated and non-fungicides were demonstrated to cause dermal sensitization in some human volunteers.

9.2.4.2.2 Ziram and thiocarbamates in workers harvesting Bosc pears, 1990

An episode of dermatitis associated with ziram in workers picking Bosc pears was reported to the Washington Department of Health (DOH) September, 1990. The cases occurred despite compliance with the existing re-entry interval, which required waiting until the spray dried in the treated field. The rashes involved arms, hands, faces, and abdomens. The only workers affected were those picking Bosc pears. No rashes appeared in other workers harvesting D'Anjou pears and golden delicious apples. The findings were consistent with an interaction between skin abrasions that commonly occur while picking the Bosc variety and the exposure to the thiocarbamate fungicide. Half-life data for ziram on pears was estimated at approximately 9 days based on dissipation studies from California, New York, and Washington. The re-entry interval was lengthened to 14 days, representing approximately 11/2 half-lives, for most formulations of ziram on Bosc pears.

REFERENCES

1. **United States Department of Agriculture,** 1997 Census of agriculture. Available online at: www.usda.gov, 1999
2. **O'Malley, M. A.,** Skin reactions to pesticides, *Occup. Med.*, 12, 327, 1997.
3. **Morrissey, B.,** Acute pesticide toxicity: Washington case studies, Yakima, Washington, 2000.
4. **Sealey, J.,** Research in fungicides results in protection for fruit packers and sorters against dermatitis, *Occup. Health*, 12, 139, 1952.

9.2.5 Oregon*

In 1997, Oregon had 5,285,659 acres of cropland with approximately 37% under irrigation. Grain and silage crops (wheat, barley, oats, and silage) accounted for approximately 40% of the cropland, and vegetables and orchard crops for approximately 5%. Total agricultural product sales for 1997 amounted to 3 billion dollars, with nursery and greenhouse crops accounting for 22.8%; livestock production for 28.8%; corn, barley, oats, and silage for 16.5%; fruits, nuts, and berries for 10.4%; vegetables for 7.2%; and other crops for 7.8%. After Washington, Oregon was the second leading U.S. producer of hops, with 13,572,000 pounds in 1997.[1]

9.2.5.1 Case Reports

In 1977, Brancaccio reported the case of a pesticide sprayer who sought treatment for a recurring wrist and hand dermatitis, with patchy areas of depigmentation, associated with handling the herbicide carbyne. The patient had used a medicated cream containing ortho-phenylphenol, but had no reaction to that material on patch testing. Testing with carbyne diluted 1:10 in mineral provoked a 2+ response at 48 h and 1 week, but did not produce depigmentation at the patch site.[2]

9.2.5.1.1 Surveillance information

The Oregon Health Division (OHD) has been conducting surveillance of pesticide illness since the late 1980s. In 1994, OHD investigated 53 pesticide-related incidents, affecting 101 separate individuals — 60% sought medical care and 2 were hospitalized. Approximately 60 of these exposures were work related, involving 13 persons in the agriculture, nursery, or forestry industries. Additional cases were found in the Oregon workers' compensation system, indicating that fewer than 20% of cases were reported directly to OHD.

In the years 1995 and 1996, 45 people were reported to have dermatologic or eye symptoms in response to pesticide exposure in Oregon; 9 reports were work-related (exposed in agricultural work or in nurseries) and 36 were non-occupational (in or around the home, or on roadside right-of-ways). Of the 45 reports, one case was confirmed — a non-occupational exposure to copper sulfate resulting in chemical burns. Nine reports were classified as probable, including two reports of contact dermatitis (one related to demeton-s-methyl), one of urticaria (diazinon), two chemical burns (demeton-s-methyl, sulfur), two of photodermatitis (one related to ferbam), and two reports of itchy skin related to the herbicide sulfometuron methyl. The other two probable reports concerned eye irritation or conjunctivitis. The remaining 35 reports included 6 possible cases, one unlikely, 3 insufficient information, and the dermal or eye problems reported by 25 individuals were classified as unrelated to pesticide exposure.

For the cumulative period from 1988 to 1994, there were 63 confirmed or probable cases of pesticide exposure resulting in dermatologic or eye symptoms. The most commonly reported active ingredients among these case reports are shown in Table 9.11. Twenty-nine other active ingredients were reported one or two times each as the first chemical product in the database.

* Data supplied courtesy of Catherine Thomsen, Oregon Health Division.

TABLE 9.11
Oregon Pesticide Dermatitis Cases 1988–1994; Most Commonly Reported Agents

Pyrethrins	14
Chlorpyrifos	6
Propetamphos	5
Chlorothalonil	5
Glyphosate	4
Malathion	4

9.2.5.1.2 Plants vs. pesticides as a source of agricultural skin disease

A non-pesticide-related dermatitis has also been encountered by surveillance program staff, an acute bullous dermatitis associated with exposure to Mayweed *(Anthemis cotula).* Outbreak of dermaitis secondary to this weed has been reported in California as well (see Section 9.2.4.1). The California outbreak included several workers who were hospitalized for nursing care of extensive burns of the lower extremities and a case of erythema multiforme.[3] Dermatitis in the hops industry has also been a particular problem — sometimes attributed to the effect of the hops plants (*Humulus lupulus*) per se and sometimes attributed to the effect of residue on the plants, particularly propargite. Investigations have not been conclusive. Mitchell and Rook's extensive review of the literature on botanical dermatology includes a section describing the irritant effects of the hop plant, citing literature as long ago as 1693. The pattern of lesions included the hands, face, and legs. The irritation occurs as the hop-cones are removed from the stalks.[4]

Data courtesy of Catherine Thomsen, Oregon Department of Health Services. Note that active ingredients identified here are only the most prominent of the first chemical listed in the Oregon database, and may not be the cause of the dermal symptoms.

REFERENCES

1. **United States Department of Agriculture,** 1997 Census of agriculture. Available online at: www.usda.gov, 1999.
2. **Brancaccio, R. R. and Chamales, M. H.,** Contact dermatitis and depigmentation produced by the herbicide Carbyne, *Contact Dermatitis,* 3, 108, 1977.
3. **O'Malley, M. and Barba, R.,** Bullous dermatitis in field workers associated with exposure to Mayweed, *Am. J. Contact Dermatitis,* 1, 34, 1990.
4. *Humulus lupulus* in *Botanical Dermatology: Plants and Plant Products Injurious to the Skin,* Mitchell, J. and Rook, A., Eds., Greengrass, Vancouver, 1979, 154.

9.2.6 Arizona

According to the 1997 census of agriculture, Arizona agricultural production involves approximately 1.3 million acres of farmland, about 80% under irrigation, divided between production of wheat, barley, cotton, vegetables, and orchard crops. Total agricultural sales for 1997 amounted to $1,903,408,000, with cattle, dairy, poultry, and other livestock products accounting for 35.7%; vegetables for 20.9%; cotton for 15.7%; grain (wheat, barley, oats, and corn), hay, and silage for 13.5%; nursery and greenhouse crops for 6.9%; fruit, nuts, and berries for 6.2%; and other crops for 1.1%.[1]

9.2.6.1 Surveillance

Arizona operates a surveillance program that depends upon telephone reports to the state department of health services and field investigations by the state department of agriculture. Data for 1995 included 48 telephone reports regarding cases occurring in California, and an additional 30 cases involving workers employed in Holtville, California, who sought treatment in Yuma (approximately 50 miles away from the site of the exposure).[2] In 1996, 157 telephone reports were received and tended to involve complaints of systemic illness or odor, with few complaints of dermatitis.

During 1996, department of agriculture investigators were contacted by members of a crew concerned about white residue and a strong odor in an orange grove near Yuma, where they had been assigned to work. Workers reported headaches, itching, and burning eyes and nose. Application records indicated that the orchard had been treated 17 days earlier with dimethoate and sulfur. Foliage samples demonstrated levels of dimethoate ranging from 7.5 to 24.0 ppm, but no analysis was conducted for sulfur residue. There was no former medical evaluation and it could not be determined whether there were frank cases of dermatitis according to information from the state registry, because the crew foreman declined to take them for medical evaluation. The department of agriculture investigation noted that the incident occurred despite apparent compliance with the application interval. California experience with monitoring sulfur residues has indicated that their dissipation is erratic and that this may have been a factor in the occurrence of the outbreak. It is obvious that no definite conclusions can be drawn in the absence of residue data confirming the presence of sulfur at the time the workers entered the treated field.*

REFERENCES

1. **United States Department of Agriculture,** 1997 Census of agriculture. Available online at: www.usda.gov, 1999.
2. **Arizona Department of Health Services**, Pesticide Poisoning Surveillance: Annual Report 1995, Office of Environmental Health, Arizona Department of Health Services. Available online at: http://www.hs.state.az.us/edc/oeh/pestcide.htm.

9.2.7 Texas

Texas has 227,000 farms; agricultural employment includes 142,000 self-employed, 30,000 unpaid workers (family members), and 55,000 wage workers.* Of 20,258,700 acres planted in 1997, 12,600,000 acres were planted with grain (barley, corn, oats, rice, rye, wheat, and sorghum.), 5,500,000 million acres with cotton, and 1,018,300 acres with other crops including dry beans, peanuts, soybeans, sugarbeets, sunflowers, sweet potatoes, and vegetables. Total cash sales for 1997 amounted to 13 billion dollars, with cattle and other livestock production accounting for 68.8%, cotton for 10.7%, grain and silage crops (corn, sorghum, wheat, barley, etc.) for 12.3%, nursery and greenhouse crops for 3.5%, vegetables for 1.8%, fruits, nuts, and berries for 0.6%, and other crops for 1.8%.

* Data courtesy of the Arizona Department of Health Services, Office of Environmental Health, Ernest Arvizu.

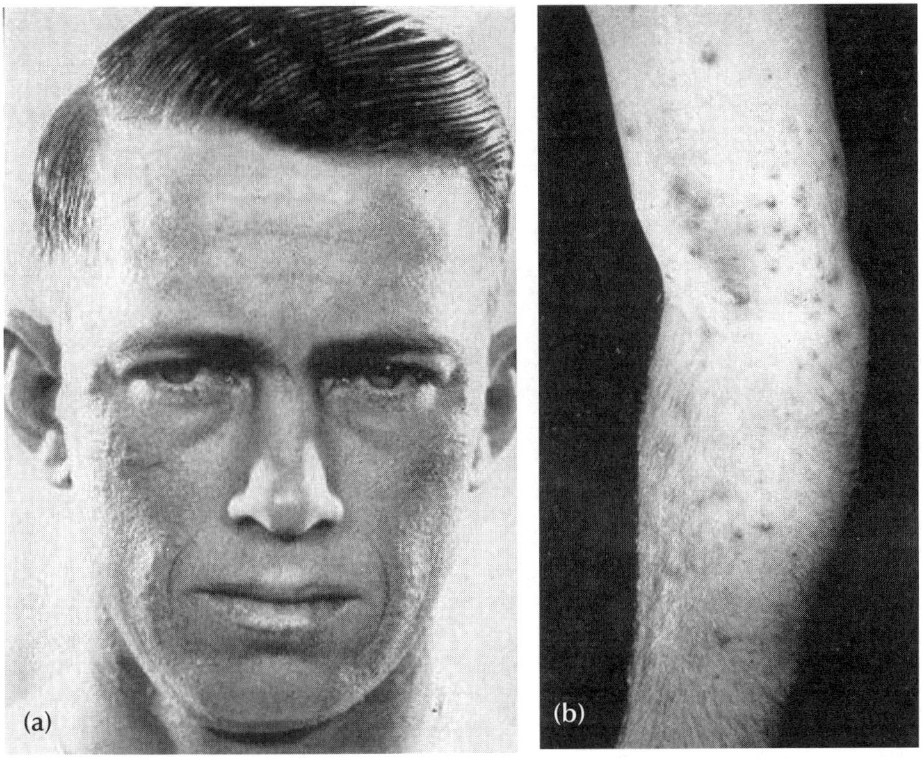

Figure 9.14 (a) Facial chloracne and (b) folliculitis associated with production of DDT. (Reported by Bowen, Archives of Dermatology, 1956. Reproduced with permission from the American Medical Association.)

9.2.7.1 Case Reports

Several cases of chloracne were described in workers employed in the manufacture of DDT in a plant near Houston in 1954.[1] The cases began with pruritic erythema of the face and forearms, followed by the development of folliculitis of the legs and comedones and inflammatory follicular papules on the face and upper extremities (Figure 9.14a and b). The acne later developed into large indurated, tender papulopustules. The index case, as well as several co-workers, had exposure to process stream materials including trichlorobenzene and benzene hexachloride, as well as finished DDT in a packaging line. Although treated with topical remedies, X-rays, and acne surgery, the acne did not improve until the workers were removed from work in the DDT process.

9.2.7.2 Surveillance Information

From 1986-1994, the Texas Department of Health confirmed 247 cases of occupational pesticide poisoning (Figure 9.15), believed to significantly underestimate the actual number of cases.* In a 1989-1990 study, hospital and physicians were sur-

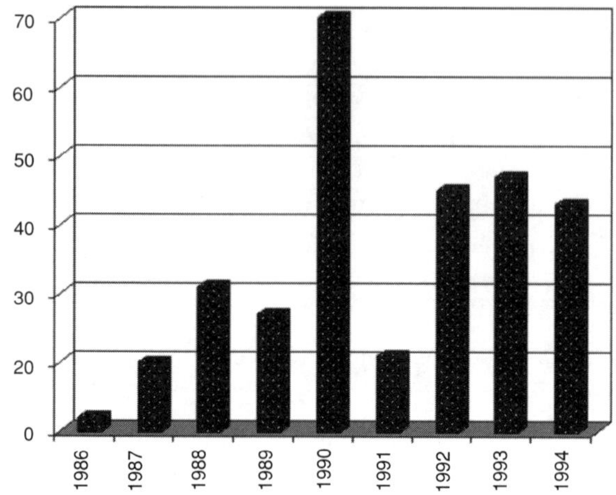

Figure 9.15 Texas pesticide surveillance data, 1986-1994.

veyed in a nine county area in the high southern Texas plains. It was found that only 20% of the cases were reported.

The available Texas data did not include a detailed assessment of dermatitis cases, but did include a noteworthy report on a fatal crop duster accident that involved paraquat burns and possibly systemic poisoning which were suspected to have been the ultimate cause of death:

> In October 1993, a crop duster crashed in a small rural Texas county. The plane had been applying paraquat dichloride for cotton defoliation. The pilot was burned by a small fire which occurred upon impact. The paraquat he had been applying had covered him causing splash-type chemical burns over a large portion of his body. Despite his injuries, the pilot was able to walk away from the crash. A witness drove the pilot to a nearby hospital where the pesticide was washed off. He was then transferred to the burn unit of a second hospital. After a preliminary examination, no evidence of acute pathology was found. The pilot remained responsive throughout this examination. Over the next five days, he became increasingly hemodynamically unstable, his respiratory status worsened, and his body temperature fell (all symptoms compatible with paraquat poisoning). On day seven following the crash, his blood pressure dropped rapidly. All efforts to resuscitate him failed. The death certificate listed chemical burns and paraquat exposure as the cause of death.

REFERENCES

1. **Bowen, S.S. and Moursund, M.P.,** Chloracne in the manufacture of DDT, *Arch. Dermatol.*, 75, 743-746, 1957.

* http://www.tdh.state.tx.us/epidemiology/eetd.html.

9.2.8 Midwestern and Eastern U.S.

Although there has been minimal effort made in pesticide illness surveillance in these regions of North America, there have been published case reports and epidemiologic studies. Investigations by the National Institute for Occupational Safety and Health (NIOSH) involving pesticides have also involved many of the pesticide manufacturing operations in the region, which account for the majority of total North American pesticide production.

9.2.8.1 Crop Production

Agricultural production in the middle western U.S. involves predominantly the mechanized production of grain and the production of livestock. The patterns of farm ownership and employment also differ markedly from the in western states which produce a large variety of specialty crops. A similar pattern is found in Canada (see Section 9.3). Data for the state of Iowa, for example, whose pilot surveillance program is discussed below, indicate that 1997 cash sales totaled $11,947,894,000, with corn and other grains accounting for 27.9%; soybeans for 23.1%; hogs, pigs, cattle, dairy, and other livestock products for 48.3%; and nursery crops, vegetables, and fruit crops for less than 1%. In Wisconsin, for 1997, agricultural cash sales totaled $5,579,861,000, with dairy products accounting for 49.2%; other livestock for 21.3%; corn, silage crops, and other grains for 13.1%; soybeans for 4.6%; and other crops for 9.1%.[1]

The eastern and southeastern sections of the U.S. show a wider variation in types of agricultural production, as illustrated by data from New York and South Carolina, whose surveillance programs are discussed below. New York had cash sales of $ 2,834,512,000 in 1997, with dairy products accounting for 51.5%; other livestock for 13.1%; corn, silage, and grain crops for 8.6%; fruits and berries for 6.5%; nursery crops for 10.3%; and other crops for 10%. 1997 agricultural cash sales for South Carolina totaled $1,588,173,000, with poultry and other livestock accounting for 50.2%; tobacco for 12.9%; nursery crops for 9.1%; cotton for 8.5%; soybeans, vegetables, fruits, and other crops for 11%; and corn, silage crops, oats, barley, and other grains for 8.3%. Florida had total cash sales of $6,004,554,000 for 1997, with fruits, vegetables, and nursery crops accounting for 67%. Dairy, poultry, and cattle were the predominant livestock products, together accounting for 17% of total sales.[1]

9.2.8.2 Case Reports

In 1929 a case of allergic contact dermatitis due to pyrethrum was reported in a New York City apartment dweller (Figure 9.16)[2] and allergic contact reactions associated with the manufacture of pyrethrum insecticide were investigated by the dermatologic component of the public health service in 1934.[3] The public health service also described dermatitis associated with alpha naphthyl isothiocyanate in 1939.[4]

Novel case reports from the region have included contact urticaria following exposure to sodium pentachlorophenate,[5] toxic epidermolysis in residents returning to houses fumigated with acrylonitrile,[6] and chloracne related to dioxins, but not phenoxy herbicides, in a chemist who synthesized brominated as well as chlorinated dioxins.[7] Reports also included suspected cases of dermatitis mediated by irritation due to

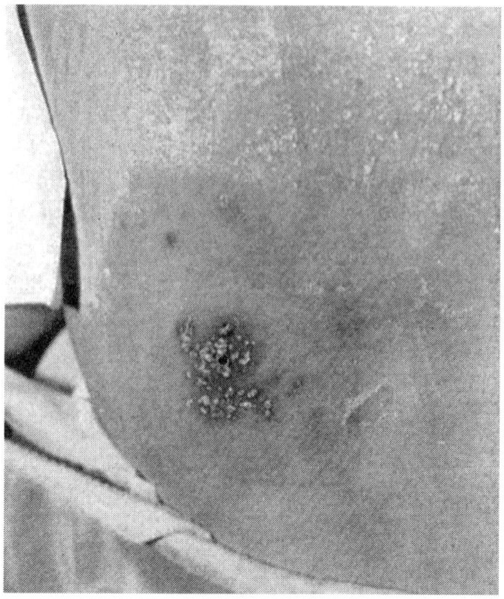

FIGURE 9.16 Patch reaction to pyrethrum powder. (Sulzberger, *JAMA*, 1930. Reproduced with permission of the American Medical Association.)

diquat,[8] naled,[9] and glutaraldehyde.[10] Allergic contact dermatitis case reports include compounds used on grain crops such as allidochlor (Randox®); (see Figure 9.17),[11] trifluralin and benefin,[12] atrazine,[13] and acetanilides.[14] Non-occupational contact dermatitis due to antimicrobial compounds, such as thimerosal,[15] has also been reported.

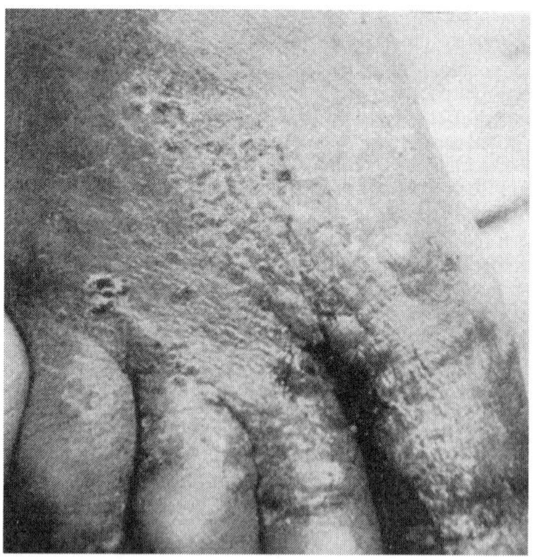

FIGURE 9.17 Dermatitis associated with allidochlor (Randox). (Spencer, *JAMA*, 1966. Reproduced with permission of the American Medical Association.)

TABLE 9.12
Pesticide Cases Reported to the State of Iowa Department of Public Health Pilot Surveillance Program — 1989 and 1990

Pesticide	Number of cases
EPTC (Eradicane®)	3
Terbufos (Counter®)	1
Malathion	1
Furadan	1
Unspecified organophosphate/carbamate	8
Chlordane	1
Unspecified organochlorine	1
2-4,D	1
Atrazine	1
Alachlor (Lasso®)	1
Dicamba (Banve1®)	1
Trifluralin (Treflan®)	3
Bifenzoquat methylsulfate (Avenge®)	1
Bromoxy nitrile (Buctril®)	1
Chloropicrin fumigant (Quasar®)	1
Alachlor/atrazine mixture (Lariat®)	1
Unspecified	3

Note: Counts are more than 26 as for some cases, multiple exposures were reported.

9.2.8.3 Surveillance

The region is not covered by ongoing surveillance programs, but pilot surveillance programs have been conducted in Wisconsin,[16] Iowa,[17] and Florida.[18] The Iowa pilot survey, based upon the Sentinel Event Model,[19] identified 26 cases in a 15-month period, principally involving cholinesterase inhibiting insecticides (e.g., terbufos, used for corn root worm) and herbicides (EPTC, alachlor, and bifenzoquat methyl sulfate) used in production of cereal crops (Table 9.12). Cases were described as including cutaneous and respiratory exposure, but dermatitis cases were not identified, per se. The Wisconsin pilot program, covering the period for 1979-1980 was based upon a combination of workers' compensation sources and pesticide incident reports from the Department of Agriculture, Trade and Consumer Protection. Like the Iowa data, the report did not specifically delineate cases of pesticide-associated skin conditions. In Florida, a pilot surveillance program evaluating agricultural injuries in the state, based upon workers' compensation reports filed between 1987 and 1989, documented 117 occupational diseases, including 40 cases of organophosphate poisoning from a single re-entry violation episode and 82 total pesticide illnesses. Skin cases accounted for 36 (30.8%) of all occupational illnesses, but the individual compounds involved were not described. A surveillance study conducted in South Carolina from 1983-1987 described 312 hospitalized pesticide poisoning cases,[20] noting that the data source utilized precluded evaluation of most cases of "allergy, dermatoses, or outpatient treatment." Based upon an earlier surveillance report,[21] the authors estimated approximately 15 outpatient cases for each hospitalizedcase.

The only active pesticide illness surveillance program in the eastern U.S. currently functions in the state of New York, based principally upon laboratory reporting of identified cases of cholinesterase depression and information reported by poison

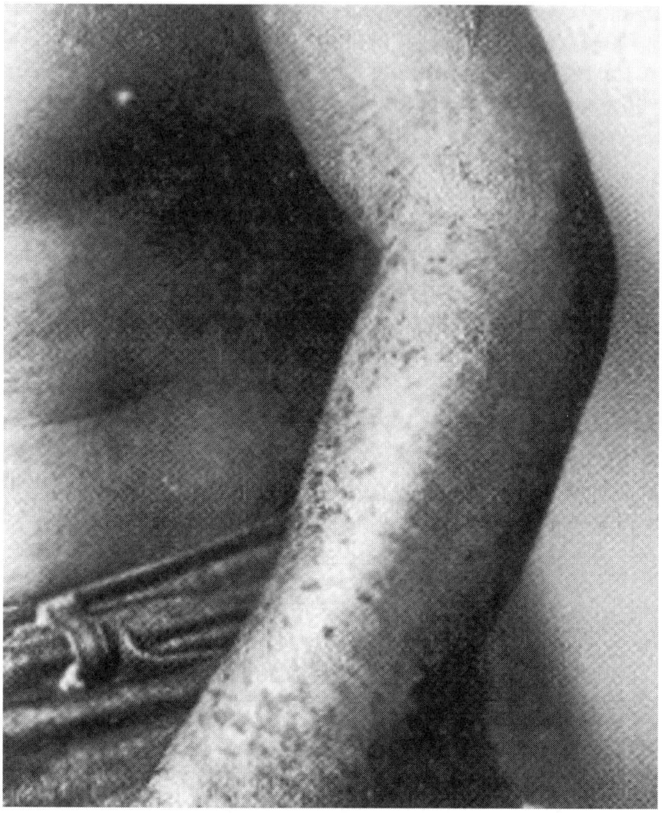

FIGURE 9.18 Dermatitis associated with exposure to residue of anilazine in tomato harvesters. (Schuman, 1980, reproduced with permission of *Lancet*.)

control. For the year 1995, 278 total reports were received and 205 for the year 1996, with 17-18% involving unspecified effects on the skin.[22] Published reports regarding pesticide illness surveillance from South Carolina have described hospitalized poisoning cases, principally due to organophosphates, carbamates, organochlorines, and synthetic pyrethroids, but did not include any description of dermatitis or other conditions seen in ambulatory care settings.[20]

9.2.8.4 Outbreak, Cohort, Incidence, and Prevalence Studies

9.2.8.4.1 Creosote

In 1935, Shambaugh [23] reported on a case of metastatic lip cancer in a fisherman with a six-year work history of handling nets impregnated with creosote tar. In doing net repairs, this patient apparently had a habit of holding a tar-smeared needle in one corner of his mouth, and developed a small hard nodule at the site of contact. After the death of this patient, Shambaugh conducted a survey of 141 lip cancer cases at one hospital and identified 4 additional fishermen and identified 3 additional cases of fishermen from ad hoc sources. In investigation of net repair and net tarring businesses, he found cases of tar warts of the hands and forearms, but no cases of skin cancer.

Jonas[24] reported an examination survey of 2700 creosote handlers employed as carpenters, roofers, and wood treatment workers; 450 (16.7%) had mild (erythema on the neck and face, followed by secondary pigmentary changes noticeable in Caucasian workers) or severe (intense burning and itching, intense bronze secondary pigmentation, and desquamation) burns and 3% also had keratitis.

9.2.8.4.2 Dermatitis associated with anilazine

The outbreak associated with anilazine reported by Schuman in 1980 is discussed in Section 8.3.7. A representative case is illustrated in Figure 9.18.

9.2.8.4.3 Dog groomers

The New Jersery Department of Environmental Protection in Trenton performed a study evaluating the prevalence of symptoms associated with insecticide use among animal groomers. About one third of the survey respondents reported symptoms compatible with pesticide exposure.[25] The symptoms evaluated in the study questionnaire included skin complaints, but the study protocol did not have an adequate means of differentiating among the many possible causes of the nonspecific symptoms listed as "pesticide-associated."

9.2.8.4.4 Floral shop workers

In 1990, Thiboutot[26] reported on a study of 57 floral shop workers from Pennsylvania, identifying a history of hand dermatitis in 15 (26%) the preceding year. Sixteen subjects volunteered for patch testing with the North American Contact Dermatitis Group Standard and Perfume Trays, a series of eight commercial pesticides antigens and 20 plant allergens. Four reacted to the antigen tuliposide A (found in Alstroemeria, Narcissus and other plants). A reaction to difolatan (Captafol) was the only pesticide reaction noted, the relevance of which is unknown. A California study reported by Saiz, in 1992,[27] identified low levels of fungicide (chlorothalonil, vinclozolin, and captan) and insecticide residues (chlorpyrifos, fluvalinate, diazinon, and permethrin) in 100% of 92 hand wash samples obtained from a sample of florists studied in February and March of 1991, indicating exposure to a broad range of pesticides. Conclusions from the Pennsylvania study about possible sensitivity to pesticides were limited because of the narrow range of antigens tested.

9.2.8.4.5 Chloracne

Cases of chloracne related to the production of 3,4-dichloroaniline and related herbicides (e.g., methazole – see Figure 9.19) were described by Taylor.[28,29] Related cases were noted in a U.S. Public Health Service survey of workers at a plant producing the herbicide propanil (also discussed in Section 8.5.2), the carbamate insecticide methomyl, and 3,4-dichloroaniline.[30] In addition to skin complaints, workers also suffered from methemoglobinemia due to 3,4-dichloroaniline, a compound used in synthesis of propanil. Of the 102 workers, 11% had been hospitalized for illness related to chemical exposures, including both methomyl poisoning and methemoglobinemia. Seventeen (61%) of 28 production workers with exposure to

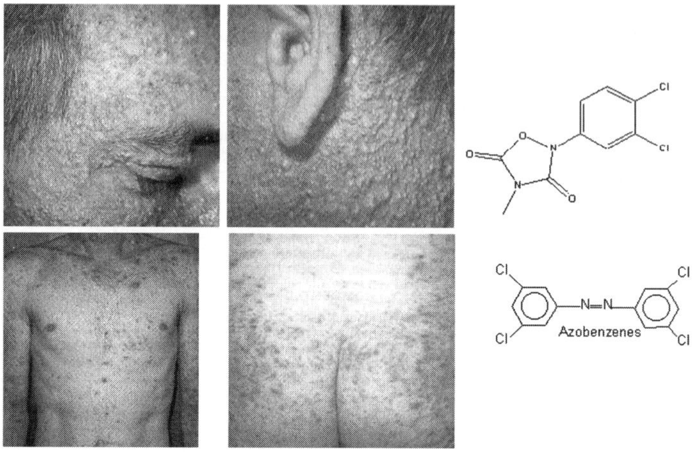

FIGURE 9.19 Chloracne from 3,3', 4,4'-tetrachloroazoxybenzene and 3,3',4,4'-tetrachloroazobenzene, associated with the production of methazole. (Reported by Taylor, *Archives of Dermatology*, 1977. Reproduced with permission from the American Medical Association.)

dichloroaniline and propanil (3'4'-dichloropropionanilide) had chloracne. The ability of both compounds to produce this condition was verified in the rabbit ear test. Investigations conducted by Taylor identified the chloracnegen as an azobenzene compound, 3,3',4,4'-tetrachloroazoxybenzene.[29]

An additional cluster of chloracne cases was described by Deeken,[31] associated with the production of the herbicide dichlobenil (Casoron®) in a factory near Kansas City. Although the factory was included in the NIOSH dioxin registry, the cases associated with the dichlobenil process were distinct from those associated with the production of phenoxy herbicides. Nevertheless, the specific contaminant responsible for the chloracne in the dichlobenil production process was never described.

The bulk of chloracne cases reported in North America have been associated with the production of phenoxy herbicides and trichlorophenol (see Section 8.5.3). The NIOSH Dioxin Registry was created in order to evaluate the risk of soft tissue sarcoma, lymphoma, and liver cancer among workers exposed to tetra-, hexa-, and hepta-chlorinated dibenzo-*p*-dioxins in the course of manufacturing and formulating pentachlorophenol, 2,4,5-trichlorophenol (TCP), 2,4,5-trichlorophenoxyacetic acid (2,4,5-T), and 2,4,5-T acid esters. A total of 7329 potentially exposed workers were identified from the personnel records of the 14 plants in the registry and 4668 (63.7%) had some type of medical record for review. The amount of detailed medical information available was not uniform for each plant and each worker in the study. For plants 1, 7, 8, 9, 10, 11, and 12 periodic medical surveillance was conducted at varying intervals for most workers exposed to phenoxy herbicides and chlorophenols. In plants 2, 3, 4, 5, 6, 13, and 14 such surveillance was minimal or altogether lacking. This variation in medical surveillance is reflected in the percentage of workers in each of the groups who had medical records available to review. In the 7 plants with periodic examinations 3999 (80.9%) of 4941 workers had medical records available; in the remaining plants only 335 (14.0%) of 2388 workers had records available for review.

TABLE 9.13
Cumulative Incidence of Chloracne by Plant in the NIOSH Dioxin Registry

Plant	Definite/ probable chloracne cases	%	Possible chloracne cases	%	Non-cases	%	Missing records	%	Total employed
1	101	21.4	6	1.2	182	38.6	182	38.6	471
2	1	1.3	6	7.7	15	19.2	56	71.8	78
3	2	0.3	12	1.6	53	7.1	683	91.1	750
4	25	6.6	19	5.0	222	58.0	114	30.0	380
5	2	2.4	0	0.0	67	78.0	16	18.0	85
6	8	4.3	0	0.0	1	0.5	176	95.0	185
7	54	5.3	23	2.2	677	65.9	298	29.0	1,027
8	259	44.6	42	6.7	248	42.7	32	5.5	581
9	222	10.1	37	1.7	1,750	79.6	190	8.6	2,199
10	71	21.8	1	0.3	130	40.0	123	37.9	325
11	4	2.1	1	0.5	99	54.4	78	42.9	182
12	11	7.1	1	0.6	60	38.5	84	53.8	156
13	0	0.0	8	4.3	140	75.7	37	20.0	185
14	0	0.0	0	0.0	88	12.1	637	87.9	725
Total	**780**	**10.6**	**156**	**2.1**	**3,732**	**54.8**	**2,069**	**31.1**	**7,329**

The NIOSH registry review identified 780 cases of definite and probable chloracne, accounting for 10.6% of the cohort as a whole, and 15.8% of those workers with available medical records. Measured as a percentage of the entire workforce, the cumulative incidence of cases ranged from 0.0% in plants 13 and 14 to 44.6% in plant 8. In addition to differences from plant to plant there was considerable variation within each plant in the distribution of cases across time (Table 9.13). These observed variations in the distribution of chloracne give valuable clues regarding the exposure of the cohort to dioxins, reflecting a variety of changes in production processes, production levels, and the occurrence of major process accidents.

The portion of the dioxin cohort involved in the 1949 process accident was the subject of two separate examination surveys. Moses[32] found persistent chloracne in 52% of 226 workers examined in a 1979 study. In this study chloracne was associated with abnormal gamma-glutamyl transpeptidase (GGT) and higher mean GGT was found in those with chloracne, compared to those without. Neurological examination showed a statistically significant higher prevalence of abnormal sensory findings in those with chloracne as was an increased prevalence of self-reported sexual dysfunction and decreased libido. No differences were found between those with and without chloracne in serum cholesterol, total urinary porphyrins, or in reproductive outcome. A portion of the same group cohort was also examined by Suskind.[33] The population consisted of two cohorts: 204 clearly exposed and 163 not exposed. The proportion of those with persistent chloracne was 55.7%, similar to the survey conducted by Moses. In the Suskind survey, chloracne was associated with actinic elastosis of the skin, and with the history of gastrointestinal tract ulcer.

An examination survey of a separate portion of the cohort was conducted by NIOSH in 1987.[34] Comparison was made with an unexposed control group in interviews, examinations, and laboratory tests. The latter included lipid adjusted serum TCDD levels. This test showed a mean value of 220 pg per gram of lipid (range = not detected to be 3400 pg per gram of lipid) in the cohort members and referents, 7 pg per gram of lipid (range not detected to be 20 pg per gram of lipid), $p < 0.001$) in the comparison group. Chloracne at the time of the survey was associated with the highest levels of 2,3,7,8-TCDD. Abnormal (GGT), the reproductive hormones serum testosterone, luteinizing, and follicle-stimulating hormones, and abnormal high density lipoprotein concentration, counts of CD3/Ta1 cells (helper lymphocytes), and fasting serum glucose levels also showed significant correlation with TCDD levels. The range of serum TCDD found in the NIOSH study was similar to that seen in other dioxin exposed production workers elsewhere in the world. A study conducted by Neuberger and Landvoigt[35] showed a median TCDD level of 340 pg/g (range 98-659 pg/g) in 9 subjects with chloracne in a German cohort compared to a median level of 18, (range 8-28 pg/g) in 4 in-plant referents and in 17 external referents (median 16, range less than 5-23 pg/g).

REFERENCES

1. **United States Department of Agriculture,** 1997 Census of agriculture. Available online at: www.usda.gov, 1999.
2. **Sulzberger, M. and Weinberger, C.,** Dermatitis due to insect powder, *JAMA*, 95, 111, 1930.
3. **Schwartz, L.,** Skin hazards in American industry: Dermatitis due to pyrethrum contained in an insecticide, *Pub. Health Bull.*, Part I, 1934.
4. **Reports, P. H.,** Dermatitis caused by a new insecticide, *Pub. Health Rep.*, 54, 1426, 1939.
5. **Kentor, P.,** Urticaria from contact with pentachlorophenate [letter], *JAMA*, 256, 3350, 1986.
6. **Radimer, G., Davis, J., and Ackerman, A.,** Fumigant-induced toxic epidermal necrolysis, *Arch. Dermatol.*, 110, 103, 1974.
7. **Schecter, A. and Ryan, J.,** Persistent brominated and chlorinated dioxin blood levels in a chemist 35 years after dioxin exposure, *J. Occup. Med.*, 34, 702, 1992.
8. **Wood, T. E., Edgar, H., and Salcedo, J.,** Recovery from inhalation of diquat aerosol, *Chest*, 70, 774, 1976.
9. **Mick, D. L., Gartin, T. D., and Long, K. R.,** A case report: occupational exposure to the insecticide naled, *J. Iowa Med. Soc.*, 60, 395, 1970.
10. **Lyon, T. C.,** Allergic contact dermatitis due to Cidex. Report of a case, *Oral Surg. Oral Med. Oral Pathol.*, 32, 895, 1971.
11. **Spencer, M. C.,** Herbicide dermatitis, *JAMA*, 198, 1307, 1966.
12. **Pentel, M. T., Andreozzi, R. J., and Marks, J. G., Jr.,** Allergic contact dermatitis from the herbicides trifluralin and benefin, *J. Am. Acad. Dermatol.*, 31, 1057, 1994.
13. **Schlicher, J. E. and Beat, V. B.,** Dermatitis resulting from herbicide use — a case study, *J. Iowa Med. Soc.*, 62, 419, 1972.
14. **Iden, D. L. and Schroeter, A. L.,** Allergic contact dermatitis to herbicides [letter], *Arch. Dermatol.*, 113, 983, 1977.
15. **Stolman, L. P. and Sands, E.,** Contact-lens hand [letter], *N Engl. J. Med.*, 311, 1521, 1984.

16. **Weiss, H. B.,** Human exposures to pesticides. A 1979-1980 report of the Wisconsin Division of Health, *Wis. Med. J.*, 80, 12, 1981.
17. **State of Iowa Department of Public Health**, Draft Summary Report: Non-infectious Illness Reporting in Iowa: Pesticides and Lead, 1989-1990, State of Iowa Department of Public Health.
18. **Becker, W.**, An analysis of serious occupational diseases and poisoning to agricultural workers in Florida during a three-year period, 1987-1989, 1991, University of Florida.
19. **Maizlish, N., Rudolph, L., and Dervin, K. T.,** The surveillance of work-related pesticide illness: an application of the Sentinel Event Notification System for Occupational Risks (SENSOR), *Am. J. Pub. Health*, 85, 806, 1995.
20. **Schuman, S., Whitlock, M., Caldwell, S., and Horton, P.,** Update on hospitalized pesticide poisonings in South Carolina, 1983-1987, *J. S. C. Med. Assoc.*, 85, 62, 1989.
21. **Caldwell, S. and Watson, M.,** Hospital survey of acute pesticide poisoning in South Carolina 1971-1973, *J. S. C. Med. Assoc.*, 71, 249, 1975.
22. **New York State Department of Health,** Pesticide Registry Report 1995-1996.
23. **Shambaugh, P.,** Tar cancer of the lip in fishermen, *JAMA*, 104, 2326, 1935.
24. **Jonas, A.,** Creosote burns, *J. Ind. Hyg. Toxicol.*, 25, 418, 1943.
25. **Bukowski, J., Brown, C., Korn, L., and Meyer, L.,** Prevalence of and potential risk factors for symptoms associated with insecticide use among animal groomers, *J. Occup. Environ. Med.*, 38, 528, 1996.
26. **Thiboutot, D. M., Hamory, B. H., and Marks, J. G., Jr.,** Dermatoses among floral shop workers, *J. Am. Acad. Dermatol.*, 22, 54, 1990.
27. **Saiz, S., Blewett, T., Haskell, D., and Margetich, S.**, Florist Hand Exposure to Pesticides, Sacramento, CA, 1992, California Department of Pesticide Regulation.
28. **Taylor, J. S., Wuthrich, R. C., Lloyd, K. M., and Poland, A.,** Chloracne from manufacture of a new herbicide, *Arch. Dermatol.*, 113, 616, 1977.
29. **Taylor, J. S. and Lloyd, K. M.,** Chloracne From 3,3',4,4'-tetrachloroazoxybenzene And 3,3',4,4'-tetrachloroazobenzene update and review, *Chlorinated Dioxins and Related Compounds*, 1982.
30. **Morse, D. L., Baker, E. L., Jr., Kimbrough, R. D., and Wisseman, C. L. D.,** Propanil-chloracne and methomyl toxicity in workers of a pesticide manufacturing plant, *Clin. Toxicol.*, 15, 13, 1979.
31. **Deeken, J. H.,** Chloracne induced by 2,6-dichlorobenzonitrile, *Arch. Dermatol.*, 109, 245, 1974.
32. **Moses, M., Lilis, R., Crow, K. D., Thornton, J., Fischbein, A., Anderson, H. A., and Selikoff, I. J.,** Health status of workers with past exposure to 2,3,7,8-tetrachlorodibenzo-*p*-dioxin in the manufacture of 2,4,5- trichlorophenoxyacetic acid: comparison of findings with and without chloracne, *Am. J. Ind. Med.*, 5, 161, 1984.
33. **Suskind, R. R. and Hertzberg, V. S.,** Human health effects of 2,4,5-T and its toxic contaminants, *JAMA*, 251, 2372, 1984.
34. **Sweeney, M. H., Calvert, G. M., Egeland, G. A., Fingerhut, M. A., Halperin, W. E., and Piacitelli, L. A.,** Review and update of the results of the NIOSH medical study of workers exposed to chemicals contaminated with 2,3,7,8-tetrachlorodibenzodioxin, *Teratog. Carcinog. Mutagen*, 17, 241, 1997.
35. **Neuberger, M., Landvoigt, W., and Derntl, F.,** Blood levels of 2,3,7,8-tetrachlorodibenzo-*p*-dioxin in chemical workers after chloracne and in comparison groups, *Int. Arch. Occup. Environ. Health*, 63, 325, 1991.

9.3 CANADA

FAO data demonstrate that grain, livestock products, and oil crops are the principal Canadian agricultural products[1] and that herbicides account for the bulk of pesticide use (Figure 9.20). Trade data indicate that most of pesticides used are imported (Figure 9.21).

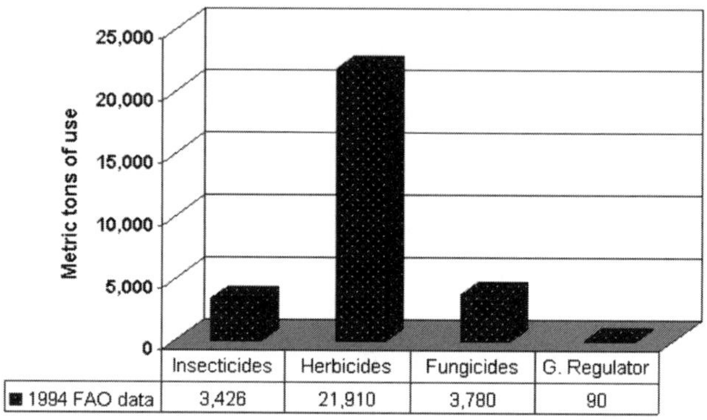

FIGURE 9.20 1994 pesticide use data from UN FAOSTAT database.

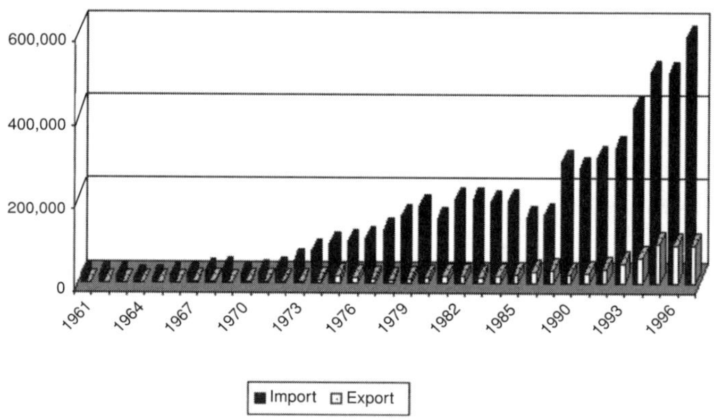

FIGURE 9.21 Canadian pesticide trade data, FAO, 1961-1997.

9.3.1 Epidemiologic Information

Limited information is available about the occurrence of pesticide skin conditions in Canada. In 1978, Newhouse[2] reported on a case of paraquat ulceration in a 39-year-old fruit farmer's wife, initially identified by the dermatology service of a hospital in Hamilton, Ontario. The patient developed systemic poisoning and died of respiratory failure 8 weeks after initial exposure to paraquat. Hogan,[3] in 1985, described two cases of allergic contact dermatitis (Figures 9.22a,b, and c) associated with the chlorocarbanilate herbicide barban (Carbyne®). The material was confirmed as a potent sensitizer in subsequent tests on guinea pigs.

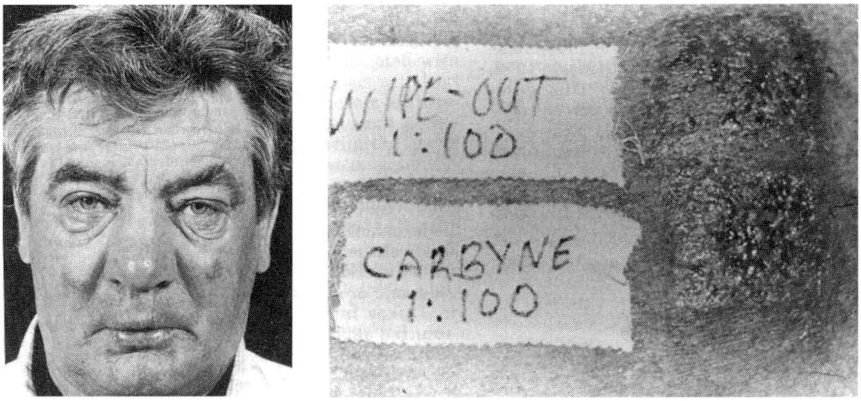

FIGURE 9.22 Allergic contact dermatitis from Rarban (Carbyne®). (Reprinted with permission from Adams, *Occupational Skin Disease*, 2nd Edition, W.B. Saunders.)

Other occupational skin problems appear to have occurred frequently among Canadian farmers (Figures 9.23a,b, and c). The importance of livestock agriculture in the country is emphasized by reports from Saskatchewan describing extensive case series of dermatophytoses associated with raising cattle.[4,5] From provincial cancer registry data, which includes non-melanoma skin cancers, Hogan also described the occurrence of occupational skin cancers in the same population of Saskatchewan farmers.[6]

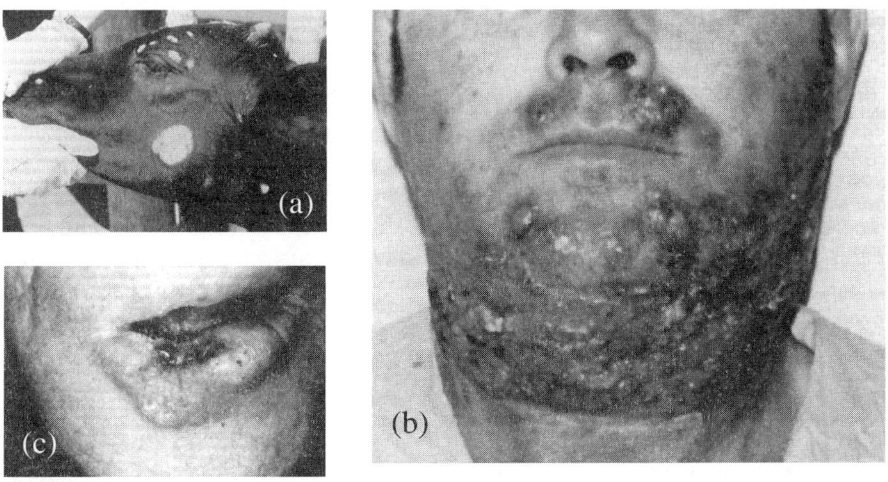

FIGURE 9.23 Non-pesticide skin problems in Canadian agriculture include fungal infections associated with handling infected livestock, and non-melanoma skin cancer. (a) *T. verrucosum* infection in a young cow; the organism may cause severe skin infection in man. (b) Excessive inflammatory skin lesions on face and neck due to skin infection with *T. verrucosum*. (c) Non-melanoma skin cancer in a Saskatchewan farmer. (Reprinted from Dosman and Cockroft, *Principles of Health and Safety in Agriculture*, CRC Press, 1989.)

REFERENCES

1. **FAOSTAT,** Canadian production data for primary livestock, primary crops, and crops by area harvested, 1997. Available online at: apps.fao.org, 1999.
2. **Newhouse, M., McEvoy, D., and Rosenthal, D.,** Percutaneous paraquat absorption. An association with cutaneous lesions and respiratory failure, *Arch. Dermatol.*, 114, 1516, 1978.
3. **Hogan, D. J. and Lane, P. R.,** Allergic contact dermatitis due to a herbicide (barban), *Can. Med. Assoc. J.*, 132, 387, 1985.
4. **Armstrong, K. R. and Post, K.,** *Principles of Health and Safety in Agriculture,* Dosman, J. A. and Cockcroft, D. W., Eds., CRC Press, Boca Raton, FL, 1989, 288.
5. **Hogan, D. J. and Martin, W. T.,** *Principles of Health and Safety in Agriculture,* Dosman, J. A. and Cockcroft, D. W., Eds., CRC Press, Boca Raton, FL, 1989, 294.
6. **Hogan, D., Lane, P., Gran, L., and Wong, D.,** Risk factors for squamous cell carcinoma of the skin in Saskatchewan, Canada, *J. Dermatol. Sci.*, 1, 97, 1990.

9.4 EUROPE

European case reports and case series, prevalence studies, and reports on outbreak investigations account for a substantial portion of the world literature on the skin effects of pesticides. The presentation of this large array of information is organized here by region of Europe. In each region, crop production, pesticide use, and trade data are reviewed based upon data from the FAO. Emphasis is given to cases involving novel chemicals or dermatoses, followed by discussion of compounds reported in multiple geographic areas, and a review of available prevalence, incidence, surveillance, and outbreak data.

9.4.1 U.K. and Ireland

Based upon 1998 FAO data, the principal agricultural products for the U.K. and Ireland include grain, root crops, dairy, and livestock.[1] FAO statistics also indicate that herbicides account for approximately two thirds of the pesticide use in the U.K. and Ireland for 1997.[2] Trade data (Figures 9.24a and b) indicate that both countries are net importers of pesticides, but also demonstrate a substantial export trade. Pesticide compounds associated with cases or outbreaks of dermatitis reported in the U.K. are shown in Figure 9.25.

9.4.1.1 Case Reports

Cases affecting agriculture workers included allergic dermatitis associated with dusting sulfur in a professional gardener;[3] contact allergy to dithianone and to lindane in a female nursery worker, who was also sensitive to primula;[4] and photoallergy associated with a formulation of glyphosate[5] that proved related to 1,2-benzisothiazoline-3-one, a preservative in the end-use formulation, rather than to the active ingredient.[6] Other compounds associated with agricultural contact dermatitis in the U.K. and Ireland included paraquat[7] and pyrethrum extract.[8]

The experimental synthesis or factory production of pesticides accounted for several cases, including dermatitis secondary to captafol in two agricultural chemists; contact

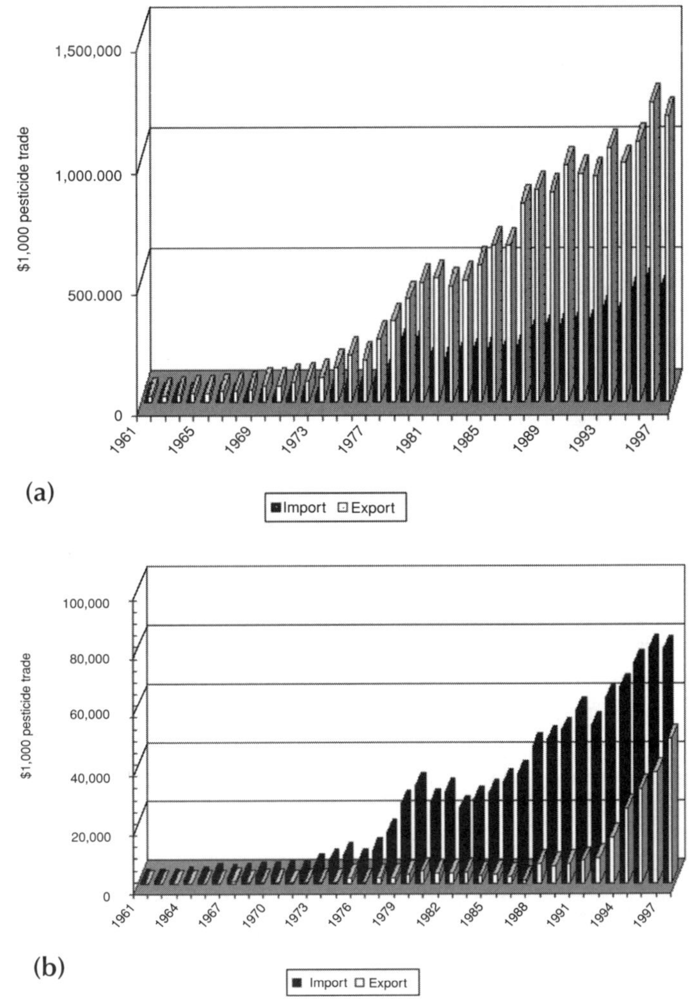

FIGURE 9.24 Pesticide trade data for the U.K. and Ireland. (a) The U.K. (b) Ireland. (UN FAO database)

allergy to the fungicide bupirimate in a formulation plant worker;[9] allergy to the chlorothalonil analog TFTPN in a synthetic chemist;[10] to the pesticide synthetic intermediates tetrachloronitrobenzene[11] and ethoxymethylenemalononitrile, an analog of malathion.[12]

Cases related to non-agricultural end-use included allergy to fentichlor in a maintenance worker exposed to a 40% concentrate[13] and dermatitis in paper mill workers associated with the fungicide/slimicide compounds bis-1,4-bromoacetoxy-2-butene and 2,3-dichloro-4-bromotetrahydrothiophene-1,1-dioxide.[14] Warin also recorded a case of facial dermatitis due to dazomet, used as a slimicide, in an English paper mill worker. [Warin, 1992 #1963]

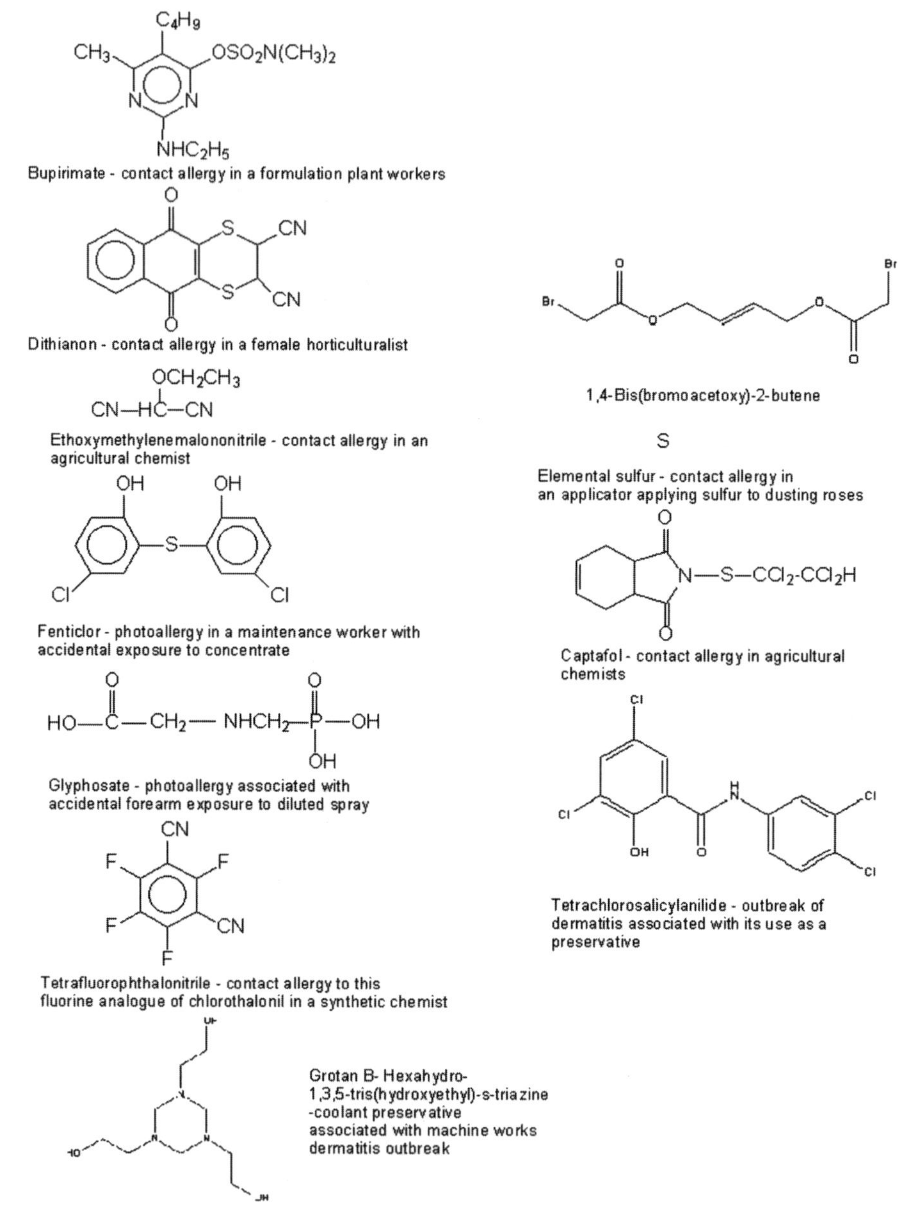

FIGURE 9.25 Pesticide contact dermatitis.

9.4.1.2 Prevalence, Incidence, Surveillance, and Outbreak Data

Epidemiologic reports have included outbreaks of dermatitis due to chlorosalicylanilide containing soaps in an English foundry,[15, 16] and an outbreak due to 2-chlorobenzoyl-chloride-azine (CCA), an intermediate in an acaricide synthesis.[17] As in other jurisdictions (see Section 8.5.3), the English manufacture of 2,4,5-trichlorophenol was associated with a process explosion and a subsequent outbreak of chloracne.[18]

In 1976, Keczkes and Brown[19] reported the results of an investigation into a cluster of dermatitis cases in a machine shop; 29 men sought treatment for dermatitis at the plant dispensary during the outbreak, 19 working with coolants. Patch tests included the standard European series, several machine coolant ingredients (Solvac® 1534, Kutwell® 40, and Grotan® 0.2%), a barrier cream, and a liquid soap. One patient had a + reaction to chromate, of no clear relevance to the outbreak; 4 ++ reactions to Grotan (Hexahydro, 1,3,5,tris (2-hydroxyethyl) triazine) while others were negative. Allergic reactions to this newly identified hazard were therefore common in this population, but did not explain all of the outbreak associated cases.

Surveillance reports on pesticide illnesses in the U.K. have not been routinely published. In 1973, Hearn[20] reviewed poisoning incidents reported to the Pesticides Branch of Ministry of Agriculture, Fisheries, and Food from 1952 to 1971. Out of a total of 250 recorded pesticide effects, 54 (21.6%) were dermatitis and 18 (7.1%) were chemical burns. The chemical burns were principally due to non-pesticidal compounds such as sulfuric acid, phenol, and hydrofluoric acid. The compounds associated with dermatitis included paraquat, organomercurials, diquat, thiocarbamates (thiram, maneb, zineb, and metam sodium), phenoxy herbicides (MCPA and 2,4-D), diesel oil, DNBP, DDT/lindane/phenylmercury spray, demeton methyl, dicamba/TBA, dinitrocresol, and sodium arsenite.

REFERENCES

1. **FAOSTAT,** Production data for primary livestock, primary crops, and crops by area harvested, 1998, for United Kingdom and Ireland. Available online at: apps.fao.org, 1999.
2. **FAOSTAT,** Pesticide Use Data, 1997, for United Kingdom and Ireland. Available online at: apps.fao.org, 1999.
3. **Wilkinson, D.,** Sulphur sensitivity, *Contact Dermatitis*, 1, 58, 1975.
4. **Calnan, C.,** Dithianone sensitivity, *Contact Dermatitis Newsletter*, 6, 119, 1969.
5. **Hindson, C. and Diffey, B.,** Phototoxicity of glyphosate in a weedkiller, *Contact Dermatitis*, 10, 51, 1984.
6. **Hindson, T. C. and Diffey, B. L.,** Phototoxicity of a weedkiller: A correction, *Contact Dermatitis*, 11, 260, 1984.
7. **Sharvill, D.,** Reaction to paraquat, *Contact Dermatitis Newletter*, 9, 210, 1971.
8. **Mitchell, J. C., Dupuis, G., and Towers, G. H.,** Allergic contact dermatitis from pyrethrum (Chrysanthemum spp.). The roles of pyrethrosin, a sesquiterpene lactone, and of pyrethrin II, *Br. J. Dermatol.*, 86, 568, 1972.
9. **McFadden, J. P., Kinoulty, M., and Rycroft, R. J.,** Allergic contact dermatitis from the fungicide bupirimate, *Contact Dermatitis*, 28, 47, 1993.
10. **Carmichael, A. J., Foulds, I. S., and Sadhra, S.,** Allergic contact dermatitis from tetrafluoroterephthalonitrile and TFX diamine, *Contact Dermatitis*, 20, 233, 1989.
11. **Cotterill, J. A.,** Contact dermatitis following exposure to tetrachloronitrobenzene, *Contact Dermatitis*, 7, 353, 1981.
12. **Wakelin, S. H., Price, A. E., Basketter, D. A., and Rycroft, R. J.,** Allergic contact dermatitis from ethoxymethylenemalononitrile in an agrochemical chemist, *Contact Dermatitis*, 38, 237, 1998.

13. **Norris, P. G., Hawk, J. L., and White, I. R.,** Photoallergic contact dermatitis from fentichlor, *Contact Dermatitis*, 18, 318, 1988.
14. **Rycroft, R. J. and Calnan, C. D.,** Dermatitis from slimicides in a paper mill, *Contact Dermatitis*, 6, 435, 1980.
15. **Jones, D.,** Investigations of an outbreak of photodermatitis confined to one shop in a large factory, *Br. J. Ind. Med.*, 19, 100, 1962.
16. **Wilkinson, D.,** Photodermatitis due to tetrachlorsalicylanilide, *Br. J. Derm.*, 73, 213, 1961.
17. **Pryce, D. W.,** Allergic contact dermatitis from 2-chlorobenzoyl chloride azine, *Contact Dermatitis*, 16, 285, 1987.
18. **Martin, J. V.,** Lipid abnormalities in workers exposed to dioxin, *Br. J. Ind. Med.*, 41, 254, 1984.
19. **Keczkes, K. and Brown, P.,** Hexahydro, 1,3,5,tris (2-hydroxyethyl) triazine, a new bacteriocidal agent as a cause of allergic contact dermatitis, *Contact Dermatitis*, 2, 92, 1976.
20. **Hearn, C. E.,** A review of agricultural pesticide incidents in man in England and Wales, 1952-71, *Br. J. Ind. Med.*, 30, 253, 1973.

9.4.2 France/Belgium/The Netherlands

Based upon FAO data, the principal agricultural products in France, Belgium, and the Netherlands include grain, root crops, pulses (legumes), vegetables, fruit, milk, and meat.[1] Although grain and livestock account for the majority of agricultural commodities produced, the quantity of fruit and vegetables produced in the regions far exceeds the quantities produced in Ireland and the U.K.

FAO data on production by metric tons and by hectares harvested do not give a detailed account of the Netherlands' production of horticultural crops. Data from The Netherlands' Ministry of Agriculture, Nature Management, and Fisheries for 1995 indicate that arable crops accounted for only 8.8% of the cash value of Dutch agricultural products, whereas horticultural products accounted for approximately 36% and products derived from livestock for the remainder.[2]

The overall quantity of pesticides reported used in the region exceeded the quantity reported used in England and Ireland, with fungicides and herbicides accounting for the majority (Figure 9.26). Trade data demonstrate approximately equal quantities of pesticide imports and exports (Figure 9.27). Pesticide compounds associated with cases or outbreaks of dermatitis reported in the Belgium, the Netherlands, and France are shown in Figure 9.28.

9.4.2.1 Case Reports and Case Series

Cases associated with cultivation of arable land in the region included an allergically mediated case of depigmentation in a grain miller following exposure to dinoterbe (2-*t*-butyl-4,6-dinitrophenol);[3] nail damage from exposure to concentrated solution of dinitro-ortho-cresol prior to application to fruit trees (Figure 9.29);[4] contact allergy, accompanied by secondary depigmentation, associated with handling the acetanilide herbicide alachlor;[5] and seven cases of contact dermatitis following contact with the thiadiazine fumigant dazomet.[6] Other compounds associated with agricultural allergic contact dermatitis included the animal feed additive furazolidone[7] and the herbicide phenmediphan.[8] Two

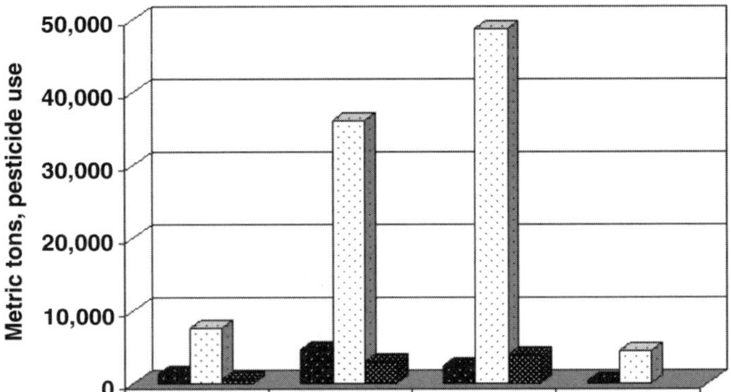

	INSECTICIDES	HERBICIDES	FUNGICIDES	G. REGULATORS
Belgium-Luxembourg	1,222	4,724	2,403	392
France	7,603	36,052	48,624	4,500
Netherlands	495	3,013	3,925	

FIGURE 9.26 1995-1996 pesticide use data for Belgium-Luxembourg, France, and The Netherlands.

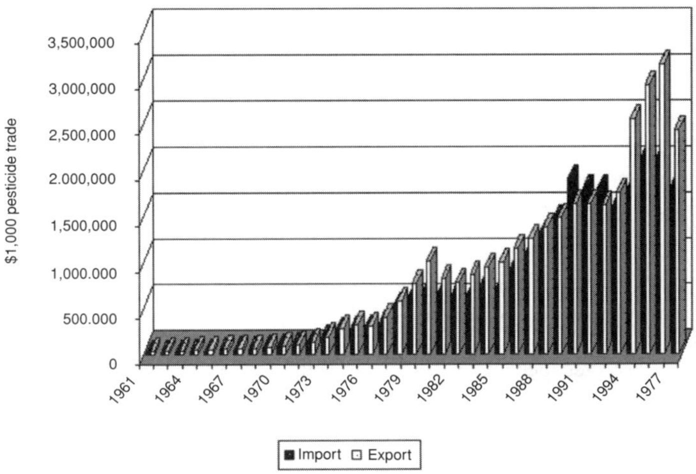

FIGURE 9.27 Pesticide trade data for Belgium-Luxembourg, The Netherlands, and France. (UN FAOSTAT database)

unusual cases of skin irritation from paraquat, accompanied by systemic toxicity, were also reported. Although one of the cases involved mistaken application of paraquat to the skin as a scabicide, the other occurred in a gardener applying both paraquat and diquat.[9]

Cases associated with the horticultural industries in the Netherlands included contact dermatitis due to the insecticides methiocarb,[10] methomyl,[11] omethoate, and dimethoate.[12] Allergic contact dermatitis was also reported due to fungicides used

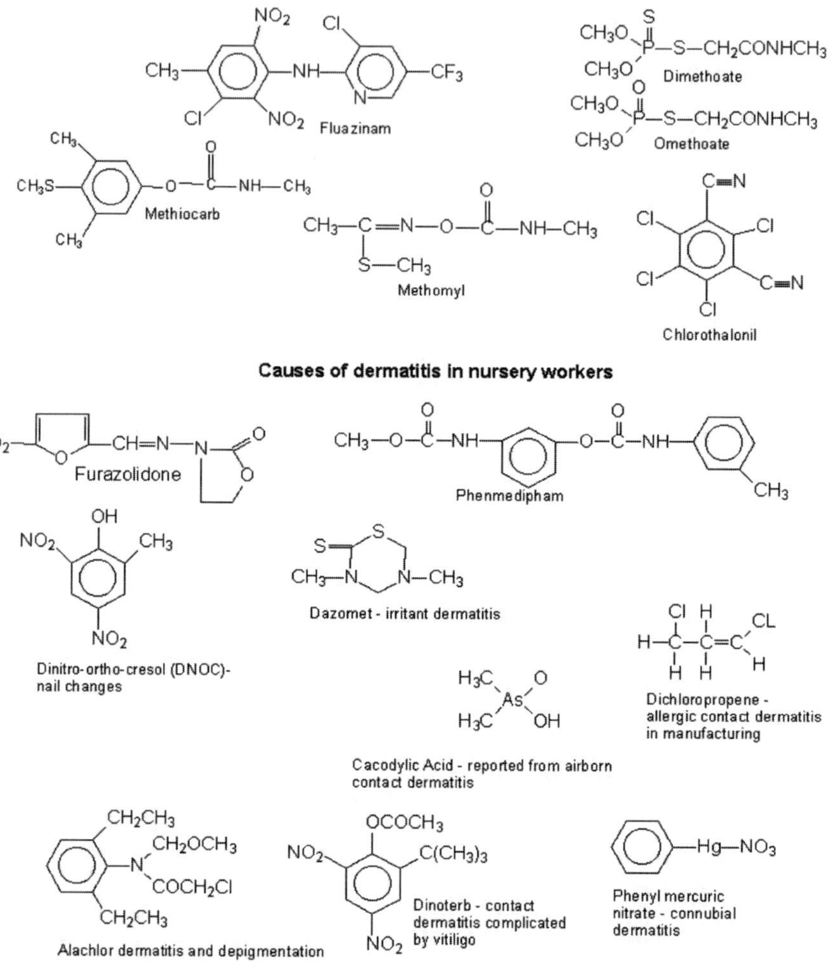

FIGURE 9.28 Selected pesticides associated with contact dermatitis in Belgium-Luxembourg, France, and The Netherlands.

in floriculture such as fluazinam,[13] chlorothalonil,[14] benomyl,[15] and Plondrel.[16] Cases of allergic contact dermatitis secondary to residues of the fungicide maneb on office plants[17] and cut flowers were also reported.[18]

Cases in pesticide laboratory and manufacturing workers included airborne contact with cacodylic acid,[19] allergic contact dermatitis due to the chemical intermediate tetrachloroacetophenone, used in organophosphate synthesis,[20] the D-D (dichloropropane/dichloropropene) fumigant mixture (Figure 9.30),[21, 22] and to methanesulfonyl (mesyl)-chloride and mesyl isopropyllactate.[23]

Compounds associated with allergic contact dermatitis following non-agricultural use included the antimicrobials thimerosal,[24] 1,2-benzisothiazolin-3-one,[25] DNCB,[26] and phenylmercuric nitrate used as a spermicide.[27]

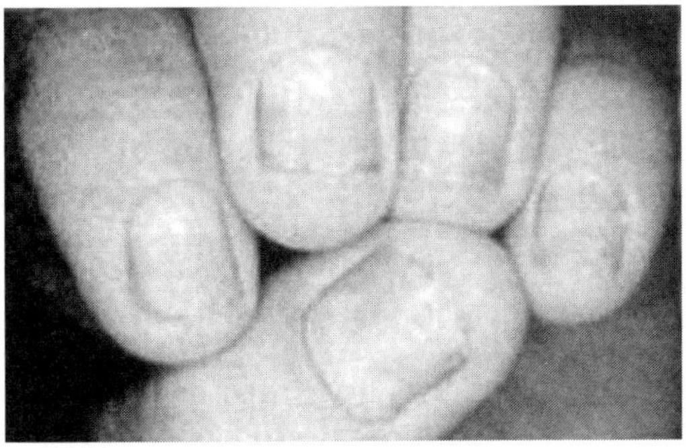

FIGURE 9.29 Mees lines in nails 2 months after exposure to DNOC. (Baran, *Archives of Dermatology*, 1974. Reprinted with permission from the American Medical Association.)

9.4.2.2 Prevalence, Incidence, Surveillance, and Outbreak Data

9.4.2.2.1 Chloracne

As discussed in Section 8.5.3, outbreaks of chloracne associated with the production of trichlorophenol and its derivatives were reported in the Netherlands[28] and following pentachlorophenol exposure in Belgium.[29] In an evaluation of the latter population, examination of 88 volunteer workers demonstrated a correlation between the presence of chloracne and the elevation of serum triglycerides and a decreased total bilirubin and lactate-dehydrogenase compared to controls.

9.4.2.2.2 Plants vs. pesticides as the cause of dermatitis in horticulture

An outbreak of dermatitis thought to be due to chlorothalonil residues was described in The Netherlands in 1992.[30] However, the dermatitis proved to be psoralen-related phototoxicity, similar to those reported in the U.S., reported by investigators from Centers for Disease Control.[31,32]

A study of the prevalence of dermatitis in 103 Dutch bulb growers compared to 49 control subjects was reported in 1993.[33] Contact dermatitis of the hands was equally common in the two groups (11 and 10%). However, minor signs of dermatitis were

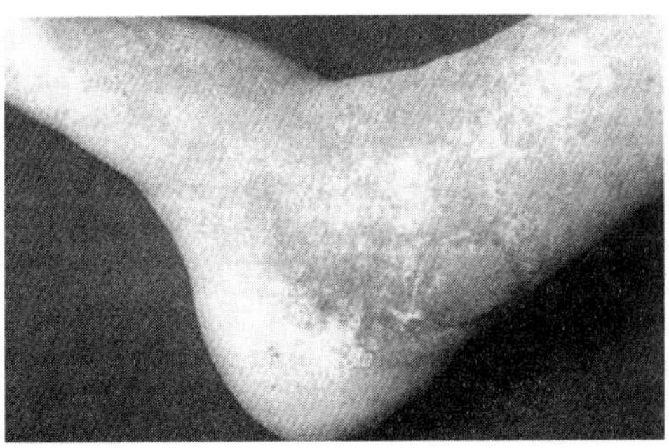

FIGURE 9.30 Dermal reaction to D-D fumigant. (Bousema, 1991. Reprinted with permission from *Contact Dermatitis*, Munskgaard.)

seen more often in bulb growers (30 vs. 8%, $p < 0.05$). Frequent exposures among growers included irritant sap from narcissus plants, hyacinth dust, and pesticides. Contact sensitization was suspected in 19 growers and 3 controls. On testing, contact sensitization to pesticides existed in probably 10 growers, and to flower-bulb extracts in 4 growers. Reactions to propachlor were among the most frequent, but were suspected to be due to irritant rather than allergic mechanisms. An exposure prevalence study in a similar population demonstrated that the most frequent pesticides encountered included 1,3-dichloropropene, primicarb, zineb, maneb, and captan.[34]

9.4.2.2.3 Surveillance

Surveillance reports were not routinely reported from Belgium, France, or the Netherlands. However, a 1968 report from France[35] described a series of 9864 cases of skin allergies, akin to a surveillance report. The reported cases included 690 (14%) associated with fertilizers and pesticides (categorized as insecticides, insecticidal solvents, oils, carbides and pesticidal carriers, helicides and molluscicides, fungicides, bactericides, herbicides, and rodenticides. A 1993 report[36] described an experimental network for toxicology surveillance in French agriculture based upon physician reports to the national agricultural insurance fund. Between 1991 and 1993, the program collected 111 observations from 22 French jurisdictions. These included 25.9% involving mucocutaneous symptoms described as irritations, burns, and allergic reactions. The individual compounds involved were not described, but insecticides were noted as the most frequent cause of reported events in the overall data set. The total absence of protective measures was noted as the probable underlying factor in 72% of the cases and the use of inappropriate protective measures in 20%.

REFERENCES

1. **FAOSTAT,** Production data for primary livestock, primary crops, and crops by area harvested, 1998, for Belgium, France and Netherlands. Available online at: apps.fao.org, 1999
2. **Ministry of Agriculture, N.M.a.F.,** Facts and figures 1997/1998 – Netherlands Agriculture. Available online at: http://www.minlnv.nl/international/stat/f-agricult1.shtml, 1999.
3. **Sabouraud, S., Testud, F., Rogerie, M. J., Descotes, J., and Evreux, J. C.,** Occupational depigmentation from dinoterbe, *Contact Dermatitis*, 36, 227, 1997.
4. **Baran, R.,** Letter: Nail damage caused by weed killers and insecticides, *Arch. Dermatol.*, 110, 467, 1974.
5. **Ducombs, G., Felix, B., Duboscq, M. F., and Geniaux, M.,** Allergic contact dermatitis and depigmentation due to a herbicide alachlor, Ninth International Symposium on Contact Dermatitis, Stockholm, Sweden, May, 23, 271, 1990.
6. **Garnier, R., Prince, C., Reygagne, A., Azoyan, P., Dally, S., and Efthymiou, M. L.,** Contact dermatitis from dazomet: Seven cases, *Arch. Mal. Prof. Med. Trav.*, 54, 649, 1993.
7. **de Groot, A. and Conemans, J.,** Contact allergy to furazolidone, *Contact Dermatitis*, 22, 202, 1990.
8. **Nater, J. P. and Grosfeld, J. C.,** Allergic contact dermatitis from Betanal (phenmedipham), *Contact Dermatitis*, 5, 59, 1979.

9. **Garnier, R., Chataigner, D., Efthymiou, M. L., Moraillon, I., and Bramary, F.,** Paraquat poisoning by skin absorption: report of two cases, *Vet. Hum. Toxicol.*, 36, 313, 1994.
10. **Willems, P. W., Geursen-Reitsma, A. M., and van Joost, T.,** Allergic contact dermatitis due to methiocarb (Mesurol), *Contact Dermatitis*, 36, 270, 1997.
11. **Bruynzeel, D. P.,** Contact sensitivity to lannate, *Contact Dermatitis*, 25, 60, 1991.
12. **Haenen, C., de Moor, A., and Dooms-Goossens, A.,** Contact dermatitis caused by the insecticides omethoate and dimethoate, *Contact Dermatitis*, 35, 54, 1996.
13. **Van Ginkel, C. J. and Sabapathy, N. N.,** Allergic contact dermatitis from the newly introduced fungicide fluazinam, *Contact Dermatitis*, 32, 160, 1995.
14. **Bruynzeel, D. P. and van Ketel, W. G.,** Contact dermatitis due to chlorothalonil in floriculture, *Contact Dermatitis*, 14, 67, 1986.
15. **van Ketel, W. G.,** Sensitivity to the pesticide benomyl, *Contact Dermatitis*, 2, 290, 1976.
16. **van Ketel, W. G.,** Active sensitization by 0,0-diethyl-phthalimido-phosphothioate (Plondrel), *Contact Dermatitis*, 3, 51, 1977.
17. **Nater, J. P., Terpstra, H., and Bleumink, E.,** Allergic contact sensitization to the fungicide Maneb, *Contact Dermatitis*, 5, 24, 1979.
18. **van Ketel, W. G.,** Thiuram-mix, *Contact Dermatitis*, 2, 232, 1976.
19. **Bourrain, J. L., Morin, C., Beani, J. C., and Amblard, P.,** Airborne contact dermatitis from cacodylic acid, *Contact Dermatitis*, 38, 364, 1998.
20. **van Joost, T. and Wiemer, G. R.,** Contact dermatitis from tetrachloroacetophenone (TCAP) in an insecticide plant, *Contact Dermatitis*, 25, 66, 1991.
21. **Bousema, M. T., Wiemer, G. R., and van Joost, T.,** A classic case of sensitization to DD-95, *Contact Dermatitis*, 24, 132, 1991.
22. **Nater, J. P. and Gooskens, V. H. J.,** Occupational dermatosis due to a soil fumigant, *Contact Dermatitis*, 2, 227, 1976.
23. **van Joost, T., Schulten-Lukassen, M., Stolz, E., Smit, A. F., and Ribbens, P. H.,** Occupational contact dermatitis to mesylisopropyllactate, *Contact Dermatitis*, 10, 187, 1984.
24. **van Ketel, W. G.,** Sensitization to mercury from mercurochrome, *Contact Dermatitis*, 6, 499, 1980.
25. **Damstra, R., van Vloten, W., and van Ginkel, C.,** Allergic contact dermatitis from the preservative (1,2-BIT; Proxel): A case report, its prevalence in those occupationally at risk and in the general dermatological population, and its relationship to allergy to its analogue Kathon CG, *Contact Dermatitis*, 27, 105, 1992.
26. **Malten, K.,** DNCB in cooling water, *Contact Dermatitis Newsletter*, 15, 1974.
27. **Bonnetblanc, J. M. and Delrous, J. L.,** Connubial dermatitis from phenylmercuric nitrate, *Contact Dermatitis*, 34, 367, 1996.
28. **Holmstedt, B.,** Prolegomena to Seveso, *Arch. Toxicol.*, 44, 211, 1980.
29. **Baxter, R. A.,** Biochemical study of pentachlorophenol workers, *Ann. Occup. Hyg.*, 28, 429, 1984.
30. **Groot, B. J., Belinfante-van Gelder, M. E., and Jans, H. W.,** An epidemic of dermatitis caused by blanched celery, *Ned. Tijdschr. Geneeskd.*, 136, 1210, 1992.
31. **Berkley, S. F., Hightower, A. W., Beier, R. C., Fleming, D. W., Brokopp, C. D., Ivie, G. W., and Broome, C. V.,** Dermatitis in grocery workers associated with high natural concentrations of furanocoumarins in celery, *Ann. Intern. Med.*, 105, 351, 1986.
32. **Seligman, P. J., Mathias, C. G., O'Malley, M. A., Beier, R. C., Fehrs, L. J., Serrill, W. S., and Halperin, W. E.,** Phytophotodermatitis from celery among grocery store workers, *Arch. Dermatol.*, 123, 1478, 1987.

33. **Bruynzeel, D. P., de Boer, E. M., Brouwer, E. J., de Wolff, F. A., and de Haan, P.,** Dermatitis in bulb growers, *Contact Dermatitis*, 29, 11, 1993.
34. **Brouwer, E. J., Brouwer, D. H., Bruynzeel, D. P., Emmen, H. H., van Hemmen, J. J., Hooisma, J., Jonkman, E. J., Ruijten, M., Salle, H. J. A., Sjardin, W., Verberk, M. M., Vermeulen, N. P. E., de Weerd, A. W., van Welie, R. T. H., et al.,** Health in Relation to Occupational Exposure to Pesticides in the Dutch Flower Bulb Culture. Part 1. General Study Design. Results, Conclusions and Recommendations, Labour Inspectorate, Ministry of Social Affairs and Employment, The Hague, The Netherlands, Report No. S144, 1, 17, 1992.
35. **Vacher Vallet, G. J.,** Allergies attributed to pesticides and fertilizers (ten years of partial observation in France), *Arch. Mal. Prof. Med. Trav.*, 29, 336, 1968.
36. **Morel, A., Delemotte, B., Nastorg, G., Portos, J., and Bonderf, J.**, Reseau Experimental De Toxicovigilance En Agriculture, Nice, France, 1993.

9.4.3 Scandinavia (Denmark, Finland, Norway, Sweden)

FAO data for 1998 indicate that agriculture in Scandinavia is dominated by grain, root crop, and livestock production.[1] Of the countries in the region, consumption of pesticides appears highest in Denmark and Sweden, with herbicides and fungicides accounting for almost of all of the reported use.[2] Trade data demonstrate net excess of imports over exports. Although FAO trade data are based upon dollar value of material sold rather than actual use, the data suggest overall increasing levels of use (Figures 9.31).The compounds associated with cases or outbreaks of dermatitis in Scandinavia are shown in Figure 9.32.

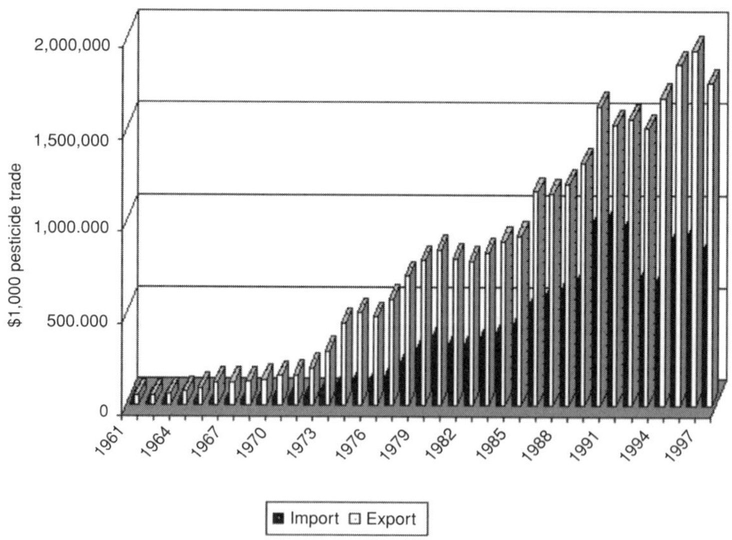

FIGURE 9.31 Scandinavian pesticide trade data. (UN FAOSTAT database)

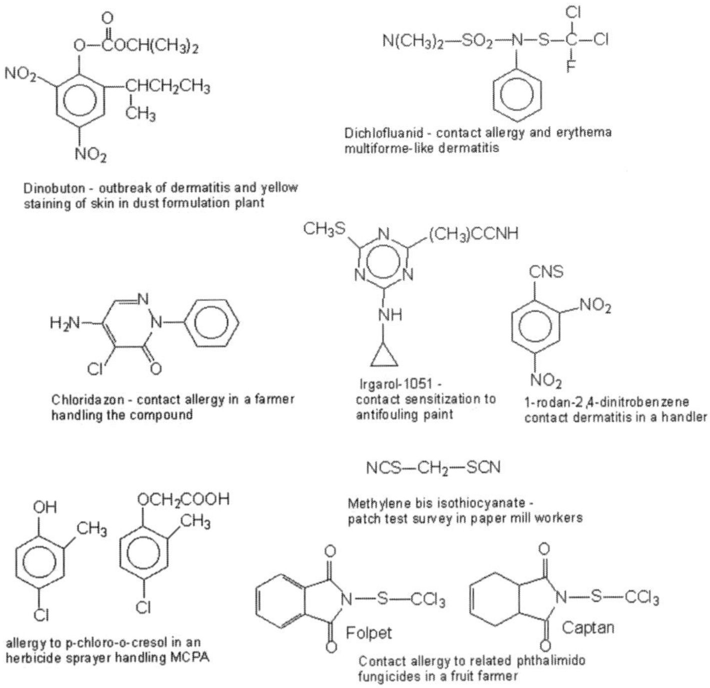

FIGURE 9.32 Pesticide contact dermatitis in Scandinavia.

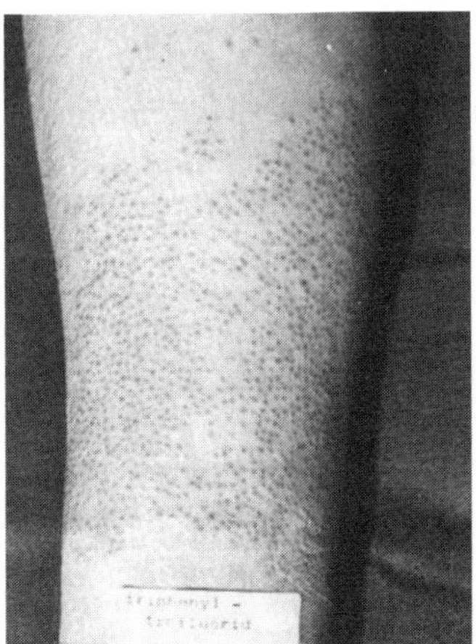

FIGURE 9.33 Follicular dermatitis associated with triphenyltin fluoride used as an antifouling paint. (With permission from *Contact Dermatitis*.)

9.4.3.1 Case Reports and Case Series

Cases associated with Scandinavian agriculture included two cases of reported contact dermatitis due to the insecticide methomyl in Danish nursery workers,[3] a case of allergic contact dermatitis associated with handling the thiocyanate insecticide rodannonitrobenzene,[4] a case of parathion poisoning accompanied by an erysipeloid-like eruption of the left index finger,[5] contact dermatitis due to the *Botritis* fungicide dichlofluanid,[6, 7] and the herbicides chloridazon[8] and *p*-chloro-*o*-cresol[9] and the phthalimido fungicides captan and phaltan.[4]

The use of chlorothalonil as a wood preservative and antimicrobial compound was associated with a controversial series of contact dermatitis cases.[10-13] Other antimicrobials associated with dermatitis included 2-*tert*-butylamino-4-cyclopropylamino-6-methylthio-1,3,5-triazine,[14] 2-n-octyl-4-isothiazolin-3-one,[15] thiuram (under the trade name Nobecutam),[16] and triphenyltin fluoride used in anti-fouling paint (Figure 9.33).[17]

9.4.3.2 Outbreak Investigations, Prevalence, Incidence, and Surveillance Data

9.4.3.2.1 Dinobuton

In 1973, a chemical works in northern Sweden began the manufacture of a new miticide dust containing dinobuton. The dust generated spread like a "carpet over the entire factory." Employees shortly complained of itching of the face, neck, and arms and yellow stains of the hair and nails. Several also reported fatigue, weakness, headache, alcohol intolerance, and alopecia. No reactions were found in groups of 10 controls tested to concentrations varying from 0.6 to 40% and the authors concluded that the compound did not cause significant irritation. Although positive responses were found in at least two of the workers, the report did not describe either the total number tested or the total number with positive patch reactions.[18]

9.4.3.2.2 Methylene-bis-thiocyanate

The prevalence of allergic reactions to the preservative methylene-bis-thiocyanate was evaluated in 1987 in 34 paper mill workers with intermittent contact to low concentrations of the preservative-slimicide used in paper and board manufacturing.[19] Non-irritant test concentrations were not established initially, and 3 workers reacted to 0.1% concentration of the formulated slimicide, and the reactions were felt to be irritant in nature. No workers had definite allergic reactions.

9.4.3.2.3 Finnish farmers

Among Finnish farmers, involved predominantly in livestock agriculture, the most frequent causes of occupational hand disorders were studied by Susitaival, in 1996, using compensation data and questionnaire studies.[20, 21] Hazards identified included cow dander, disinfectants and detergents, wet work, and rubber chemicals. Zoonotic diseases, especially *Trichophyton* infections, were also frequent.

9.4.3.2.4 Danish greenhouse workers

In 1986, a questionnaire survey was conducted of greenhouse workers handling potted plants and flowers; 23% reported contact dermatitis and 11% reported urticaria.[22] A 1998 questionnaire study by Paulsen[23] estimated that the lifetime prevalence of dermatitis among 1958 Danish gardeners and greenhouse workers was 19.6%. A subsequent dermatologic study[24] evaluated the causes of dermatitis among 253 gardeners and greenhouse workers (including 183 diagnosed who had occupational eczema on a skin examination) using patch tests with the standard allergen series, a gardeners series, a Compositae series, a pesticides series, and additional allergens based on the subjects' case histories. Overall, 120 subjects (48%) had at least one positive patch test reaction. The most frequent positive reactions included nickel sulfate, Compositae species, thimerosal, and captan. Irritant reactions to leaves or extracts were also frequently observed.

9.4.3.2.5 Mushroom workers

A study by Larsen[25] evaluated the prevalence of contact allergy to benomyl in 62 pickers/former pickers from a mushroom farm provoked by observation of 2 index cases of contact allergy. No additional cases were found by testing among the group of 60 additional present and former mushroom pickers.

9.4.3.2.6 Forestry workers

In 1995, Kolmodin-Hedman[26] evaluated the frequency of symptoms following exposure of forestry workers to permethrin in a unique double-blind study. Workers planted conifers either sprayed with 1.5% active ingredient or sprayed only with the vehicle and solvent. Among the a few symptomatic workers, ocular and nasal irritation were the most common complaints. In a separate exposure study and symptom survey, [27] 139 planters reported skin and upper respiratory irritation symptoms associated with exposure to pyrethroids: 73% for fenvalerate, 63% for permethrin (trans/cis 75/25) and 33% for permethrin (trans/cis 60/40).

9.4.3.2.7 Creosote workers

In 1992, Karlehagen[28] reported a study of cancer incidence among 922 Norwegian and Swedish men employed in work with creosote treated wood products. Although total cancer risk was not increased, the incidence rates for lip cancer, non-melanoma skin cancer, and malignant lymphoma were elevated in the study group relative to the population background. The standardized incident ratios (SIRs) were 2.50, 2.37, and 1.9, but elevated SIRs for all 3 outcomes were based on fewer than 10 reported cases. Most of the increased risk was found to occur in workers having 20 or more years latency. The study was unique in concentrating on creosote exposed workers and also unique in being able to study non-melanoma skin cancers tracked in Scandinavian cancer registries, but not in the U.S. and in many other parts of the world.

REFERENCES

1. **FAOSTAT,** Production data for primary livestock, primary crops, and crops by area harvested, 1998, for Denmark, Finland, Norway and Sweden. Available online at: apps.fao.org, 1999.
2. **FAOSTAT,** Pesticide Use Data, 1996, for Denmark, Finland, Norway and Sweden. Available online at: apps.fao.org, 1999.
3. **Paulsen, E., Andersen, K. E., and Skoven, I. G.,** Concomitant sensitization to lannate and Gerbera, *Contact Dermatitis*, 37, 128, 1997.

4. **Fregert, S.,** Allergic contact dermatitis from the pesticides captan and phaltan, *Contact Dermatitis Newsletter*, 2, 28, 1967.
5. **Svindland, H. B.,** Subacute parathion poisoning with erysipeloid-like lesion, *Contact Dermatitis*, 7, 177, 1981.
6. **Hansson, C. and Wallengren, J.,** Allergic contact dermatitis from dichlofluanid, *Contact Dermatitis*, 32, 116, 1995.
7. **Bjorkner, B., Bruze, M., and Gruvberger, B.,** Sensitization to dichlofluanide, Ninth International Symposium on Contact Dermatitis, Stockholm, Sweden, May, 23, 1990.
8. **Bruze, M. and Fregert, S.,** Allergic contact dermatitis to chloridazon, *Contact Dermatitis*, 8, 427, 1982.
9. **Fregert, S.,** Allergic contact dermatitis from *p*-chloro-*o*-cresol in a pesticide, *Contact Dermatitis Newsletter*, 3, 46, 1968.
10. **Bach, B. and Pedersen, N.,** Contact dermatitis from a wood preservative containing tetrachloroisophthalonitrile, *Contact Dermatitis*, 6, 142, 1980.
11. **Eilrich, G. L. and Chelsky, M.,** Facial dermatitis caused by chlorothalonil in a paint [letter; comment], *Contact Dermatitis*, 25, 141, 1991.
12. **Meding, B.,** Contact dermatitis from tetrachloroisophthalonitrile in paint, *Contact Dermatitis*, 15, 187, 1986.
13. **Johnsson, M., Buhagen, M., Leira, H. L., and Solvang, S.,** Fungicide-induced contact dermatitis, *Contact Dermatitis*, 9, 285, 1983.
14. **Andersson, T.,** Occupational allergic contact dermatitis from 2-tert-butylamino-4-cyclopropylamino-6-methylthio-1,3,5-triazine, *Contact Dermatitis*, 36, 272, 1997.
15. **Thormann, J.,** Contact dermatitis to a new fungicide, 2-n-octyl-4-isothiazolin-3-one, *Contact Dermatitis*, 8, 204, 1982.
16. **Pock-Steen, B.,** Contact allergy to Nobecutan, *Contact Dermatitis*, 18, 52, 1988.
17. **Andersen, K. E. and Petri, M.,** Occupational irritant contact folliculitis associated with triphenyl tin fluoride (TPTF) exposure, *Contact Dermatitis*, 8, 173, 1982.
18. **Wahlberg, J.,** Yellow staining of hair and nails and contact sensitivity to dinobuton, *Contact Dermatitis Newsletter*, 16, 481, 1974.
19. **Jappinen, P. A.,** Patch tests with methylene-bis-thiocyanate in paper mill workers, *Contact Dermatitis*, 16, 233, 1987.
20. **Susitaival, P.,** *Medical Sciences*, Kuopio University, Finland, 1996, 93.
21. **Susitaival, P., Husman, L., Hollmen, A., and Horsmanheimo, M.,** Dermatoses determined in a population of farmers in a questionnaire-based clinical study including methodology validation., *Scand. J. Work Environ. Health*, 21, 30, 1995.
22. **Lander, F., Skytthe, A., and Jeune, B.,** Working environment and health of gardeners, *Arbejdsmiljofondet*, *Vesterbrogade*, 69, 1620.
23. **Paulsen, E., Sogaard, J., and Andersen, K. E.,** Occupational eczema in gardeners and greenhouse workers in the county of Funen, *Ugeskr. Laeger.*, 160, 6972, 1998.
24. **Paulsen, E.,** Occupational dermatitis in Danish gardeners and greenhouse workers (II). Etiological factors, *Contact Dermatitis*, 38, 14, 1998.
25. **Larsen, A. I., Larsen, A., Jepsen, J. R., and Jorgensen, R.,** Contact allergy to the fungicide benomyl?, *Contact Dermatitis*, 22, 278, 1990.
26. **Kolmodin-Hedman, B., Akerblom, M., Flato, S., and Alex, G.,** Symptoms in forestry workers handling conifer plants treated with permethrin, *Bull. Environ. Contam. Toxicol.*, 55, 487, 1995.
27. **Kolmodin-Hedman, B., Swensson, A., and Akerblom, M.,** Occupational exposure to some synthetic pyrethroids (permethrin and fenvalerate), *Arch. Toxicol.*, 50, 27, 1982.
28. **Karlehagen, S., Andersen, A., and Ohlson, C.,** Cancer incidence among creosote-exposed workers, *Scand. J. Work. Environ. Health*, 18, 26, 1992.

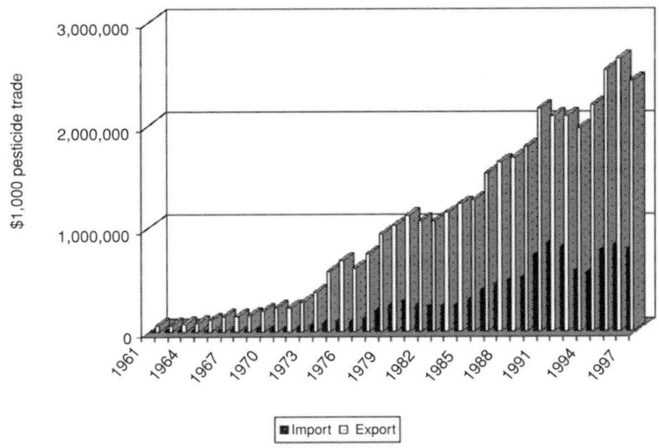

FIGURE 9.34 Pesticide trade data for North Central Europe.

9.4.4 North Central Europe (Austria, Germany, Switzerland)

FAO data indicate that agricultural production in Austria, Germany, and Switzerland is dominated by grain, root crops, and livestock production.[1] Total pesticide use exceeds that in the neighboring countries of Scandinavia,[2] with fungicides and herbicides accounting for the majority of use. Trade data demonstrated a net export of pesticides from the region, steadily increasing in economic value between the 1960s and the 1990s (Figure 9.34). The compounds associated with cases or outbreaks of dermatitis in Austria, Germany, and Switzerland are shown in Figure 9.35.

9.4.4.1 Case Reports and Case Series

Cases associated with agricultural production included the simultaneous sensitization to several compounds (parathion, methyl parathion, azinphos methyl, metathion or fenitrothion, metasystox, pyrethrum and EDTA, accompanied by *de novo* sensitization to fenticlor) in a vintner,[3] the food additive olaquindox (both contact allergy and photoallergy);[4] eczema associated with a mixture of 2,4-D and 2,4,5-T,[5] allergy and photoallergy to phenmedipham;[6] contact allergy to dimethoate;[7] contact allergy to the thiocarbamate fungicides mancozeb, metiram, maneb, nabam, propineb, and zineb;[8] contact dermatitis associated with benomyl;[9] urticaria in a gardener associated with handling of captan;[10] allergy to propachlor in a 29-year-old male farmworker;[11] and with metam sodium in the production of potatoes and other root crops.[12,13]

Cases associated with the laboratory handling or factory production of pesticides included allergy in three laboratory workers to the novel acaricide compounds 4-chloro-2-(alpha-hydroxy-beta, beta,beta-trichloroethyl)-phenol and 4,6-dichloro-2-(alpha-hydroxy-beta, beta,beta-trichloroethyl)-phenol,[14] and chloracne and a variety of systemic symptoms associated with trichlorophenol (herbicide) production.[15]

The non-occupational use of pesticides accounted for cases of contact dermatitis associated with exposure to chlorothalonil used as a wood preservative,[16] irritant dermatitis due to tributyltinoxide,[17] contact dermatitis associated with the use of

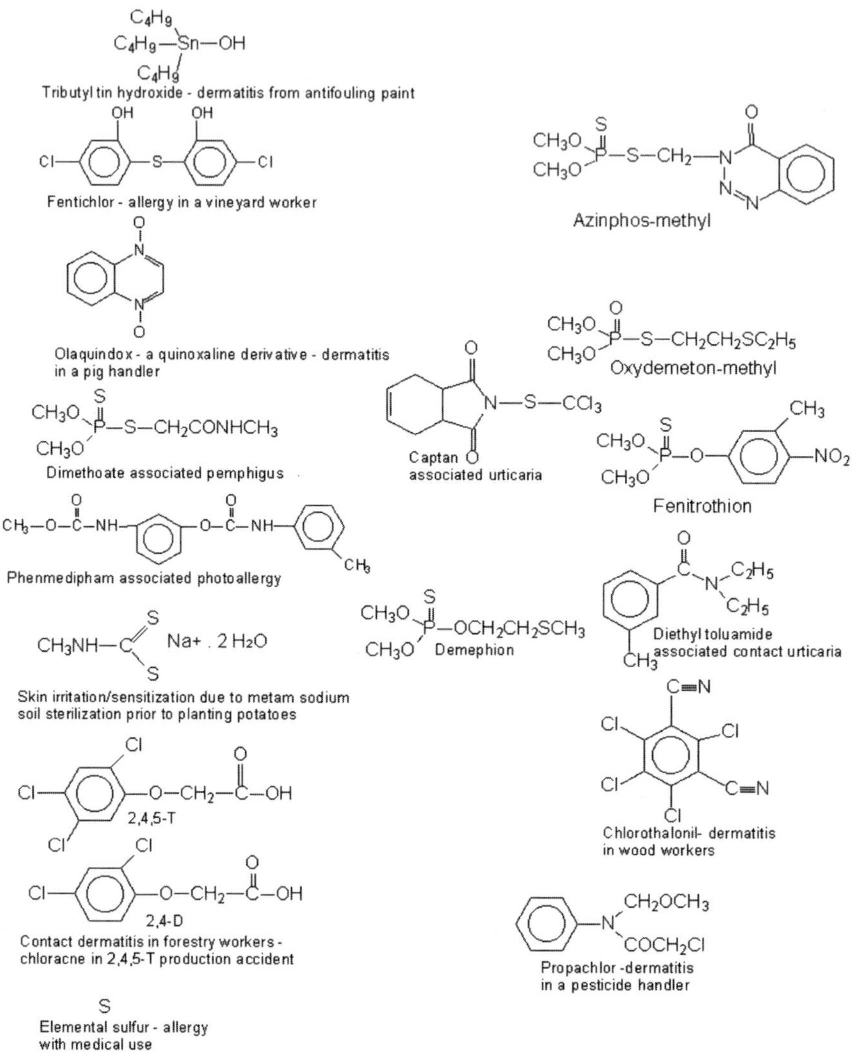

FIGURE 9.35 Contact dermatitis and other skin conditions associated with pesticide in Germany.

dichlorvos in structural pest control,[18] with medical use of sulfur,[19] irritant dermatitis and systemic toxicity from deliberate ingestion of paraquat[20] irritant, and possibly allergic dermatitis associated and urticaria associated with diethyltoluamide.[21] Muller-Mohnssen[22] reported a case of persistent contact dermatitis secondary to non-occupational exposure to a mixture of pyrethrin and resmethrin (Figure 9.36).

Reports associating pesticides with dermatoses usually considered of endogenous origin, included lupus erythematosus[23] and pemphigus possibly associated with exposure to dimethoate.[24]

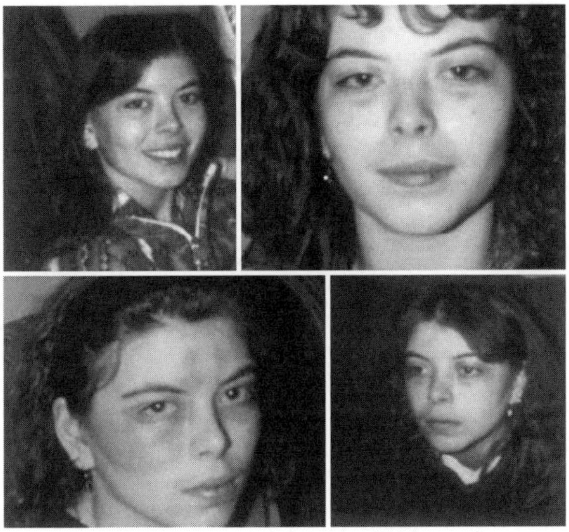

FIGURE 9.36 Persistent contact dermatitis over a several month period following direct accidental exposure to a mixture of pyrethrin and bioresmethin. (Muller-Mohnssen, *Toxicology Letters*, 107, 161, 1999. With permission from the Scientific and Medical Division, Elsevier Science.)

9.4.4.2 Prevalence, Incidence, Surveillance, and Outbreak Data

9.4.4.2.1 Skin reactions to PCP

In 1979, Klaschka et al.[25] described a study of skin reactions to 1 and 5% pure PCP, technical PCP, and PCP-Na (sodium salt) performed on 100 selected PCP-exposed individuals and a group of unselected control individuals. The PCP in some cases caused contact dermatitis felt to be irritant rather than allergic in nature.

9.4.4.2.2 Chloracne

In 1982, Thiess and Link[26] conducted a mortality study on a cohort of 74 workers exposed to tetrachlorodibenzodioxin as a result of an accident in the BASF AG factory in Ludwigshafen, Federal Republic of Germany in 1953; 70 were directly involved in the accident, 4 were involved in the cleanup, and 66 (89.1%) suffered from chloracne or severe dermatitis. The small cohort demonstrated a modest excess of gastrointestinal neoplasms (stomach cancer — 0.61-0.70 expected vs. 3 observed [crude chi square = 9.6]; colon cancer — 1 observed vs. 0.29-0.42 expected).

9.4.4.2.3 Dermatitis in billposters

An outbreak investigation in 1986 evaluated sudden onset irritant contact dermatitis in a group of 30 billposters after the introduction of two new pastes.[27] Possible causes of the outbreak included adhesive resins (acetone-formaldehyde), pentachlorophenol present as an impurity, acetone solvents, or the alkalinity of one paste.

9.4.4.2.4 Surveillance study

In 1989 Jung[28] described a series of 20 dermatitis cases occurring in 1985-1986 in gardeners, florists, vegetable and fruit tree farmers, and manufacturing workers. The hands, face, and neck were the most common site affected. The most common plant allergies involved primula and chrysanthemum. Pesticides associated with skin reactions in this group included thiocarbamates (maneb, zineb, mancozeb, propineb, thiram, and metham-Na), demephion, dimethoate, propachlor, benomyl, chloralhydrate with chloralmethylhalbacetal, carbaryl, dinocap, captan, phenyl-Hg-acetate; and trichlodinitrobenzene.

REFERENCES

1. **FAOSTAT,** Production data for primary livestock, primary crops, and crops by area harvested, 1998, for Austria, Germany, Switzerland. Available online at: apps.fao.org, 1999.
2. **FAOSTAT,** Pesticide Use Data , 1998, for Austria, Germany, and Switzerland. Available online at: apps.fao.org, 1999.
3. **Pevny, I.,** Pesticide allergy — allergic contact eczema of a vintner (author's transl.), *Derm. Beruf. Umwelt.*, 28, 186, 1980.
4. **Dunkel, F., Elsner, P., Pevny, I., and Burg, G.,** Olaquindox-induced photoallergic contact dermatitis and persistent light reaction in a pig breeder; Ninth International Symposium on Contact Dermatitis, Stockholm, Sweden, May 17-19, 1990, *Contact Dermatitis*, 23, 283, 1990.
5. **Jung, V. H. and Wolf, F.,** Kontaktezekzeme durch das Herbizid Selest 100 in der Forstwirstschaft, *Dtsch. Gesundheitwes.*, 30, 1540, 1975.
6. **Koch, P. and Bahmer, F. A.,** Photoallergic dermatitis caused by the herbicide phenmedipham, *Derm. Beruf. Umwelt.*, 37, 203, 1989.
7. **Pambor, M. and Bloch, Y.,** Dimethoate and dithiocarbamate as occupational contact allergens in a female agrotechnician, *Dermatol. Monatsschr.*, 171, 401, 1985.
8. **Koch, P.,** Occupational allergic contact dermatitis and airborne contact dermatitis from 5 fungicides in a vineyard worker. Cross-reactions between fungicides of the dithiocarbamate group?, *Contact Dermatitis*, 34, 324, 1996.
9. **Kuhne, G., Heise, H., Plottke, B., and Puskeiler, T.,** Dermatitis after Benlate contact, *Z. Gesamte Hyg.*, 31, 710, 1985.
10. **Croy, I.,** Etiology of urticaria due to Captan-based antifungal agents, case report, *Z. Gesamte Hyg.*, 19, 710, 1973.
11. **Schubert, H.,** Allergic contact dermatitis due to propachlor (author's transl.), *Dermatol. Monatsschr.*, 165, 495, 1979.
12. **Wolff, F. and Jung, H. D.,** Acute contact dermatitis after contact with nematin, *Z. Gesamte Hyg.*, 16, 423, 1970.
13. **Jung, H. D. and Wolff, F.,** Nematin (Vapam) contact dermatitis in agricultural workers, *Dtsch. Gesundheitswes.*, 25, 495, 1970.
14. **Dehne, H. and Wozniak, K. D.,** Toxic and sensitizing properties of substituted aryltrichloromethylcarbinols, *Z. Gesamte Hyg.*, 19, 535, 1973.
15. **Kleu, G. and Goltz, R.,** Late and permanent lesions following chronic occupational exposure to chlorophenol compounds. Catamnestic, neurologic-psychiatric and psychologic studies, *Med. Klin.*, 66, 53, 1971.

16. **Spindeldreier, A. and Deichmann, B.,** Contact dermatitis against a wood preservative with a new fungicidal agent (author's transl.), *Derm. Beruf. Umwelt.*, 28, 88, 1980.
17. **Gammeltoft, M.,** Tributyltinoxide is not allergenic, *Contact Dermatitis*, 4, 238, 1978.
18. **Stoermer, D.,** Occupational contact dermatitis caused by dichlorvos pesticide, *Dermatol. Monatsschr.*, 171, 245, 1985.
19. **Schneider, H.,** Schwefelallergie [Sulfur allergy], *Hautartz*, 29, 340, 1978.
20. **Zilker, T., Fogt, F., and von Clarmann, M.,** Kein Parkinson syndrom nach akuter Paquatintoxikation, *Klin. Wochenschr.*, 66, 1988.
21. **von Mayenburg, J. and Rakoski, J.,** Contact urticaria to diethyltoluamide, *Contact Dermatitis*, 9, 171, 1983.
22. **Muller-Mohnssen, H.,** Chronic sequelae and irreversible injuries following acute pyrethroid intoxication, *Toxicol. Lett.*, 107, 161–175, 1999.
23. **Beer, K. R., Lorincz, A. L., Medenica, M. M., Albertini, J., Baron, J., Drinkard, L., and Swartz, T.,** Insecticide-induced lupus erythematosus, *Int. J. Dermatol.*, 33, 860, 1994.
24. **Tsankov, N., Dimitrowa, J., Obreschkowa, E., and Lasarowa, A.,** Induced pemphigus caused by the pesticide phosphamide, *Z. Hautkr.*, 62, 196, 1987.
25. **Klaschka, F., von Nieding, G., and Walter, R.,** Skin sensitization to wood preservatives containing pentachlorophenol (PGP), *Zentralbl. Arbeitsmed. Arbeitsschutz. Prophyl*, 29, 150, 1979.
26. **Thiess, A. M. and Link, R.,** Mortality study of persons exposed to dioxin in a trichlorophenol-process accident that occurred in the BASF AG on November 17, 1953, *Am. J. Ind. Med.*, 3, 179, 1982.
27. **Lindemayr, H. and Pospischil, E.,** Occupational eczema in billposters, *Wien. Klin. Wochenschr.*, 98, 689, 1986.
28. **Jung, H. D., Honemann, W., Kloth, C., Lubbe, D., Pambor, M., Quednow, C., Ratz, K. H., Rothe, A., and Tarnick, M.,** Contact eczema caused by pesticides in East Germany, *Dermatol. Monatsschr.*, 175, 203, 1989.

9.4.5 Southern Europe (Portugal, Spain, Italy, Greece)

FAO data for southern Europe demonstrate a diverse variety of agricultural production, with the output of fruits and vegetables approximately equal to the production of cereal crops and livestock. [1] Pesticide use data[2] are limited by absence of information on Spanish pesticide use in the FAO database. Nevetheless, import data suggest a level of pesticide use in Spain just slightly lower than reported use for Italy.[3] Trade data for the region overall demonstrate a considerable excess of imports over exports (Figure 9.37). Case reports, epidemiologic studies and outbreak investigations described below are summarized in Figure 9.38. In addition, a viewpoint on Spanish experiences with pesticide skin conditions is given by Conde-Salazar in Section 9.4.5.3.

9.4.5.1 Case Reports and Case Series

Cases associated with agricultural use of pesticides in southern Europe included erythema multiforme and contact dermatitis in agricultural warehouse workers associated with the insecticide dimethoate,[4] erythroderma[5] and contact dermatitis[6-9] associated with thiocarbamates (maneb, mancozeb, and zineb) and phthalimido fungi-

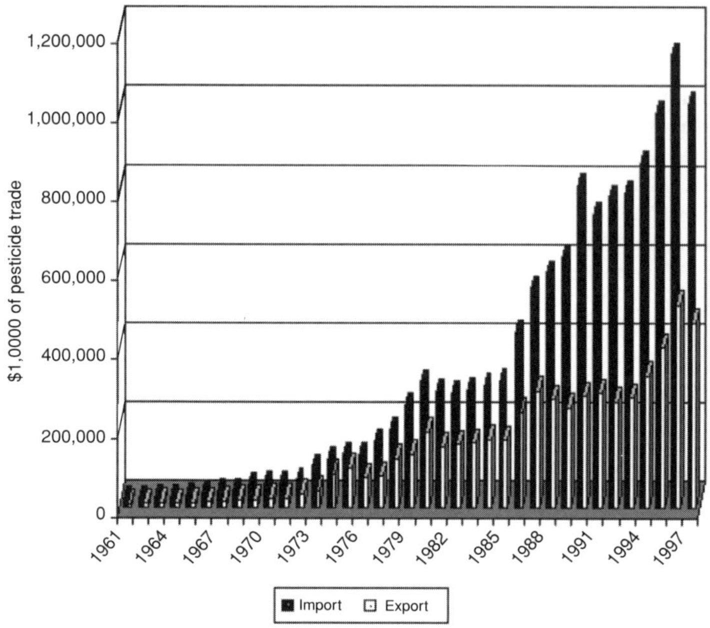

FIGURE 9.37 Pesticide trade data for Southern Europe.

cides (captafol, captan, and folpet).[6] Novel reports included a case of acne associated with exposure to paraquat,[10] and pellagra-like dermatitis and vitiligo associated with mancozeb.[11] Pizzini et al.[12] reported a case of secondary childhood exposure to an an abandoned container of paraquat resulting in "strawberry tongue" and severe contact burns of the lower extremities.[12]

Non-agricultural pesticide cases included a 35-year-old man handling concentrated (19%) triphenyltin who developed persistent urticaria, elevated liver transaminases, and abdominal pain, after spilling the compound on his unprotected arms;[13] and a masseuse with contact allergy to povidone-iodine.[14] Other reports included allergy to 1,2-dichloropropane used in paint,[15] occurring on the fingers and hands of painters or metalworkers in the engineering industry from 1985 to 1988. Spettoli[16] also describes an interesting case of mixed allergy to captafol, pyrethrum, and sesquiterpene lactones in a 65-year-old man who gardened as a hobby.

9.4.5.2 Prevalence, Incidence, Surveillance, and Outbreak Data

9.4.5.2.1 Monferrato region of Italy, 1960

A descriptive investigation conducted in 1960 by Coricciata[17] evaluated the annual occurrence of dermatitis in the Monferrato agricultural region of Italy; 80 cases were observed in 1958-1959 and associated the problem with the introduction of dithiocarbamate zinc in the region.

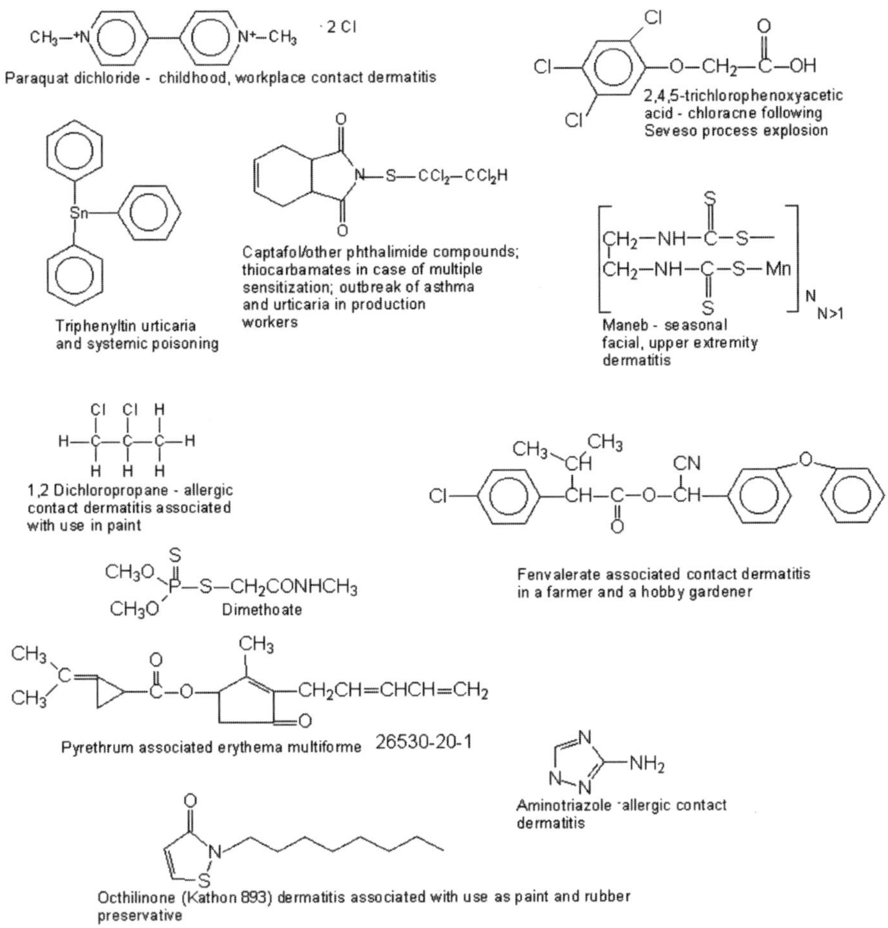

FIGURE 9.38 Pesticide-related skin conditions in Southern Europe.

9.4.5.2.2 Difolatan dermatitis, urticaria, and asthma in Spanish difolatan formulation plant

In 1975, Camarasa[18] investigated outbreaks of dermatitis in a Spanish packing operation and found severe dermatitis in 7 of 41 workers. Symptoms included widespread pruritus, morbilliform and urticarial eruptions, and marked asthmatic dyspnea. These symptoms forced several workers to stop work. Four patients had +++ epicutaneous allergic response to 1% concentration of orthodifolatan in petrolatum.

9.4.5.2.3 Standard patch test antigens in Spanish agricultural and nonagricultural workers, 1984

Garcia-Perez[19] described the frequency of reactions to the standard European patch test series in a group of agricultural workers with dermatitis (n = 34) compared with a control group (n = 244). The study group showed significantly higher frequent sensitivity to contact allergens such as paraphenylenediamine, balsam of Peru, neomycin, mercury, *p*-aminoazobenzene, diphenylguanidine, and cobalt-chloride. Nickel, parabens, wood tars, and naphthyl mix exhibited significantly lower frequencies as contact allergens in the study group.

9.4.5.2.4 *Italian nursery workers and poinsettia dermatitis*

A case of eczema was identified in an Italian flower nursery in 1985.[20] This was followed by a small prevalence study, involving 30 individuals working in 3 nurseries and 30 individuals from other occupations. All were patch tested with a standard series of allergens, pesticides, fertilizers, extracts of poinsettia plant parts, and poinsettia latex. Three female florists were positive to the poinsettia latex and two also showed positive results to aqueous extracts of the leaves and latex. The observed patch reactions were biopsied, which demonstrated spongiosis and lymphocytic infiltration, suggestive of an allergic reaction. Tests in the control subjects were negative.

9.4.5.2.5 *Patch test reaction in agricultural and nonagricultural workers, Italy*

Lisi reported in 1986[21] on the prevalence of patch test reactions to a pesticide test battery in 50 agricultural workers and 150 non-agricultural workers. A total of 24 positive reactions were observed, principally involving captan, difolatan, and folpet. Reactions to thiram were seen in four nonagricultural workers without a prior history of handling pesticides. Reactions were recorded about twice as frequently among current or former agricultural workers, a finding of borderline statistical significance. Positive reactions occurred about nine times more frequently among workers with a history of hand dermatitis.

9.4.5.2.6 *Italian patch test series and risk factors for positive reactions*

In 1987 Lisi[22] also reported the prevalence of positive pesticide patch tests in 652 subjects seen at a patch test clinic. The test battery included a battery of 12 fungicides, 12 insecticides, 8 herbicides, and 4 miscellaneous pesticides. Allergic reactions to 1 or more pesticides occurred in 35 subjects (5.4%), including 24 that appeared relevant to a history of contact dermatitis. The positive pesticide patch reactions occurred among 17 (7.6%) of the 223 current or former agricultural workers, and among 18 (4.2%) of the 429 subjects who never worked in agriculture, a difference that is of borderline statistical significance using an uncorrected chi square test ($p = 0.07$).[23] The allergic reactions occurred principally to the phthalimide compounds (captan, difolatan, and folpet), but irritant reactions were most commonly due to the organotin compound fentin hydroxide. Thiram was positive only in four non-agricultural workers who had never had any contact with pesticides.

9.4.5.2.7 *Prevalence of common skin conditions among agricultural and nonagricultural workers*

Between 1990 and 1992, Cellini and Offidani[24] studied 426 agricultural workers, authorized to use pesticides, together with a control group of 100 age and sex matched nonagriculture workers. Common skin conditions, vitiligo, psoriasis, and seborrheic dermatitis, had a similar prevalence in the two groups. Irritant plant dermatoses (2.5%), episodes of acute systemic intoxication from pesticides (6.8%), onychopathies due to chronic trauma (16%), contact dermatitis of the hands (12%), chilblains (7%), and Favre-Racouchot syndrome (2.5%) showed a higher prevalence in the study of agricultural workers.

9.4.5.2.8 Patch test reactions in feed mill workers

In 1990 Staffa et al.[25] described the prevalence of patch test reactions in a group of 204 feed mill workers from 15 separate northern Italian feed mills. No external control group was tested, but workers with skin complaints were compared to other workers. There were 16 cases of irritant and 12 cases of allergic contact dermatitis in the entire group, a prevalence of 13.7%. For workers employed more than 10 years, contact dermatitis complaints on questionnaire were more prevalent than for workers employed less than 10 years, but no difference was observed in the rate of positive patch tests. Sensitizers identified included antimicrobial agents, anti-fungal compounds, anti-parasitics, antioxidants, emulsifiers, dyes, and cobalt (used in feed as a trace element). Compounds not previously identified as sensitizers included indigo carmine, monensin sodium, thiabendazole, methylchlorpindol, and amprolium hydrochloride.

9.4.5.2.9 Patch test reactions to pyrethroids

In a 1992 report, Lisi[26] described a series of 230 patients, including 82 agricultural workers, tested with 1, 2, or 5% allethrin, cypermethrin, deltamethrin, fenothrin, fenvalerate, permethrin, and resmethrin. Two non-atopic patients showed irritant reactions, both caused by resmethrin. Two patients with positive reactions to fenvalerate, a farmer and a hobby gardener, had chronic contact dermatitis of the hands. An apparent allergic reactions to cypermethrin was not considered clinically relevant.

9.4.5.2.10 Seveso studies

The Seveso accident and similar accidents associated with production of 2,4,5-T are discussed in Section 8.5.3. Among reported dioxin episodes, the July 10, 1976, explosion at Seveso was unique in the extent of environmental contamination it produced and the occurrence of large numbers of chloracne cases outside of the work force at the affected plant.[27] Effects on vegetation were observed three days after the accident and animal deaths and acne lesions among children were observed two weeks after the accident.[28] Apart from approximately 200 cases of chloracne, few acute health effects were definitely established. Follow-up monitoring studies demonstrated serum dioxin (TCDD) levels ranging up to 56,000 ppt.[29] Geometric mean values (53.2 ppt) more modestly exceeded the 3.9 ppt background level.[30] Epidemiologic studies have documented an excess mortality from cardiovascular and respiratory diseases, a possible excess of diabetes cases, and an increased occurrence of GI, lymphatic, and hematopoietic cancers.[31] Reported reproductive effects associated with the accident have included an alteration of the sex ratio in children born to parents residing in the most heavily contaminated area of the community after April 1977 and prior to 1984.[29]

TABLE 9.14
Pesticide Dermatosis Reported In Spain

Pesticide	Clinical picture	No. patients	Ref.
Difolatan 1% pet.	Respiratory symptoms, urticaria, dermatitis	3	Camarasa[18]
Thiuram 1%	Dermatitis	7	Garcia-Perez et al.[19]
Vondozeb 0.5% pet.	Dermatitis	1	Chacon et al.[32]
Zineb 1% pet.	Dermatitis	3	Grimalt[33]
Dimethoate (++), Mancozeb (IR), dinocap (IR)	Dermatitis	1	Conde-Salazar[34]
Paraquat	Dermatitis, nail disease	1	Botella et al.[35]
Herbicides	Chloracne	5	Londono[36]

Note: Pet.= petrolatum, IR= irritant.

9.4.5.3 Spanish Experience with Pesticide Dermatoses

The King's order number 3349/83 (November 30, 1983) is the legislative basis for pesticide regulation in Spain. Although it mentions the importance of the dermal absorption route as cause of acute clinical toxic manifestations, it does not discuss the importance of acute or chronic dermatological signs and symptoms.

There are no official statistics in Spain regarding the effects of pesticides on people exposed to them, especially on agricultural workers. However, in Spain there are 2.1 million farmers, some 32% of the active working population.[28] As in other countries, knowledge of the pesticide dermatosis (PD) derives from clinical case reports, clinical studies of groups of patients, and from statistical compilations of patch test clinic reports. Below, we discuss the available literature regarding pesticide dermatoses in Spain.

Pesticide-related skin diseases in Spain are summarized by Garcia-Perez et al.[16] and by Conde-Salazar[29] both of these reviews are considered by them as a work of the Spanish Contact Dermatitis Research Group (GEIDC). During an occupational dermatology course held in Madrid (1987, Spain), Vilaplana[30] presented a paper entitled "Pesticides: present and future." He summarized knowledge about the pesticides in general and recommended a battery of patch tests.

Contact dermatitis is the most frequently reported pesticide-associated skin condition reported in the Spanish dermatological literature, with irritant contact dermatitis accounting for 75% and the allergic type 25% (Table 9.14). Published reports included 21 cases, 14 (76%) allergic, 5 were from chloracne, 2 irritant dermatitis and, one a nail disease. Pesticides involved were fungicides, herbicides, and insecticides. Fungicides involved were difolatan, thiuram, and carbamates (zineb, vondozeb). As noted in reports from other countries, herbicides were related to produce chloracne.

Conclusions: Pesticide dermatosis may be underreported in Spain. More research is necessary to establish the incidence and prevalence of skin conditions related to pesticides in the country and to document typical conditions of exposure. It is also necessary to develop a uniform protocol to assess pesticide dermatosis including a pesticide patch test series. A summary of the Spanish experience is presented in Table 9.14.

REFERENCES

1. **FAOSTAT,** Production data for primary livestock, primary crops, and crops by area harvested, 1998, for Greece, Italy, Spain, and Portugal. Available online at: apps.fao.org, 1999.
2. **FAOSTAT,** Pesticide Use Data , 1998, for Greece, Italy, and Portugal. Available online at: apps.fao.org, 1999.
3. **FAOSTAT,** Pesticide import and export data for Spain and Italy, 1961-1997. Available online at: apps.fao.org, 1999.
4. **Schena, D. and Barba, A.,** Erythema-multiforme-like contact dermatitis from dimethoate, *Contact Dermatitis*, 27, 116, 1992.
5. **Babini, G.,** On a case of erythroderma caused by zinc dithiocarbamate, *Arch. Ital. Dermatol. Venereol. Sessuol.*, 34, 230, 1966.
6. **Peluso, A. M., Tardio, M., Adamo, F., and Venturo, N.,** Multiple sensitization due to bis-dithiocarbamate and thiophthalimide pesticides, *Contact Dermatitis*, 25, 327, 1991.
7. **Piraccini, B. M., Cameli, N., Peluso, A. M., and Tardio, M.,** A case of allergic contact dermatitis due to the pesticide maneb, *Contact Dermatitis*, 24, 381, 1991.
8. **Crippa, M., Misquith, L., Lonati, A., and Pasolini, G.,** Dyshidrotic eczema and sensitization to dithiocarbamates in a florist, *Contact Dermatitis*, 23, 203, 1990.
9. **Manuzzi, P., Borrello, P., Misciali, C., and Guerra, L.,** Contact dermatitis due to Ziram and Maneb, *Contact Dermatitis*, 19, 148, 1988.
10. **Angelo, C., Ruatti, P., and Comba, P.,** A case of acne in a subject exposed to paraquat, *Med. Lav.*, 77, 247, 1986.
11. **Lisi, P. and Caraffini, S.,** Pellagroid dermatitis from mancozeb with vitiligo, *Contact Dermatitis*, 13, 124, 1985.
12. **Villa, L., Pizzini, L., Vigano, G., Ferioli, A., Maroni, M., Ruggeri, R., Barlassina, C., Vannini, P., and Salacrist, L.,** Paraquat-induced acute dermatitis in a child after playing with a discarded container, *Med. Lav.*, 86, 563, 1995.
13. **Colosio, C., Tomasini, M., Cairoli, S., Foa, V., Minoia, C., Marinovich, M., and Galli, C. L.,** Occupational triphenyltin acetate poisoning: a case report, *Br. J. Ind. Med.*, 48, 136, 1991.
14. **Tosti, A., Vincenzi, C., Bardazzi, F., and Mariani, R.,** Allergic contact dermatitis due to povidone-iodine, *Contact Dermatitis*, 23, 197, 1990.
15. **Baruffini, A., Cirla, A., Pisati, G., Ratti, R., and Zedda, S.,** Allergic contact dermatitis from 1,2-dichloropropane, *Contact Dermatitis*, 20, 379, 1989.
16. **Spettoli, E., Silvani, S., Lucente, P., Guerra, L., and Vincenzi, C.,** Contact dermatitis caused by sesquiterpene lactones, *Am. J. Contact Dermatitis*, 9, 49, 1998.
17. **Coricciata, L.,** Intolleranze associate agli anticrittogamici nei lavoratori agricoli, *Minerva Dermatol.*, 35, 65, 1960.
18. **Camarasa, G.,** Difolatan dermatitis, *Contact Dermatitis*, 1, 127, 1975.
19. **Garcia-Perez, A., Garcia-Bravo, B., and Beneit, J. V.,** Standard patch tests in agricultural workers, *Contact Dermatitis*, 10, 151, 1984.
20. **Santucci, B., Picardo, M., and Cristaudo, A.,** Contact dermatitis from *Euphorbia pulcherrima*, *Contact Dermatitis*, 12, 285, 1985.
21. **Lisi, P., Caraffini, S., and Assalve, D.,** A test series for pesticide dermatitis, *Contact Dermatitis*, 15, 266, 1986.
22. **Lisi, P., Caraffini, S., and Assalve, D.,** Irritation and sensitization potential of pesticides, *Contact Dermatitis*, 17, 212, 1987.

23. **Centers for Disease Control,** EPISTAT software, , 1997.
24. **Larese, F., Fiorito, A., Casasola, F., Molinari, S., Peresson, M., Barbina, P., and Negro, C.,** Sensitization to green coffee beans and work related allergic symptoms in coffee workers, *Am. J. Ind. Med.*, 34, 623, 1998.
25. **Stanek, C.,** Skin necrosis in a horse as a consequence of a disinfection accident, *Tieraerztl. Prax.*, 12, 45, 1984.
26. **Lisi, P.,** Sensitization risk of pyrethroid insecticides, *Contact Dermatitis*, 26, 349, 1992.
27. **Needham, L. L., Gerthoux, P. M., Patterson, D. G., Jr., Brambilla, P., Turner, W. E., Beretta, C., Pirkle, J. L., Colombo, L., Sampson, E. J., Tramacere, P. L., Signorini, S., Meazza, L., Carreri, V., Jackson, R. J., and Mocarelli, P.,** Serum dioxin levels in Seveso, Italy, population in 1976, *Teratog. Carcinog. Mutagen*, 17, 225, 1997.
28. **Holmstedt, B.,** Prolegomena Seveso, *Arch. Toxicol.*, 44, 211, 1980.
29. **Mocarelli, P., Needham, L. L., Marocchi, A., Patterson, D. G., Jr., Brambilla, P., Gerthoux, P. M., Meazza, L., and Carreri, V.,** Serum concentrations of 2,3,7,8-tetrachlorodibenzo-*p*-dioxin and test results from selected residents of Seveso, Italy, *J. Toxicol. Environ. Health*, 32, 357, 1991.
30. **Landi, M., Consonni, D., Patterson, D. J., Needham, L., Lucier, G., Brambilla, P., Cazzaniga, M., Mocarelli, P., Pesatori, A., Bertazzi, P., and Caporaso, N.,** 2,3,7,8-Tetrachlorodibenzo-*p*-dioxin plasma levels in Seveso 20 years after the accident, *Environ. Health Perspect.*, 106, 273, 1998.
31. **Bertazzi, P. A., Bernucci, I., Brambilla, G., Consonni, D., and Pesatori, A. C.,** The Seveso studies on early and long-term effects of dioxin exposure: a review, *Environ. Health Perspect.*, 106 Suppl. 2, 625, 1998.
32. **Chacón Espino, V. P., Valeron, M. P., and Hernandez, H. B.,** Dermatitis de Contacto por Vondozeb (Pesticida del Grupo de los Carbamatos), *Actas Dermo-Sifiologr.*, 76, 53, 1985.
33. **Grimalt, F., Romaguera, C., and Vilaplana, J.,** Dermatitis de contacto por pesticidas. Tres casos de sensibilización al zineb, *Med. Segur. Trab.*, 133, 75, 1986.
34. **Conde-Salazar, L.,** Unpublished data, 1998.
35. **Botella, R., Sastre, A., and Castells, A.,** Contact dermatitis to paraquat, *Contact Dermatitis*, 13, 123, 1985.
36. **Londoño, F.,** Acne ocupacional a proposito de 5 casos producidos por matamalezas, *Med. Cut.*, 3, 225, 1966.

9.4.6 Eastern Europe

FAO data for Eastern Europe indicate agriculture in the area is dominated by production of grain, root crops, and livestock.[1] Reported pesticide use data indicate a level of use roughly comparable to use in France, The Netherlands, and Belgium (40,000 metric tons annually).[2] Trade data indicate very minimal level of export trade (Figures 9.39a and b). Imports reached up to $700 billion in 1997. Pesticides involved in case reports, epidemiologic studies, and outbreak investigations for the region are summarized in Figure 9.40.

Cases associated with agricultural production included a case of contact pemphigus in a 63-year-old woman that appeared shortly after spraying DDT on watermelons;[3] dermatitis was associated with exposure to the insecticides thiometon[4] and lindane.[5] Contact dermatitis was also reported following contact with the herbicide

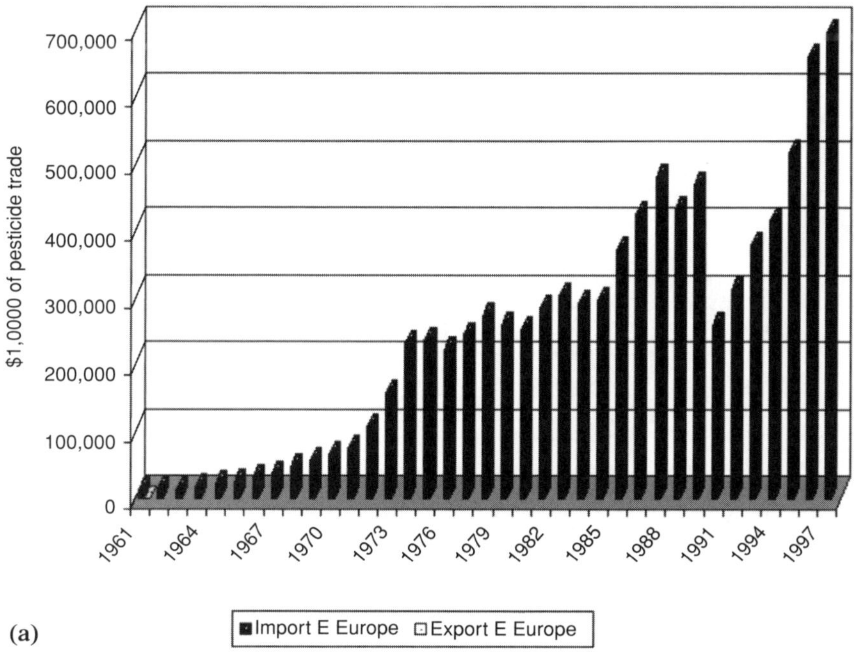

(a)

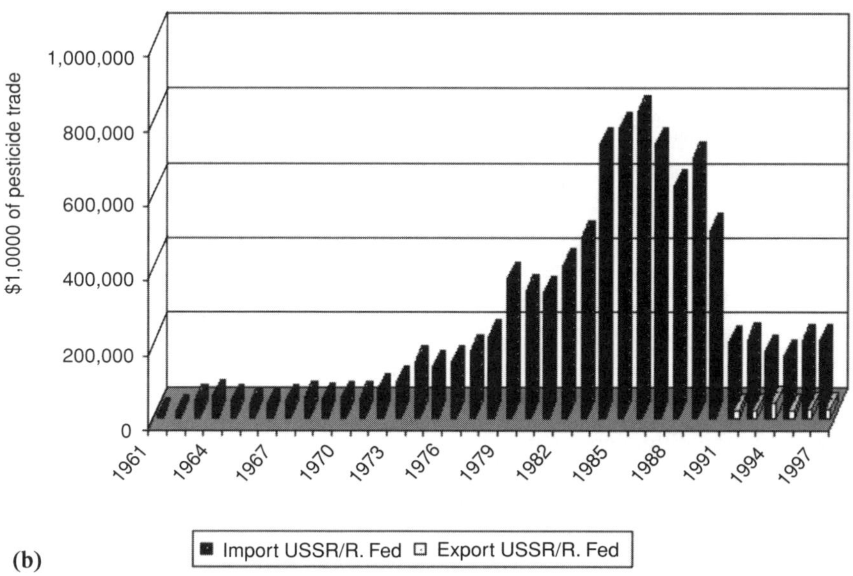

(b)

FIGURE 9.39 Pesticide trade data for Eastern Europe and the USSR/Russian Federation. (a) Eastern Europe and (b) the USSR/Russian Federation.

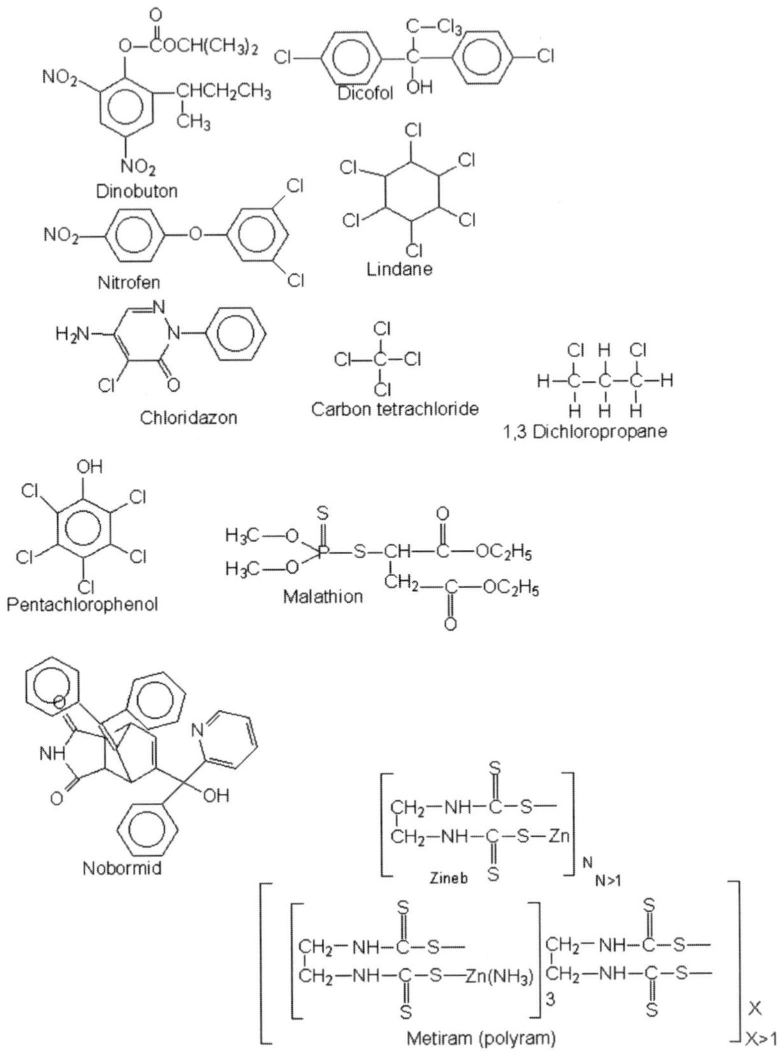

FIGURE 9.40 Reported causes of dermatitis in the USSR/Russian Federation and Eastern Europe.

nitrofen,[6] and the dithiocarbamate compound mancozeb,[7] the insecticide formithion (under the name trade name Anthio®) following direct exposure to a tractor driver from Tashkent, Uzbekistan,[8] and bullous contact dermatitis following exposure to the organophosphate defoliant butifos.[9]

Non-agricultural exposures resulted in reported episodes of contact dermatitis associated with use of dichloropropane as a preservative in an adhesive formula,[10] and irritant dermatitis of the hands and forearms in a woman mixing a solution of the rodenticide nobormid,[11] and dermatitis in a florist exposed to thiram residues on cut flowers.[12]

In Russia, the factory production of pesticides accounted for several cases affecting workers exposed to zineb and ethylene diamine, an intermediate in its production.[13] In Czech workers, intermediates in the production of the herbicide pyrazon were similarly associated with cases of dermatitis.[14]

9.4.6.1 Prevalence, Incidence, Surveillance, and Outbreak Data

9.4.6.1.1 Greenhouse workers

A 1983 Bulgarian prevalence study,[15] based upon periodic health surveillance examinations, conducted among 116 greenhouse workers in 1983 found occupational dermatoses in 15 (12.93%). Occupational contact allergic dermatitis was found in 10, with the insecticide Acrex (dinobuton) identified as the most common cause of contact sensitization.

A 1978 Russian study evaluated 26 women working in nurseries for at least 5 years,[16] who all reported being healthy prior to the beginning of their employment. Pesticide exposures included phosphamide (dimethoate), Acrex (dinobuton), kelthane (dicofol), trichlorfon, and polycarbazine (metiram); 20 had complaints of fatigue, weakness, and recurrent skin rash following the application of pesticides and 14 were diagnosed with allergic dermatitis, 5 with asthmatic bronchitis, and 2 with allergic rhinitis. Organophosphate exposure was evidenced by reduction of group mean cholinesterase activity to 60% of the level of non-exposed controls. The authors also attributed numerous other health problems to the pesticide exposure including a disturbance of regional blood flow known as "vascular dystonia," skin hyperthermia, and asymmetry of skin temperature. Laboratory effects attributed to pesticide exposure included leukopenia, eosinophilia, and relative lymphocytosis, a three-fold increase in histidase activity relative to control subjects indicated possible effect of work exposures on inducing liver enzymes, but liver cell injury (evidenced by normal transferase enzyme levels) did not seem to be occurring. Inhibition of the phytohemagglutinin and blast transformation reactions was also observed in some subjects.

9.4.6.1.2 Southern Russia and Asian republics of the former U.S.S.R.

A 1978 case series/surveillance report[17] described dermatoses in 138 farm workers of the Saratov region, in southern Russia between the Ukraine and Kazakhstan, treated over a 5-year period. The 108 men and 30 women patients treated ranged in age from 16 to 68 years old. Conditions included possibly work related eczemas in 32 patients, mycosis in 10, and photodermatosis in 27. Psoriasis, collagenosis, and other endogenous conditions prone to aggravation in the workplace accounted for the remaining cases. In tractor drivers, pesticides and fertilizers were viewed as the cause of most of the cases of allergic dermatitis and eczema.

A similar case series that amounted to a surveillance report (from the cotton growing region of Uzbekistan around Tashkent) by Karimov in 1975[18] described treatment of pesticide related skin conditions in 159 patients between the ages of 21 and 40 (38% males, 62%, yr). Occupational pesticide contact varied between 1 month and 3 years. The most common skin findings included dryness, cracking, edema, erythema, and papulovesicular eruptions and the most commonly involved compounds included BHC, seed-dressing agents (copper trichlorophenolate, thiram), and compounds used as cotton defoliants (calcium cyanamide, magnesium chlorate, and pentachlorophenol).

Karimov[18] evaluated the incidence of contact sensitivity and dermatitis in 300 cotton growers in 1972 and reported sensitization to DDT in 9 subjects and sensitization to BHC in 12 subjects. A series of 112 patients had seasonal acute or subacute dermatitis, usually of the face, neck, chest, and hands. This group, testing with 3 to 5% concentrations, demonstrated sensitization to BHC in 47(42%) and to DDT in 38 (34%).

9.4.6.1.3 Prevalence studies in Czechoslovakia

In 1975,[19] the incidence of different diseases was compared in an area of Czechoslovakia with limited use of pesticides and in another area in which the intensity of pesticide use was four times higher. Eczema, dermatitis, and urticaria were found to be reported more frequently in the high use region compared with the control area. Other findings included a higher incidence of circulatory diseases, nervous system disorders, chronic coronary insufficiency, myocarditis, myocardiodystrophy, polyneuritis, polyradiculitis, chronic gastritis and colitis, lung tuberculosis, residual pneumonia, chronic nephritis, and cystitis.

9.4.6.1.4 Pesticide production

In 1981 an examination survey of 241 workers involved in pesticide production was reported from Kharkov, in northeastern Ukraine.[20] The report reviewed guinea pig tests documenting the capacity of lindane, thiram, copper trichlorophenolate, and phenthiuram to produce allergic reactions. Examination demonstrated that 53 (22.0%) had contact dermatitis or eczema, including 25 (10.3%) with suspected allergic reactions.

Eighteen cases of dermatitis were reported during the first year (1971) of a new factory producing carbophos (malathion).[21] Industrial hygiene review of the operation estimated that workers were in contact with process chemicals 59-92% of their working hours. Other conditions noted by the author included decreased cholinesterase, hypoalbuminemia, headache, vertigo, increased irritability, nausea, trembling of the extended fingers, exaggerated tendon reflexes, disturbances in the orientation reflexes, distortion of sensitivity in the extremities, and edema.

In 1972, Elizarov[22] reported 124 cases of contact dermatitis among workers manufacturing propazine and simazine. Mild cases lasting 3 or 4 days involved pale pink erythema and slight edema. Serious cases lasting 7-10 days involved greater erythema and edema and also a vesiculopapular reaction that sometimes progressed to bullae.

In 1976[23] and 1981,[24] Czechoslovakian investigators reported on 55 workers exposed during the production of 2,4,5-sodium trichlorphenoxyacetate and related esters between 1965 and 1968. Chloracne affected the majority of workers and 11 developed porphyria cutanea tarda. Problems with elevated lipids, glucose intolerance, and biochemical and histological evidence of liver abnormalities were common and 17 developed symptoms of peripheral neuropathy.

REFERENCES

1. **FAOSTAT,** Production data for primary livestock, primary crops, and crops by area harvested, 1998, for Russian republic and former Soviet republic and Eastern Europe. Available online at: apps.fao.org, 1999.

2. **FAOSTAT,** Pesticide Use Data , 1997 and 1994, for USSR/Russian Republic and Eastern Europe. Available online at: apps.fao.org, 1999.
3. **Tsankov, N., Dimitrowa, J., Obreschkowa, E., and Lasarowa, A.,** Induced pemphigus caused by the pesticide phosphamide, *Z. Hautkr.*, 62, 196, 1987.
4. **Maric, M.,** Toxic exanthema caused by ekatin, *Med. Pregl.*, 23, 51, 1970.
5. **Kovalenko, M.,** [Allergic dermatitis caused by hexachlorane] Allergicheskii dermatit, vyzvannyi geksakhloranom, *Vestn. Dermatol. Venerol.*, 40, 69, 1966.
6. **Tsyrkunov, L. P.,** Toxicoallergic contact dermatitis due to the pesticide kelthane, *Vestn. Dermatol. Venerol.*, 0, 44, 1986.
7. **Kleibl, K. and Rackova, M.,** Cutaneous allergic reactions to dithiocarbamates, *Contact Dermatitis*, 6, 348, 1980.
8. **Karimov, A. M.,** Case of toxidermia caused by the pesticide Anthio, *Vestn. Dermatol. Venerol.*, 40, 1979.
9. **Karimov, A. M., Inagamova, D. A., and Iusupov, B.,** A case of bullous contact dermatitis caused by butifos, *Gig. Tr. Prof. Zabol.*, 57, 1988.
10. **Gryzwa, Z. and Rudzki, E.,** Dermatitis from dichloropropane, *Contact dermatitis*, 7, 151, 1981.
11. **Tsyrkunov, L. P.,** Skin lesions caused by exposure to a pesticide norbormid, *Gig. Tr. Prof. Zabol.*, 8, 46, 1989.
12. **Rudzki, E. and Napiorkowska, T.,** Dermatitis caused by the Polish fungicide Sadoplon 75, *Contact Dermatitis*, 6, 300, 1980.
13. **Tsyrkunov, L. P.,** Toxico-allergic dermatitis induced by ethylenediamine in the manufacture of the herbicide zineb, *Gig. Tr. Prof. Zabol.*, 45, 1987.
14. **Kolesar, D. and Romancik, V.,** Clinical and laboratory findings in the acute damage by precursors of the weed-killer "burex" (author's transl), *Bratisl. Lek. Listy*, 64, 132, 1975.
15. **Popov, I., Darlenski, B., and Mladenova, S.,** Occupational dermatoses among agricultural workers in green-house production of vegetables, *Higiena i zdraveopazvane*, 26, 214, 1983.
16. **Zolotnikova, G. P. and Somov, B. A.,** Role of pesticides in development of occupational dermatoses by greenhouse workers, *Vestn. Dermatol. Venerol.*, 4, 76, 1978.
17. **Dovzhanskii, S. I., Ardentov, N. N., Yakubovskaya, T., Suvorov, A. P., and Orkin, V. F.,** Dermatoses in farm workers, *Gig. Tr. Prof. Zabol.*, 7, 47, 1978.
18. **Karimov, A. M.,** Treatment of pesticide induced dermatoses, *Vestn. Dermatol. Venerol.*, 5, 64, 1975.
19. **Pol'chenko, V. I., Tkach, L. I., and Gumennyy, V. S.,** Comparative incidence of diseases in two rural zones with various intensities of pesticide use, *Vrach. Delo*, 12, 106, 1975.
20. **Zadorozhny, B. A., Petrov, B. R., Olefirenko, V. A., and Kirilko, V. A.,** Professional'nye dermatozy i ikh profilaktika v usloviyakh proizvodstva pestitsidov. [Occupational dermatoses and their control under conditions of pesticide manufacture.], *Vestn. Dermatol. Venerol.*, 6, 48, 1981.
21. **Aizenshtad, V. S., Dolmatova-Guseva, E. G., Perkhurova, V. P., Shtifel'man, A. V., and Bogomolova, L. M.,** Industrial hygiene and health status of workers in carbophos production, *Gig. Tr. Prof. Zabol.*, 15, 49, 1971.
22. **Elizarov, G. P.,** Occupational skin diseases caused by simazine and propazine, *Vestn. Dermatol. Venerol.*, 46, 27, 1972.
23. **Jirasek, L., Kalensky, J., Kubec, K., Pazderova, J., and Lukas, E.,** Chloracne, porphyria cutanea tarda, and other poisonings due to the herbicides, *Hautarzt*, 27, 328, 1976.
24. **Pazderova-Vejlupkova, J., Nemcova, M., Pickova, J., Jirasek, L., and Lukas, E.,** The development and prognosis of chronic intoxication by tetrachlorodibenzo-*p*-dioxin in men, *Arch. Environ. Health*, 36, 5, 1981.

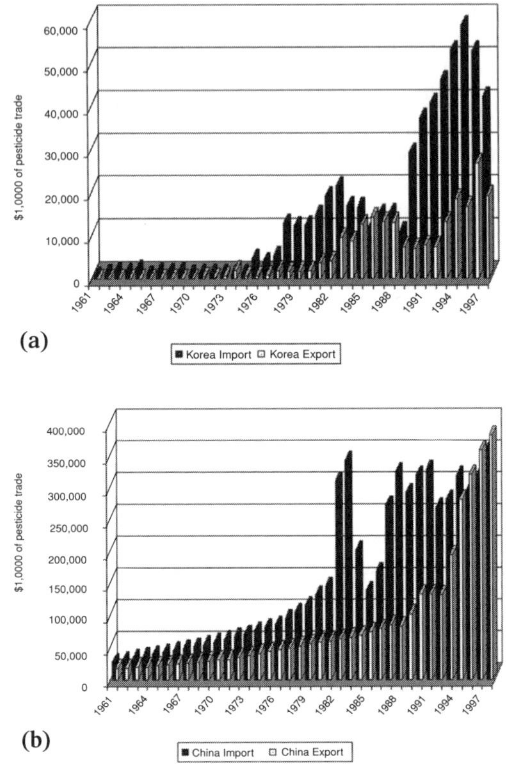

FIGURE 9.41 Korea and China pesticide trade data. (a) Korea and (b) China.

9.5 ASIA

Information on the occurrence of skin conditions associated with pesticides in Asia is sporadic. Although the effect of pesticides on the skin has been studied intensively in Japan, relatively little information from other Asian countries has been published. The sections below review regional data on China, Taiwan, and Korea; Japan; India, Malaysia, Sri Lanka, and Thailand; and Israel, Turkey, and the Middle East. As in other sections, data on agricultural production and pesticide use derive from FAO data.

9.5.1 China, Taiwan, and Korea

Grain crops dominate agricultural production in China and Korea, according to FAO figures for 1998; the principal livestock raised are pigs, chickens, beef, and ducks.[1] Taiwan and Korea have a similar pattern of agriculture. Pesticide use data are available only for Korea, with fungicides, herbicides, and plant growth regulators accounting for about 70% of the total usage.[2] Although use data were not available for China, trade data demonstrate that China has both imported and exported increasing quantities of pesticides since 1961 (Figures 9.41a and b). Case reports, epidemiologic studies, and outbreak investigations reported for the region are summarized in Figure 9.42.

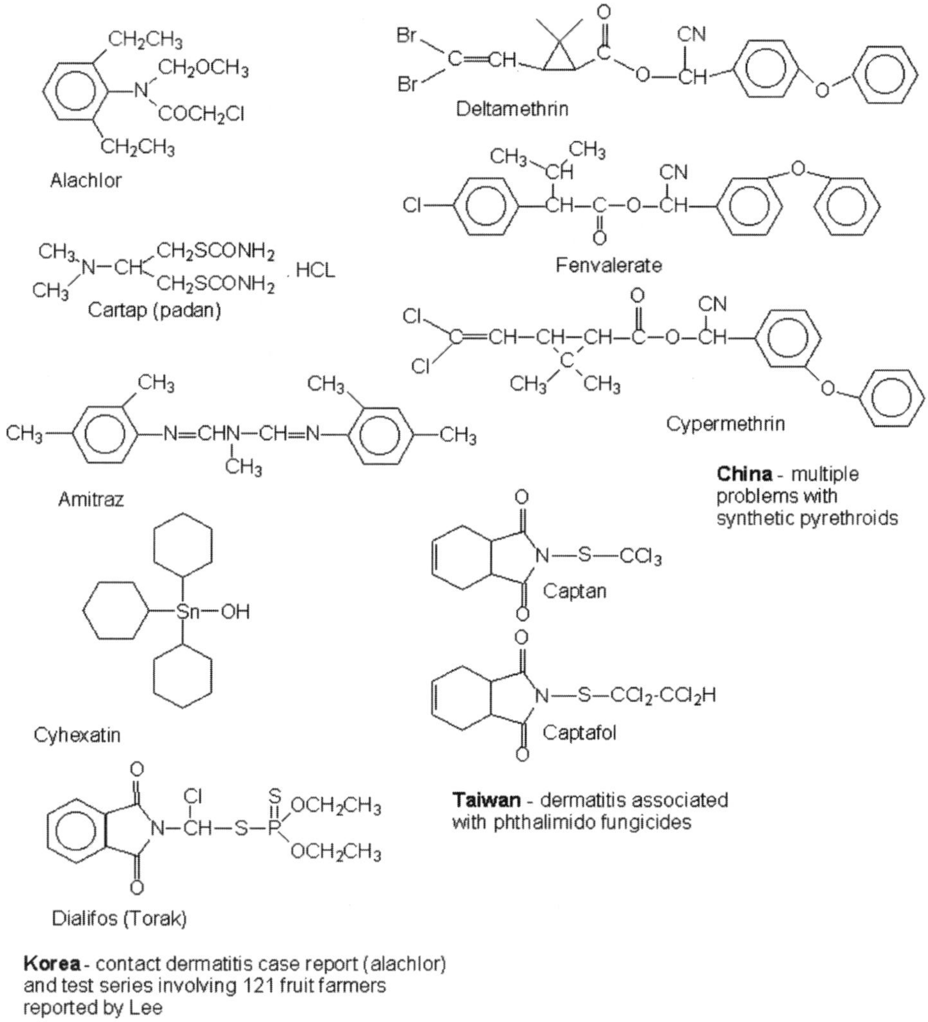

FIGURE 9.42 Pesticides associated with contact dermatitis in China, Korea, and Taiwan.

9.5.1.1 Case Reports and Case Series

In 1993, Won[3] described a case of sensitization due to alachlor in a Korean patient who had developed dermatitis after contaminating her socks with both alachlor and paraquat (Figure 9.43). Tests were positive to 0.1 and 0.2% aqueous alachlor and were negative in 10 controls, with the exception of one who developed a case of active sensitization 20 days following test application. Tests with 0.1 and 0.2% paraquat and with the European standard series were negative.[3] Although paraquat did not contribute to the allergic response experienced by this patient, it may have contributed as an irritant to the initial dermatitis.

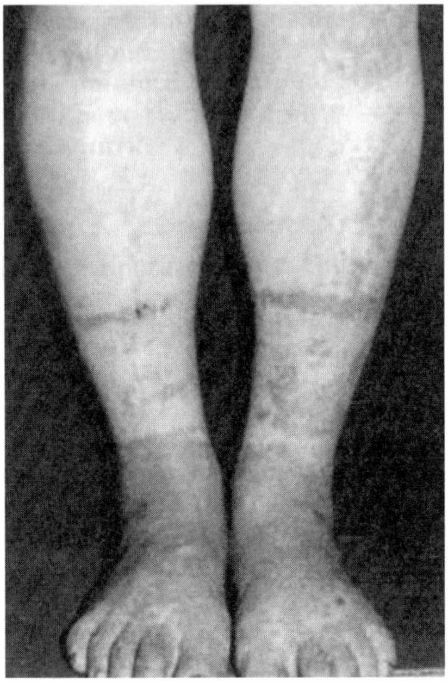

FIGURE 9.43 Dermatitis in a footwear pattern associated with exposure to alachlor and paraaquat. (Won, 1993. Reprinted with permission from *Contact Dermatitis*, Munksgaard.)

9.5.1.1.1 MATDA in rural counties near Shanghai

The fungicide MATDA (*N, N′*-methylene-*bis*-(2-amino-1,3,4-thiadiazole) was produced in China for control of *Xanthomonus oryzae* pollution of rice and estimated to save 6 million tons of rice annually. Although concerns about the compound centered on its effect as a teratogen in animal studies, the principal effect observed in workers producing and farmers handling the compound was skin irritation and dermatitis. An experimental evaluation by the Buehler method showed that the pesticide was both an irritant and a weak sensitizer. The problem was partially controlled by changing the formulation from a powder to an emulsfiable concentrate. Field studies of 704 sprayers conducted from 1980 to 1985 in Jianhu and Wujiang, two counties in eastern China near Shanghai, showed that the occurrence of dermatitis depended upon the control measures adopted. Although several tons of MATDA were applied in Wujiang and neighboring areas, dermatitis was controlled totally. Control was achieved in Jianhu county by changing the method of dipping seed and spraying MATDA seedlings.[4]

9.5.1.2 Prevalence, Incidence, Surveillance, and Outbreak Data

Most publications from China on pesticides have dealt with acute systemic poisoning[5, 6] due to organophosphates and other acute toxins. However, reports from China have included an extensive series of pyrethroid poisoning cases, often with accompanying dermal irritation.[7] Individual compounds included deltamethrin, fenvalerate, and cypermethrin; and most cases involved failure to use proper application technique or failure to wear protective equipment. Within 4-6 h after exposure, approximately half complained of abnormal facial sensation — described as burning, itching, or tingling of the skin. These were sometimes accompanied by papules on the skin that disappeared within 24 h. Some of the occupational cases had systemic symptoms, but overt systemic poisonings were more common among the 344 cases of accidental non-occupational ingestion. For the ingestion cases, skin findings were not significant. In a 1988 survey of 199 workers dividing and packaging pyrethroids, He[8] reported that up to two thirds had facial irritation and some had fine papular (miliary) dermatitis. No systemic poisoning was observed.

9.5.1.2.1 Fruit farmers in Taiwan

In 1996, Guo[9] evaluated the frequency of skin disease in fruit farmers of Taiwan by examining 122 fruit farmers who regularly handled pesticides and a comparison group of 63 printers; 30% of farmers had hand dermatitis and more than two thirds had pigmentation and thickening on the hands; 25% had dermatomycosis. The two groups had similar frequency of response to the standard antigen series, but 40% of farmers were sensitive to agricultural chemical allergens, most frequently the phthalimido compounds captofol, folpet, and captan.

9.5.1.2.2 Korean farmers

Lee[10] reported a patch test study in 1981, conducted on 121 Korean farmers with contact dermatitis, evaluating the prevalence of hypersensitivity to commonly used pesticides (difolatan, metasystox, plictran, omite, DDVP, padan, sevin, micut (amitraz), torak, dithane m-45, topsin M (thiophanate methyl), agrepto, torque, kelthane). Tests were developed by testing on the skin of healthy volunteer medical students. Positive reactions were observed for difolatan (18%), padan (4.1%), dithane m-45 (3.3%), the OP insecticide torak (2.5%), metasystox (1.7%), and plictran 1.7%. One positive test was also observed to propargite, at a 2.5% concentration, but probably represented an irritant response.

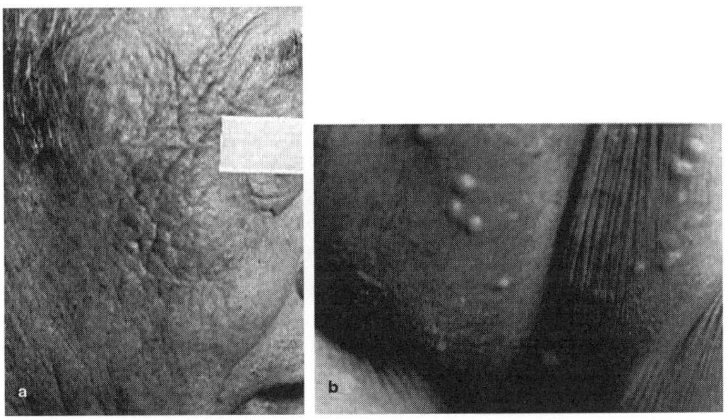

FIGURE 9.44 Chloracne in a pentachlorophenol worker from Tianjin, China. (Reprinted from the *American Journal of Industrial Medicine*, 1993, with permission from John Wiley & Sons.)

9.5.1.2.3 Chloracne

In 1990 and 1993 Cheng[11, 12] described the results of a prevalence investigation of 109 workers employed in the production of pentachlorophenol and BHC (Figure 9.44). Chloracne was found in 80 (73.4%) of the workers examined. The illness affected 20 of 21 workers employed in a trichlorobenzene tank area where the levels of dioxins and dibenzofurans exceeded 1000 ppm.. In addition to the chloracne, the subjects had higher levels of urinary porphyrins than controls.

REFERENCES

1. **FAOSTAT,** Production data for primary livestock, primary crops, and crops by area harvested, 1998, for China and Korea. Available online at: apps.fao.org, 1999.
2. **FAOSTAT,** Pesticide Use Data, 1993, for Republic of Korea. Available online at: apps.fao.org, 1999.
3. **Won, J. H., Ahn, S. K., and Kim, S. C.,** Allergic contact dermatitis from the herbicide Alachlor, *Contact Dermatitis*, 28, 38, 1993.
4. **Gu Xue, Q., Qian, C., Xue, S., Lu, Q., Wang, X., Jin, X., Yin, F., Huang, L., Lin, H., Fu, L., Li, D., and Huang, N.,** Risk assessment on using MATDA [n,n'-methylene-bis2 -amino-1,3,4-thiadiazole)] as a pesticide., *Bull. WHO Collaborating Center Occup. Health, Shanghai, Occup. Health Industrial. Modern.*, 2, 19, 1988.
5. **Xue, D., Li, Y., Jin, Z., Tao, B., and Zhou, Q.,** Trend of pesticide poisoning in recent decades in China, *Med. Lav.*, 89, 99, 1998.
6. **Sun, D., Zhou, H., and Xue, S.,** Epidemiologic survey on organophosphate-induced delayed polyneuropathy (OPIDP) among patients recovered from Methamidophos poisoning, *Med. Lav.*, 89, 123, 1998.
7. **He, F., Wang, S., Liu, L., Chen, S., Zhang, Z., and Sun, J.,** Clinical manifestations and diagnosis of acute pyrethroid poisoning, *Arch. Toxicol.*, 63, 54, 1989.
8. **He, F., Sun, J., Wu, Y., Yao, P., Wang, S., and Liu, L.,** Effects of pyrethroid insecticides on subjects engaged in packaging pyrethroids, *Br. J. Ind. Med.*, 45, 548, 1988.
9. **Guo, Y. L., Wang, B. J., Lee, C. C., and Wang, J. D.,** Prevalence of dermatoses and skin sensitisation associated with use of pesticides in fruit farmers of southern Taiwan, *Occup. Environ. Med.*, 53, 427, 1996.
10. **Lee, S., Chin, Y., Chang, W., and Kim, J.,** A study on hypersensitivity of Korean farmers to various agrochemicals. Determination of concentration for patch test of fruit-tree agrochemicals and hypersensitivity of orange orchard farmers in Che-ju Do. Korea, *Seoul Med. J.*, 22, 137, 1981.
11. **Cheng, W., Coenraads, P., and Olie, K.,** Chloracne in a pesticide factory, *Contact Dermatitis*, 23, 300, 1990.
12. **Cheng, W. N., Coenraads, P. J., Hao, Z. H., and Liu, G. F.,** A health survey of workers in the pentachlorophenol section of a chemical manufacturing plant, *Am. J. Ind. Med.*, 24, 81, 1993.

9.5.2 JAPAN

FAO data indicate that grains, root crops, fruits, and vegetables are the chief agricultural products in Japan.[1] Japanese agriculture employs intensive cultivation methods and FAO pesticide trade data suggest increasing pesticide use in the country

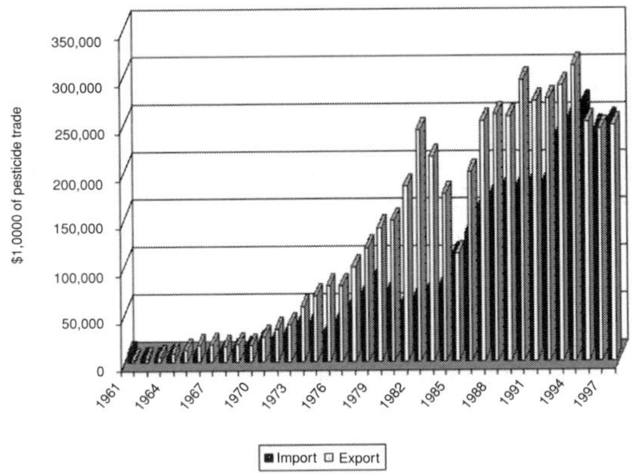

FIGURE 9.45 Japanese pesticide trade data.

from the initial point of data collection in the 1960s to the present (Figure 9.45). Current pesticide use data are not reported to the FAO database. However, a 1978 review by Wakatsuki[2] reported that the amount of pesticides applied per unit area in Japan is six times that of Europe and seven times that of the U. S. Based upon reports collected by the Japanese Association of Rural Medicine, Wakatsuki described problems stemming from heavy pesticide use including acute poisoning, dermatitis, and eye irritation. During this time period, other Japanese authors of published studies from the 1970s had similar experiences (see Sections 9.5.2.1 and 9.5.2.2).

9.5.2.1 Case Reports and Case Series

Reported cases associated with agricultural production or processing included photoallergy to the fungicide mancozeb in a pesticide handler[3] and photoallergy to chlorothalonil in a gardener.[4]

In the pesticide production industry, case reports included contact dermatitis due to tributyl tin phthalate;[5] allergy to the herbicide nitralin, along with its precursor compund 4-chloro 3,-5- dinitrophenylmethyl sulfone (DNC); and cross sensitivity to DNCB.[6] A non-agricultural case report involved an acute phosphine intoxication in a warehouseman, accompanied by a an acute dermatitis.[7]

9.5.2.2 Prevalence, Incidence, Surveillance, and Outbreak Data

Numerous case series describing the effects of pesticides on the skin of Japanese agricultural workers were published, based upon surveillance programs operated by the Japanese Association of Rural Medicine (JARM). Important national level studies and reports regarding individual outbreaks are discussed below.

Matsushita[8] reported in 1980 on 815 cases of pesticide poisoning recorded on a standard form from rural Japanase medical institutions for the Pesticide Poisoning Study Group of JARM between 1968 and 1970. Of these, 316 (38.8%) had skin lesions, including 216 (26.5%) with contact dermatitis. The compounds responsible included captafol (Difolatan), organophosphates, and chlorinated hydrocarbons; 78% of the cases occurred in pesticide applicators. In a 1985 report, additional details were given about the cases involving organophosphates. These most frequently included DDVP, salithion, sumithion, phosvel, cyanox, kilval (Vamidothion), diazinon, and malathion.[9] Sugaya[10] reported on 1376 cases identified by the JARM system between 1970 and 1977, with 879 (63.8%) occurring during spray operations, but also including numerous cases of deliberate or accidental ingestion. Of the total series, 573(41.6%) involved dermatitis.

9.5.2.2.1 Dilfolatan in tangerine orchards, 1966 and 1967

In 1968, Takamatsu[11] described work conducted in the two preceding years to evaluate the incidence of dermatitis among farmers using difolatan in the tangerine orchards of the Kumamoto prefecture (on the southern island of Kyushu). A questionnaire survey administered in 1966 to farmers and their families identified 264 cases of dermatitis (25.0% incidence). During 1967, 356 cases were identified, affecting 38% of those surveyed. The dermatitis was most often acute, beginning 1-3 days after handling the fungicide, affecting the neck, face, and forearms. After exposure ceased, the dermatitis resolved over a variable period, but few cases lasted as long as a month.

9.5.2.2.2 Propargite in orange grove applicators, 1969

In reporting an outbreak among orange farmers in the Osaka prefecture, Nishioka[12] noted "the scale and variability of problems associated with the use of pesticide increases with continuous mechanization of agriculture." Of 49 applicators handling propargite (under name BPPS), 40 (81.6%) had dermatitis, usually of acute onset. Non-specific symptoms including headache, fatigue, loss of appetite, and pharyngeal symptoms occurred in more than 50%; ocular and upper respiratory symptoms occurred in more than 20%. Experimental testing on two volunteer subjects demonstrated that the compound provoked irritation in both subjects in concentrations as low as 0.0005%.

9.5.2.2.3 Triforine

The prevalence of delayed-contact dermatitis to type allergenicity of triforine and other fungicides was evaluated among Japanese chrysanthemum growers.[13] Positive patch reactions to triforine (0.02% aqueous) were seen in 18 (17%) of 107 subjects tested and the authors reported that no reactions were seen among an undisclosed number of healthy non-agricultural subjects in preliminary testing. Among workers with reactions to triforine, 44% had a history of possibly work related contact dermatitis and 67% also had positive patch test reactions to dichlorvos (DDVP, 0.02% aqueous). The authors also evaluated cross reactions between these two compounds in the guinea pig maximization test.

9.5.2.3 Regional Studies

Regional studies indicate the breadth of interest in the investigation of skin disease in agricultural workers throughout Japan from the mid 1960s to the present. The groups studied included pesticide applicators, citrus workers, and chrysanthemum producers. The studies were principally descriptive in nature, and frequently had

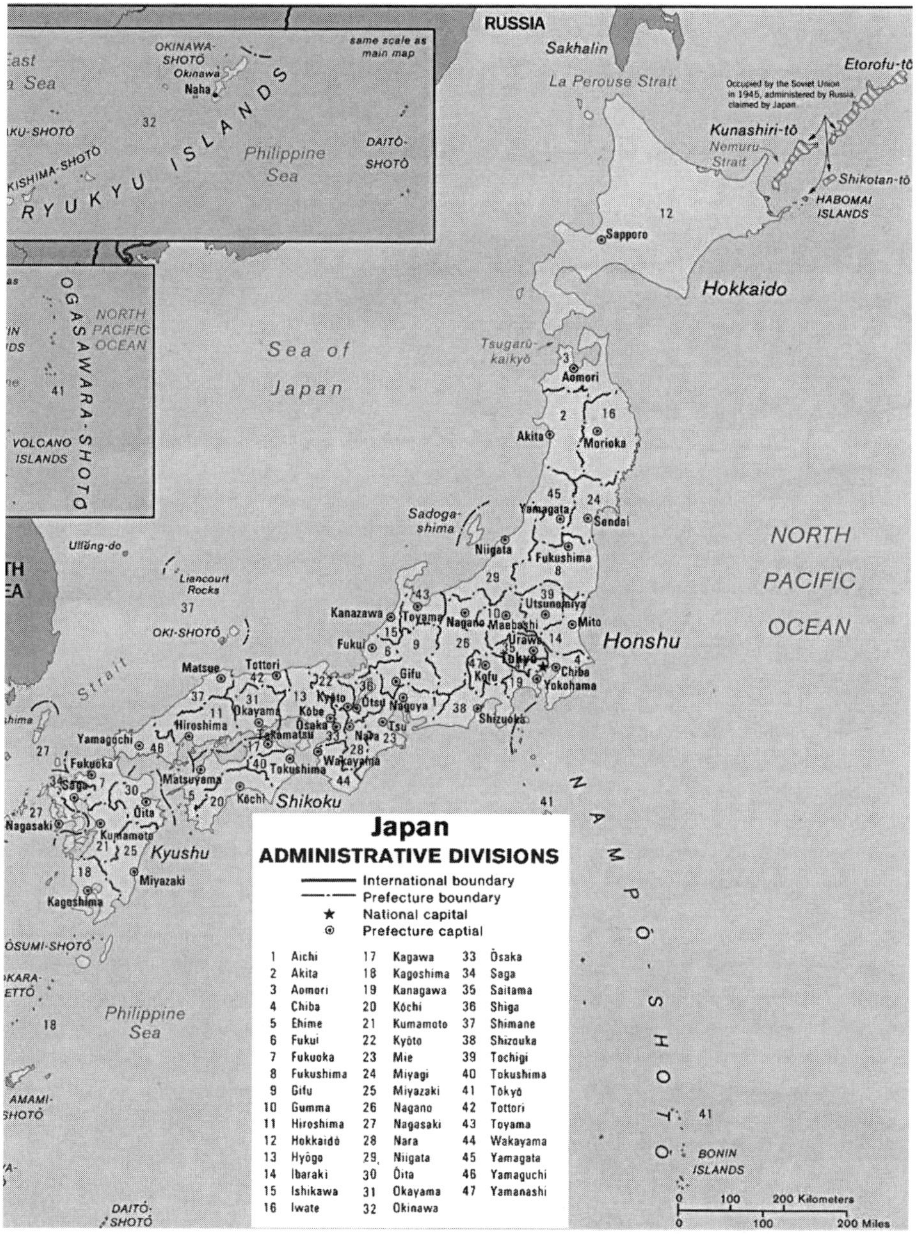

FIGURE 9.46 Japan administrative divisions. (Map courtesy of Perry-Castaneda Collection, University of Texas, Austin.)

systemic as well as topical effects. Details of selected studies are given in Table 9.15 and the affected administrative divisions (prefectures) are shown in Figure 9.46.

TABLE 9.15
Case Series and Health Surveys Reported from Local Jurisdictions (Prefectures) in Japan

Akita (the northern section of Honshu, central Japan)

Takano reported a series of pesticide intoxications, including suicide attempts and accidents, in 311 patients, farmers, aged 30-49, occurring between 1967 and 1971 in Akita prefecture, located near the northern tip of Honshu, the central island in Japan. The most frequently involved compounds included Blasticidin-S, fenitrothion, and BHC. Dermatitis cases occurred principally from residue exposure.[14]

Sugaya reported a survey of 26 pesticide sprayers from Nakayama, Akita prefecture conducted between June and August, 1974, that identified 11 cases of possibly pesticide related illness, involving OPs, OCs, and carbamates. The cases included 1 case of dermatitis and 7 cases of deformed nails.[15]

Chiba (central island of Honshu, east of Tokyo)

A health survey reported by Uchida was carried out on pesticide applicators exposed to maneb, fenitrothion, and paraquat in a pear growing and greenhouse area of Chiba prefecture from June 1974 to January 1975.[16] Occupational health problems identified in the survey included acute poisoning, dermatitis, and eye irritation and conjunctivitis, perhaps attributable to the infrequent use of protective garments. Systemic complaints considered possibly related to pesticide handling by the authors included abnormal liver function and anemia, myocardial damage, proteinuria, and glycosuria.

Ehine (southern island of Shikoku)

A survey of the citrus growing farmers in Ehime prefecture (Shikoku island, south of the central island of Honshu), Japan, revealed that 70% had previously experienced contact dermatitis due to pesticides.[17] The most frequently involved compounds were the fungicides dithianone, maneb, and manzeb. The monthly incidence rate of skin injuries from spraying pesticides varied from 0%-3.8% in May, to 26.0%-50% in July.

Fukuoka (southernmost island of Kyushu)

The health of 50 families of workers in illuminated chrysanthemum growing sheds in Fukuoka prefecture was examined in May and October, 1971, to determine any relationship between contact dermatitis and pesticide applications.[18] Illnesses included contact dermatitis due to pesticides — (dichlorvos, EPN, Estox, dithane (nabam), Lannate (methomyl), Nissol (MNFA based), and gramoxone (paraquat) — headaches, dizziness and nausea. Dermatitis was also induced by the leaves of the chrysanthemums, proven by 48 h patch tests. Non-specific systemic symptoms included susceptibility to figure, lumbago, shoulder discomfort pruritis, and stomach disorders. Serum ChE activity during the October season was siginificantly lower in man and probably related to organophosphate applications.

Kagoshima (southernmost island of Kyushu)

A field survey of 105 citrus growers revealed that 49 (46.7%) had experienced skin effects from pesticides in the past, associated with positive patch tests for maneb, carbaryl, chlorothalonil, and glyphosate.[19] An experimental study also identified putative cross reactivity between two structurally dissimilar compounds, chlorothalonil (TPN) and methidathion.

Kumamoto prefecture (southernmost island of Kyushu)

To investigate high rates of contact dermatitis occurring in prior years among 30-65% of workers applying thiophanate- methyl to stored citrus fruit, an examination survey was conducted in Kumamoto prefecture (on the southern island of Kyushu) in 1974.[20] During the survey, dermatitis occurred in 0.6-2.3% of men and 1.0-5.4 % of women applying benomyl. Specifically the pesticides that caused dermatitis were emulsified sulfur and calcium polysulfide as well as thiophanate-methyl. The symptoms are itching, redness, swelling, dry skin, and sometimes sensitized dermatitis, as well as congestion of the ocular mucosa. Results of medical examination revealed some abnormalities of hemoglobin. In patch tests, the percentage of positive reactions was about 5 for both maneb and thiophanate methyl.

TABLE 9.15
Case Series and Health Surveys Reported from Local Jurisdictions (Prefectures) in Japan (Continued)

Nomura reported medical evaluations of workers spraying thiophanate-methyl and benlate in Kumomoto prefecture for prevention of fungal diseases in stored citrus fruit. Dermatitis occurred in 30.3% of male thiophanate-methyl applicators in 1972 and in 32.3% of the women; in 48.2% of male applicators in 1973 and in 56.3% of the women; and in 52.8% of men in 1974 and in 64.8% of the women. Corresponding rates for benlate did not exceed 6% in any of the study years and study subgroups.[20]

Mie (Honshu island, central Japan, south of Tokyo)

Medical examinations were carried out on pear growers in the Mie prefecture (in central Japan, south of Tokyo) Hane area on 54 subjects during the working season (October, 1974) and on 46 subjects during the off-season (February, 1975).[21] Repeatedly applied pesticides were EPN, kitazin-P(IBP), dithane (maneb), fenitrothion, malathion, and kitathion. Dermatitis was among the most frequent finding in pesticide applicators. Anemia, reduced hematocrit, reduced total protein in blood, increased GOT and GP reduced cholnesterase, and hypertension during the October, 1974 farming season relative to the February. 1975 off-season were also described.

Nagano (Honshu island, central Japan, northwest of Tokyo)

Horiuchi reported a series of 88 cases from the general hospital of the Nagano prefecture in central Japan (Honshu) northwest of Tokyo from 1975 to 1978. Clinical findings included chemical burns, acute dermatitis, eczema, and photosensitivity. The authors reported that 10-28% of the farmers were sensitive to chlorthalonil, captafol, nicotine sulfate, dichlorvos, and maneb. Reported photosensitizers included paraquat, chlorthalonil, and dichlorvos.[22] Another series from the same institution described 142 cases of dermatitis in greenhouse workers, vegetable farmers, and other patients involving dicofol, the sulfonate compound chlorfenson (ovex), the dicofol analog, chlorfenethol, dichlorvos, fenitrothion, salithion, chlorthalonil, zineb, captafol, and paraquat.[23]

REFERENCES

1. **FAOSTAT,** Production data for primary livestock, primary crops, and crops by area harvested, 1998, for Japan. Available online at: apps.fao.org, 1999.
2. **Wakatsuki, S.,** Pesticide intoxication as an accident, present and future, *Nippon Saigai Igakkai Kaishi*, 26, 394, 1978.
3. **Higo, A., Ohtake, N., Saruwatari, K., and Kanzaki, T.,** Photoallergic contact dermatitis from mancozeb, an agricultural fungicide, *Contact Dermatitis*, 35, 183, 1996.
4. **Matsushita, S., Kanekura, T., Saruwatari, K., and Kanzaki, T.,** Photoallergic contact dermatitis due to Daconil, *Contact Dermatitis*, 35, 115, 1996.
5. **Hamanaka, S., Hamanaka, Y., and Otsuka, F.,** Phthalic acid dermatitis caused by an organostannic compound, tributyl tin phthalate, *Dermatology*, 184, 210, 1992.
6. **Nishioka, K., Asagami, C., Kurata, M., and Fujita, H.,** Sensitivity to the weed killer DNA-nitralin and cross- sensitivity to dinitrochlorobenzene, *Arch. Dermatol.*, 119, 304, 1983.
7. **Akoboshi, S., Takahashi, M., Seki, K., Kameda, S., Tashiro, M., and Harada, A.,** A case of acute phosphine toxicosis with chronic course, *Proc. Annu. Meet. Jpn. Assoc. Indust. Health*, 50, 285, 1977.
8. **Matsushita, T., Nomura, S., and Wakatsuki, T.,** Epidemiology of contact dermatitis from pesticides in Japan, *Contact Dermatitis*, 6, 255, 1980.

9. **Matsushita, T., Aoyama, K., Yoshimi, K., Fujita, Y., and Ueda, A.,** Allergic contact dermatitis from organophosphorus insecticides, *Ind. Health*, 23, 145, 1985.
10. **Sugaya, T., Wakatsuki, T., and Matsushima, S.,** A Statistical Study on Clinical Cases of Pesticide Poisoning in Rural Districts for Eight Years, Proceedings of the VII International Congress of Rural Medicine, Salt Lake City, Utah, September,, 1978.
11. **Takamatsu, M., Futatsuka, M., Arimatsu, Y., Maeda, H., and Inuzuka, T.,** Epidemiologic survey on dermatitis from a new fungicide used in tangerine orchards in Kumamoto prefecture, *Kumamoto Igakkai Zasshi*, 42, 854, 1968.
12. **Nishioka, K., Kozuka, T., and Tashiro, M.,** Agricultural miticide (BPPS) dermatitis, *Skin Res.*, 12, 15, 1970.
13. **Ueda, A., Aoyama, K., Manda, F., Ueda, T., and Kawahara, Y.,** Delayed-type allergenicity of triforine (Saprol), *Contact Dermatitis*, 31, 140, 1994.
14. **Takano, Y.,** On the clinical cases of pesticide intoxication during the last 5 years in Akita Prefecture, *Nippon Noson Igakkai Zasshi*, 21, 240, 1972.
15. **Sugaya, T., Hayashi, S., Ito, S., Suzuki, T., Kimura, M., Hayashi, M., and Kurokawa, S.,** Results of health examinations on pesticide applying farmers, *Akita Ken Noson Igaku Kenkyusho Nempo*, 4, 23, 1974.
16. **Uchida, A., Takamiya, T., and Ishige, T.,** Studies on methods of health management of persons engaged in pesticide application. Application of pesticides in Chiba Prefecture and health survey therein, Report on Studies Commissioned by the Ministry of Health and Welfare, 18, 1974.
17. **Shiwaku, K., Hirai, K., Torii, M., and Tsuboi, T.,** Skin injuries caused by pesticides among farmers in a citrus fruit area, *J. Jp. Assoc. Rural Med. May*, 32, 22, 1983.
18. **Rikimaru, T., Goto, Ezaki, H., and Takamatsu, M.,** Status of pesticide application and injuries in cultivators of illuminated chrysanthemum, *Nippon Noson Igakkai Zasshi*, 21, 236, 1972.
19. **Obama, K.,** Studies on allergic skin disease caused by pesticides in citrus growers: Field survey study and animal experiments, *Med. J. Kagoshima Univ.*, 48, 13, 1996.
20. **Nomura, S., Matsushita, T., Arimatsu, T., and Tomio, T.,** Results of health examination on persons applying organosulfur fungicides, Report on Studies Commissioned by the Ministry of Health and Welfare, *Jpn. Soc. Rural Med.*, 2, 36, 1974.
21. **Kato, T., Nishikawa, T., Yamamoto, S., Noda, H., Wada, K., Kato, H., Ogura, K., Morishita, M., Suzuki, H., and Tsutsui, J.,** Effects of pesticides on the health of pear growers in Hane area, Ueno City, Mie Prefecture, Report on Studies Commissioned by the Ministry of Health and Welfare, 27, 1974.
22. **Horiuchi, N. and Ando, S.,** Contact dermatitis due to pesticides for agricultural use, *Nippon Hifuka Gakkai Zasshi*, 90, 1980.
23. **Horiuchi, N., Ando, S., and Kambe, Y.,** Dermatitis due to pesticides for agricultural use, *Nippon Hifuka Gakkai Zasshi*, 90, 1980.

9.5.3 India, Malaysia, Sri Lanka, and Thailand

This section reviews information for India and for Malaysia, Sri Lanka, and Thailand, and other countries in the region with reported pesticide related skin conditions.

FAO data indicate that the major crops in India and other countries in the region are sugar and grain crops, pulses, oil crops, fiber, tea, coffee, and tobacco. Livestock production included 4,569,375 metric tons of meat and 1,611,000 tons of eggs.[1] Pesticide production is reflected in FAO data by increasing dollar value of pesticide exports between 1961 and 1997 and a slightly less marked increase in imported pesticides (Figures 9.47a and b). Case reports, epidemiologic studies, and outbreak investigations for the region are summarized in Figure 9.48.

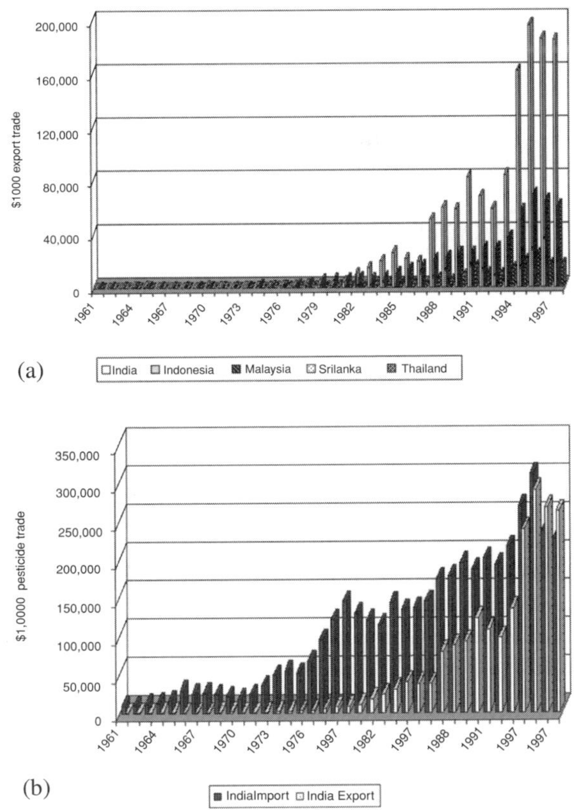

FIGURE 9.47 (**a**) Pesticide trade data for Southern Asia is dominated by (**b**) data from India.

9.5.3.1 Case Reports and Case Series

A novel case reported in 1977 involved a case of erythema multiforme associated with exposure to methyl parathion, which is used in India for indoor pest control. A few hours after spraying the material to kill bedbugs, a 25-year-old woman developed erythematous papulosquamous pattern lesions all over her body, persisting for 20 days. The eruption was not explained by any medication she was taking or any specific allergy to food. Laboratory evaluation was unremarkable except for a slight eosinophilia. The authors reported that re-exposure to a challenge dose of methyl parathion "in the form of snuff" promptly caused the lesions to recur.[2]

In 1983, Tungsanga[3] reported a bizarre accident that affected a 44-year-old Thai farmer in 1983, who accidentally used liquid Gramoxone®, containing 20% paraquat dichloride, stored in a soap container, to clean his perineum. He experienced immediate irritation symptoms and developed an ulcer at the site several days later and was admitted to a health center for medical care. Three weeks later his BUN began to rise and he developed an infiltrate on chest X-ray. Although he developed severe respiratory distress and required ventilatory support, he eventually recovered and was discharged from the hospital without apparent sequelae.

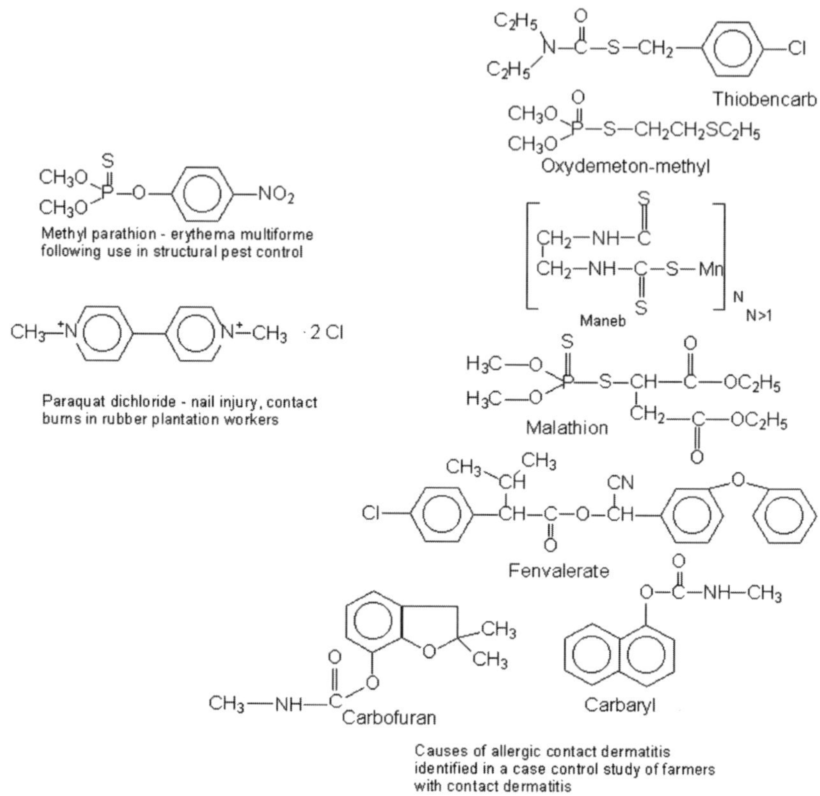

FIGURE 9.48 Southern Asia — reported causes of skin reactions to pesticides.

9.5.3.2 Prevalence, Incidence, Surveillance, and Outbreak Data

9.5.3.2.1 Cohort studies of paraquat applicators and formulators

1969 exposure study conducted by Swan[4] included clinical observations for 22 workers using backpack application equipment during the spray seasons of 1965 and 1967. The total volume sprayed for most workers exceeded 2000 gallons during a typical season lasting 12-14 weeks. In the group of 6 workers observed during 1965, one developed a chemical dermatitis of the scrotal skin during the first week of spraying and was hospitalized briefly for skin care; he suffered dermatitis of the hands and desquamation of the skin of the ear at week 4 of the study, his second week back at work. A second worker had numerous frictional lacerations during the spray period and developed septic sores and dermatitis at week eight, findings that the author did not believe were related to working with paraquat. Three episodes of skin irritation occurred in a group of six workers wearing no protective clothing during the 1967 spray season. Members of two additional groups, wearing either face masks and gloves, or gum boots and gloves, suffered similar episodes of dermatitis during trials conducted during 1967. Skin abrasions, lacerations, and dermatitis were all associated with transient appearance of paraquat in the urine (<0.010-0.32 ppm) of affected workers, but the levels did not approach those associated with acute poisoning (often >100 ppm).[5] Wearing protective clothing in most cases reduced levels of paraquat in the urine to below the limit of detection (0.01 ppm at the time of the study).

In 1979, Howard[6] studied a group of 18 U.K. and 18 Malay workers, both employed in formulating paraquat-based herbicides. Clinical records were examined, medical and occupational histories were obtained, and a clinical examination, particularly of the skin, was undertaken. Skin rashes, nail damage, and epistaxes were encountered by most workers as a result of direct contact of skin and mucous membranes with paraquat. These conditions did not persist, and no sequelae were found on the follow-up examination. In a study of Malaysian paraquat spray workers reported in 1981, it was noted that 11(40.7%) of 27 sprayers recorded one or more episodes of skin irritation of the hands, legs, or groin during the observation period.

A 1993 study by Senanayake[7] evaluated the potential respiratory effects of paraquat on 85 Sri Lankan tea plantation workers compared to two sets of controls (76 factory workers and 79 general workers), but also recorded each worker's history of skin symptoms. No differences in pulmonary function tests were noted between exposed and controls, but skin injuries were more frequent in exposed workers — being reported by 20 (23.6%) spraymen compared with a total of 21 (13%) subjects in the two control groups. Nail injuries were not seen in the factory controls, but occurred with approximately equal frequency among paraquat sprayers (10.5%) and controls identified as general workers. The authors noted that the skin damage recorded on questionnaire proved on examination to consist principally of cracks and abrasions of the feet, which may not have actually been attributable to paraquat.

9.5.3.2.2 Case control study of patch test sensitivity in Indian farmers

Sharma[8] reported a case control study, evaluating 30 Indian farmers with signs of contact dermatitis and 20 controls at the Postgraduate Institute of Medical Education and Research in Chandigarh, India (in the vicinity of Punjab). The 30 affected farmers had an age range of 28-70 years, and had dermatitis for an average of 2.5 years (range 3 months to 15 years), affecting the face, neck, and arm and lower extremities. The 20 controls were described as young healthy volunteers without dermatitis. Both groups were patch tested with a battery of 13 insecticides, 11 fungicides, and 13 herbicides. Patch test reactions consistent with allergy to one or more pesticides occurred in 11 patients but none of the controls. Reactions included thiocarbamates (maneb), and also the insecticidal (N-methyl carbamate) compounds carbofuran and carbaryl. Other positives included the insecticides malathion, oxydemeton methyl (4 patients), and fenvalerate, the antibiotic streptocycline, and the herbicide thiobencarb. The common irritants included captafol, organophosphorus compounds, and thiobencarb. The report did not explicitly discuss the clinical relevance of the observed reactions.

9.5.3.2.3 Mosquito control workers

In 1980, Jeyaratnam[9] studied 125 public health workers to evaluate the effects of fenthion used for mosquito control. Although none of the workers demonstrated depression of blood cholinesterase (ChE), symptoms in the workers included headache, sweating, pain in the eyes, impaired vision, muscle cramps, tightness of the chest, excessive salivation and nasal secretions, and dermatitis.

9.5.3.2.4 Indian coffee plantation workers

In 1990, Narahari[10] reported the results of an examination survey of workers on an Indian coffee plantation. Of 34 workers, 19 had erythema of sun-exposed, but not covered areas. Periungual telangiectasias were noted in 4 workers, and 4 also had onycholysis. An additional 15 workers had less severe nail changes. The work involved handling nitrogen/phosphorus fertilizer, spraying bordeaux mixture (copper sulfate and calcium oxide) and paraquat, and handling a 50% mixture of benzene hexachloride. Although no specific cause of the observed skin lesions was identified by the authors, the nail lesions appeared typical of those associated with paraquat.

9.5.3.2.5 Pesticide spraying and symptom frequencies in Indonesian farmers

Kishi et al. reported a season-long cohort study of 204 Indonesian farmers in Central Java, east of Jakarta, conducted by the FAO integrated pest management program, correlating work and spray schedules with the occurrence of 69 signs and symptoms. The difference in symptom frequencies between non-spray and spray periods were evaluated for each subject. Compounds handled by study participants included organophosphate and carbamate insecticides; organochlorine; pyrethroids; carbamate, thiocarbamate, phthalimido, and benzonitrile fungicides; and urea herbicides. Skin symptoms occurring with elevated frequency during the spray periods included itching skin (15%), blisters (1.5%), and red skin (0.8%). The authors contended that the elevated rates of symptoms documented the effects of "true normal use" of pesticides in developing countries.

9.5.3.2.6 Nonchemical sources of dermatitis in agricultural workers

Mahajan, Sharma, and Chakrabarti[11] reported a comparison between 30 male agricultural workers seen for dermatitis at the Department of Dermatology at a medical institute in Chandigarh, and 20 agricultural workers seen for non-dermatologic complaints at the same institution. Both groups were tested to Parthenium, Helianthus (sunflowers), Trifolium (barzeem), Xanthium, and several grain crops (barzeem, paddy rice, maize, cotton, and mustard). By clinical history, 16 patients associated their dermatitis with Parthenium, 4 with barzeem (a fodder crop), and 2 with wheat straw husk. Positive patch test reactions were obtained in 20 patients and in 1 control. Reactions were recorded for extracts of Parthenium in 20 patients (67%) and Xanthium in 12 (40%). Single patients reacted to sunflower and cotton seed extracts.[11]

REFERENCES

1. **FAOSTAT,** Pesticide Use Data , 1993, for India, Sri Lanka, Indonesia, Thailand and Malaysia. Available online at: apps.fao.org, 1999.
2. **Bhargava, R. K., Singh, V., and Soni, V.,** Erythema multiforme resulting from insecticide spray [letter], *Arch. Dermatol.*, 113, 686, 1977.
3. **Tungsanga, K., Chusilp, S., Israsena, S., and Sitprija, V.,** Paraquat poisoning: evidence of systemic toxicity after dermal exposure, *Postgrad. Med. J.*, 59, 338, 1983.
4. **Swan, A. A.,** Exposure of spray operators to paraquat, *Br. J. Ind. Med.*, 26, 322, 1969.
5. **Hayes, W. and Laws, E.,** , Academic Press, San Diego, 1990.
6. **Howard, J. K.,** A clinical survey of paraquat formulation workers, *Br. J. Ind. Med.*, 36, 220, 1979.
7. **Senanayake, N., Gurunathan, G., Hart, T., Amerasinghe, P., Babapulle, M., Ellapola, S., Udupihille, M., and Basanayake,** An epidemiological study of the health of Sri Lankan tea plantation workers associated with long term exposure to paraquat., *Br. J. Ind. Med.*, 50, 257, 1993.
8. **Sharma, V. K. and Kaur, S.,** Contact sensitization by pesticides in farmers, *Contact Dermatitis*, 23, 77, 1990.
9. **Jeyaratnam, J. and Ponnambalam, N.,** Monitoring of public health workers exposed to fenthion, in *Field Worker Exposure during Pesticide Application,* Tordoir, W. F. and Van Heemstra, E. A. H., Eds., 1980.
10. **Narahari, S., Srinivas, C., and Kelkar, S.,** LE-like erythema and periungual telangiectasia among coffee plantation workers, *Contact Dermatitis*, 22, 1990.
11. **Mahajan, V., Sharma, V., and Chakrabarti, A.,** Contact dermatitis in agricultural workers: role of common crops, fodder and weeds, *Contact Dermatitis*, 35, 373, 1996.

9.5.4 Turkey and Israel

Case reports and epidemiologic data on the skin effects of pesticides have not been reported frequently from the Middle East, although systemic illnesses have been reported from Egypt, Israel, and other countries in the region. Limited data from Israel and Turkey are discussed below.

FAO data indicate the diverse nature of agriculture in Israel and Turkey, including grains, fruits and vegetables, typical of the region as a whole.[1] Pesticides available for Turkey indicate the predominant use of herbicides;[2] only trade data are available for Israel. These indicate a substantial trade in both imports and exports, with an excess of the latter since the mid-1970s. Trade data for Turkey indicate a substantial excess of imports (Figures 9.49a and b). Case reports and epidemiologic investigations for the region are summarized in Figure 9.50.

9.5.4.1 Cases Reports and Case Series

A case series from Israel described 15 cases of first and second degree burns (Figure 9.51) to agricultural workers resulting from paraquat accompanied by minimal systemic absorption.[3] Other cases reported from Israel include a 1982 series of 10 cases of antecubital irritation (burning, erythema, and blisters to ulceration and scarring) in soldiers using a 50% diethyltoluamide based skin repellent.[4] A case similar to those reported in 1982 was reported by Amichail[5] in 1994.

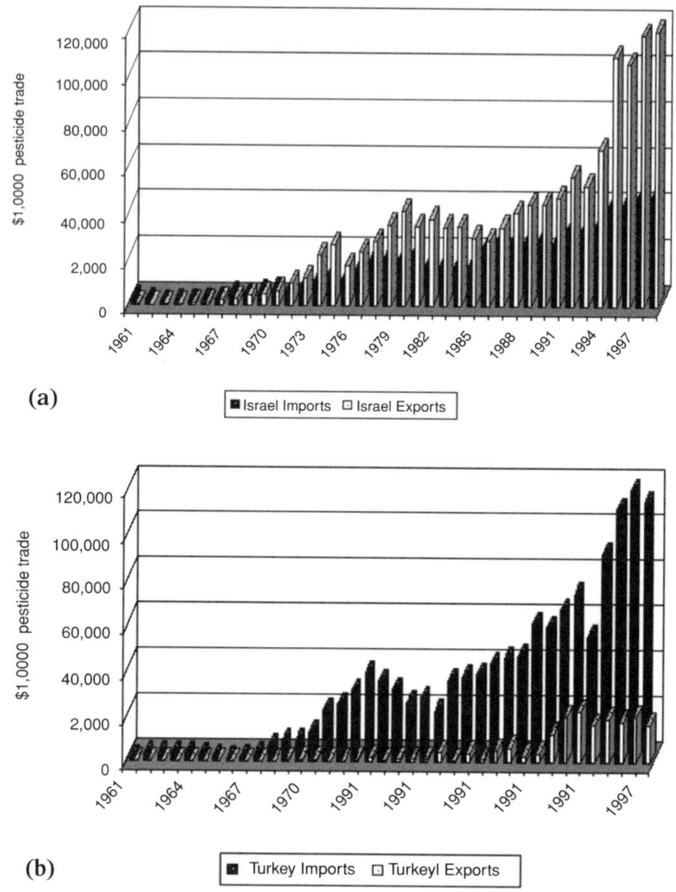

FIGURE 9.49 Israel (**a**) and Turkey (**b**) pesticide trade data.

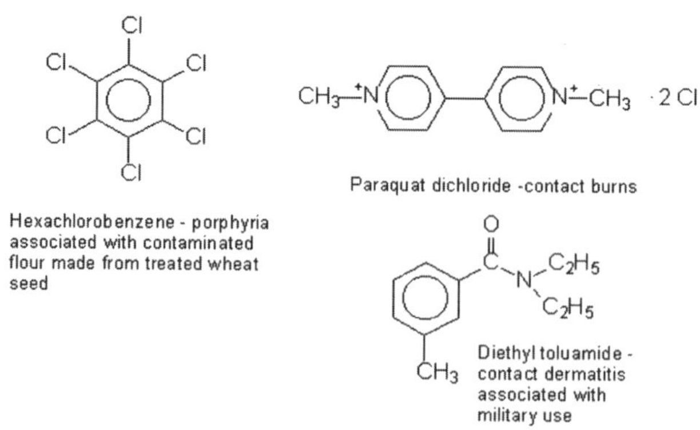

FIGURE 9.50 Pesticide skin conditions in Israel and Turkey.

9.5.4.2 Prevalence, Incidence, Surveillance, and Outbreak Data

In 1955, seed grain treated with hexachlorobenzene was mistakenly used for production of bread during a famine in southeast Turkey leading to an outbreak of toxic porphyria and involving an estimated 3000–4000 persons, with a 10% case fatality rate. Children born to women affected by the condition died from secondary HCB exposure across the placenta or through breast milk of a condition known as pink sore or "pembe yara." Many children with the sores suffered from tremor and seizures and many died from cardiorespiratory failure. The dermatologic findings in the adult and pediatric patients were as described by Peters:[6]

> Dermatologic lesions, precipitated by sunlight, tended to remit in winter and relapse during spring and summer. The skin showed bullae, milia, extensive scarring, loss of elasticity, atrophy, hyperpigmentation, and, at times, sclerodermatous change. In the acute stage, patients showed a Nikolsky's sign in which there was a lack of cohesion between the upper and lower layers of the skin, resulting in separation and ulceration with only slight pressure or trauma, and delayed healing. The children were labeled "monkey children" due to a substantial increase in pigmentation and hypertrichosis that occurred on the trunk, extremities, and especially on the face. Port-wine-colored urine having an intense fluorescence under Wood's light was characteristic in the active stages.

The illness observed included hepatic, arthritic, visceral, and neurological pathology as well as skin bullae, milia, atrophy, and hyperpigmentation. Of 161 patients reexamined 20-25 years later (between 1977 and 1981), more than half still had hyperpigmentation, scarring at site of previous bullae, hirsutism, and arthritis.[6]

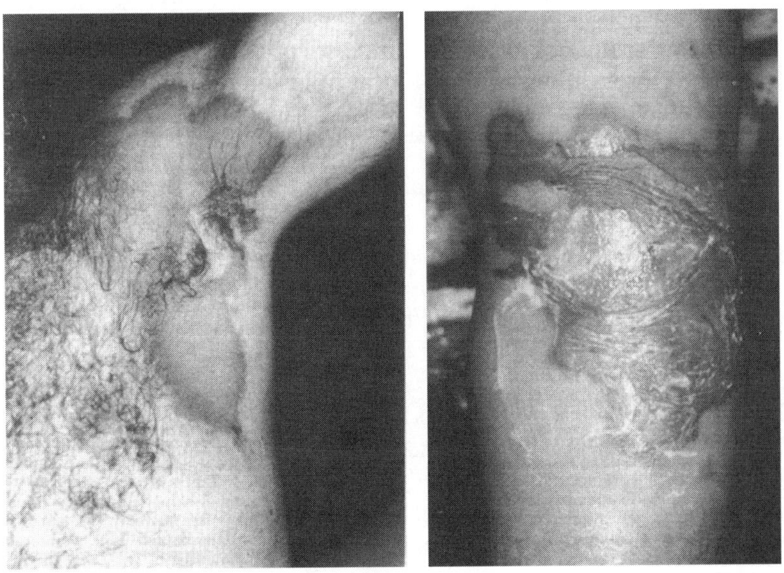

FIGURE 9.51 Paraquat burns in case series reported from Israel Poison Information Center. (Hoffer and Teitelbaum, *Human Toxicology*, 1989, with permission from Macmillan Scientific Publishing.)

REFERENCES

1. **FAOSTAT,** Production data for primary livestock, primary crops, and crops by area harvested, 1998, for Israel and Turkey. Available online at: apps.fao.org, 1999.
2. **FAOSTAT,** Pesticide Use Data , 1995, for Turkey. Available online at: apps.fao.org, 1999.
3. **Hoffer, E. and Taitelman, U.,** Exposure to paraquat through skin absorption: clinical and laboratory observations of accidental splashing on health skin of agricultural workers, *Hum. Toxicol.*, 8, 483, 1989.
4. **Reuveni, H. and Yagupsky, P.,** Diethyltoluamide-containing insect repellent: adverse effects in worldwide use, *Arch. Dermatol.*, 118, 582, 1982.
5. **Amichai, B., Lazarov, A., and Halevy, S.,** Contact dermatitis from diethyltoluamide, *Contact Dermatitis*, 30, 188, 1994.
6. **Peters, H., Gocmen, A., Cripps, D., Bryan, G., and Dogramaci, I.,** Epidemiology of hexachlorobenzene-induced porphyria in Turkey: clinical and laboratory followup after 25 years, *Arch. Neurol.*, 39, 744, 1982.

9.6 AFRICA

9.6.1 Agricultural Production and Pesticide Use Patterns

Agricultural production statistics reflect the wide variation in the types of agriculture conducted throughout the continent, including grain, root crops, sugar, vegetables, fruit, fiber, coffee and tea, pyrethrum flowers, and livestock.[1]

FAO and other researchers have also documented the complex social climate of African agriculture. Table 9.16 demonstrates, for example, co-existence of subsistence farming and cash agricultural economies side by side in many countries. At the same time, widespread urbanization has led to a shift of male employment to cities, so that women make up the majority of the agricultural workforce in many countries (Figure 9.52). London,[2] has documented a unique problem in South Africa, where a minority of farm laborers have received payments in allotments of alcohol rather than in cash, fostering a unique form of dependency among those victimized by the "DOP" payments.

The FAO use data for 1997 demonstrate that South Africa and Kenya account for the bulk of the reported pesticide use in Africa, but no data are available at all from many countries.[3] It cannot be ascertained from the data whether this actually reflects lower use (associated with a larger proportion of the work force engaged in subsistence farming), or failure to report use to the FAO system.

The limitations of official statistics have also been addressed in South Africa by surveys of local pesticide use,[4] conducted to aid in epidemiological studies of pesticide illnesses.[5] A perspective on the limitations of use data can also be obtained from trade statistics, demonstrating that Africa has been importing increasing levels of pesticides since the 1960's. Pesticide exports in 1998 were only about one fifth the level of imports (Figure 9.53). Case reports, epidemiologic investigations, and outbreak investigations for the region are summarized in Figure 9.54.

TABLE 9.16
Agricultural production in selected African countries[12]

Country	% of GDP	% of Employment	Exports	Type of Production	Crops
Burkina Faso	30	90	60	Mainly subsistence	
Cote d'Ivoire	70-80	70		Mixed export/ farming subsistence	Export: coffee, cocoa, rubber, oil palm, cotton, fruit; staple production – livestock, poultry
Ethiopia	40	85	80% of total earnings	Diverse subsistence systems	Cereals, oilseeds, livestocks
Kenya	30	75	70	.	Pesticide use heavy in export sector - fruit, vegetables, flowers, coffee
Mali	46	80			Millet, sorghum, rice, maize, cotton, groundnuts
Nigeria	30				
South Africa					Coarse grains, tobacco, rubber, coffee, beet/cane sugar; large export sector, backwards conditions – DOP system of payment
Tanzania	50-60				Tea, wattle, sugar, barley, maize
Zimbabwe	13	26	40	Modern /traditional	Cotton, groundnut, wheat, soybean, millet, sorghum, maize

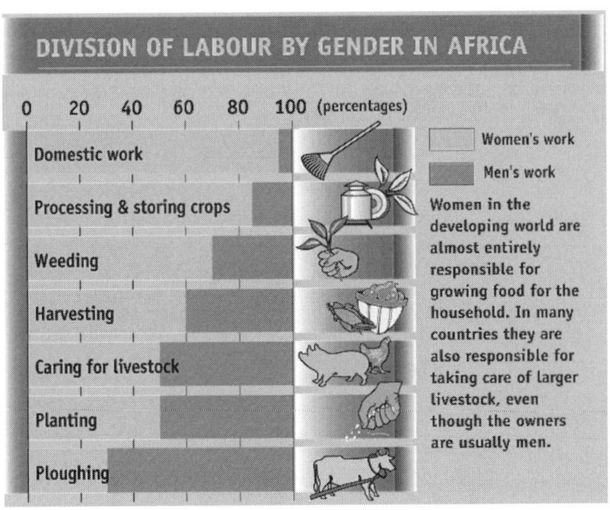

FIGURE 9.52 Distribution of work activities in African agriculture by gender. (UN FAO)

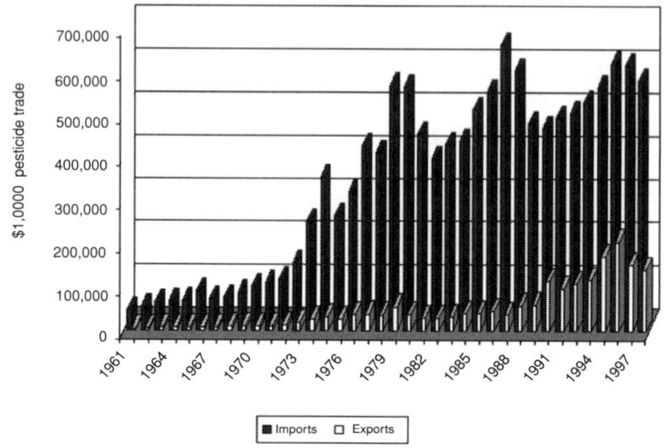

FIGURE 9.53 Pesticide trade data for Africa.

9.6.1.1 Case Reports and Case Series

A case of suspected occupational vitiligo or leukoderma of the hands and feet was observed in a 65-year-old Nigerian cocoa farmer.[6] This was considered possibly related to spraying he did with paraquat-dichloride, because the distribution of hypopigmentation corresponded to areas of skin that were uncovered while he sprayed. There was no evidence of chronic dermatitis. Partial repigmentation began

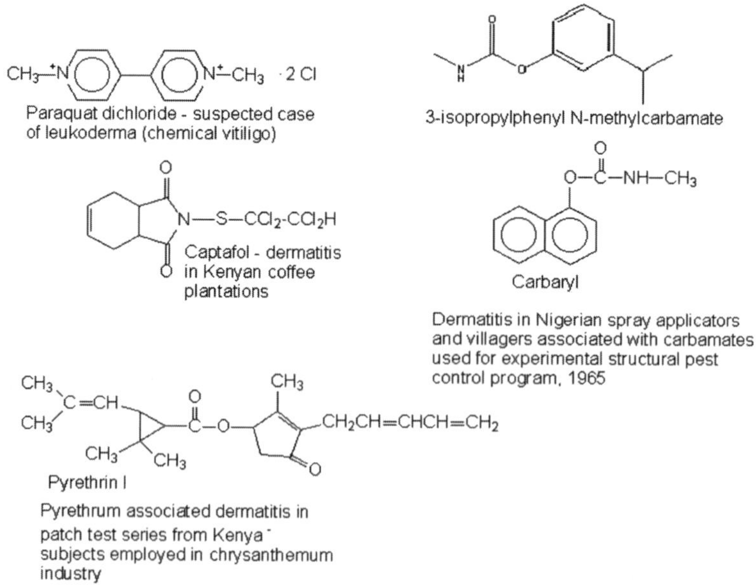

FIGURE 9.54 Reported pesticide-associated skin conditions in Africa.

6 weeks after discontinuation of paraquat spraying. The author suggested a relationship between the hypopigmentation and the patient's hypopigmentation, but could not exclude idiopathic vitiligo.

9.6.1.2 Prevalence, Incidence, Surveillance, and Outbreak Data

9.6.1.2.1 Surveillance in South Africa

Recent published reports on illness surveillance from the continent have been limited to South Africa and have concentrated on systemic poisonings in the Cape Town region.[7] This identified 14 poisoning events involving 56 hospitalized cases, with 2 events together accounting for 44 cases and only one case reported by a general practitioner. Because surveillance was focused on hospitalized poisoning, cases of dermatitis have not been included in the reports.

9.6.1.2.2 Experimental WHO field trials of insecticides for public health and structural pest control

An experimental field trial of the pesticides folithion, carbaryl, and 3-isopropylphenyl N-methylcarbamate (cumenyl methylcarbamate, CAS# 64-00-6) for public health structural pest control in 3 villages (365 people) was reported from Nigeria in 1965.[8] No adverse effects were observed during applications of folithion in either spraymen or villagers. During applications of carbaryl, a slight depression of cholinesterase was observed; the report also described a case of apparent dermatitis in a worker who accidentally splashed himself with carbaryl. During the trial of cumenyl methylcarbamate, symptoms of poisoning in applicators occurred within 2-3 h of spraying manifested by weakness, nausea, abdominal cramps, tightness in the chest, headache, blurred vision, dizziness, pinpoint pupils, profuse sweating and vomiting — typical of moderate to severe cholinesterase inhibition. There were also cases of poisoning in some villagers, including one severe poisoning in a 14-month-old child. During the first night after spraying, an itching macular hyperemic rash on uncovered areas of skin occurred in 17/34 exposed male villagers (Figure 9.55) and only one rash among the female villagers. The trial of cumenyl methylcarbamate was discontinued after nine dwellings were sprayed because of its obvious adverse effects.

9.6.1.2.3 Contact dermatitis in Kenya, 1974

A case series of patch tests was carried out in 298 consecutive patients with a clinical diagnosis of contact dermatitis at the dermatology clinic of the University of Nairobi, Kenya.[9] Positive tests were observed to standard European series antigens including metal salts (chromium, nickel, cobalt) in 50 cases (37%), rubber chemicals (IBT, PPD, thiuram) in 35 cases (26%), and plant extract products (turpentine, balsam of Peru, colophony, wool tars) in 27 (20%), but also included 3 patients with allergic reactions to pyrethrum. These were believed to derive from the Kenyan cultivation of chrysanthemums for pyrethrum extract. Three patients also had reactions to copper sulfate from its use as an agricultural spray.

Apart from the Nairobi case series, 23 cases seen in rural clinics involved ortho-difolitan (thio-tetra-hydrophthalimide), used in the coffee industry and commonly

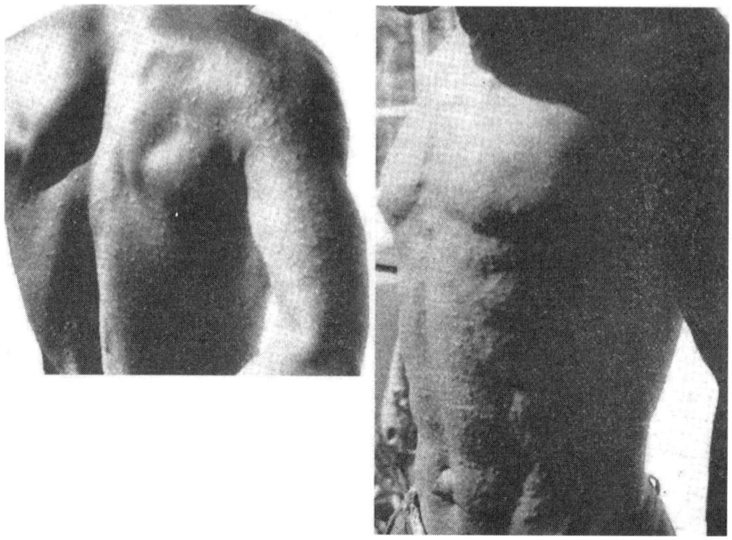

FIGURE 9.55 Dermatitis in a sprayer involved in an experimental pest control program, Nigeria. (From Vandekar, *Bulletin of the World Health Organization*, 1965, reprinted with permission.)

affecting workers employed as pickers. Cases involved severe dermatitis, conjunctivitis, stomatitis, and painful bronchitis. After the series of cases had been identified, inquiries in the affected districts revealed many additional cases. These included 13 affected workers brought on a bus to the visiting dermatologist's clinic by a rural coffee cooperative.

9.6.1.2.4 Questionnaire survey in rural Kenya, 1987

Mwanthi and Kimani[10] collected questionnaire data on the use of pesticides in 1797 homes in rural Kenya between 1987 and 1990. The survey demonstrated that 62% of homes were used for chemical storage. The residents stored chemicals in their living quarters, but 60% were unable to understand a container label. Of 133 poisoning cases seen at the main hospital in the study district, 90% were due to organophosphate pesticides. Lack of protective clothing may explain the sneezing, breathing difficulties, and contact dermatitis experienced by 66% of the participants.

9.6.1.2.5 Pyrethroid use in Tanzania

Similar to the WHO field trial described above, the use of lambda-cyhalothrin in Tanzania was evaluated following continuous use over a 6-day period in March 1990, 2.7-5.1 h/day. All had some contact related symptoms, most frequently itching and burning of the non-protected areas of the face, upper respiratory irritation and nose or throat irritation, and sneezing and coughing. Facial symptoms occurred on non-protected areas only. One resident of a treated house also complained of facial pruritus, but no other symptoms were reported from residue of the applied compound.[11]

REFERENCES

1. **FAOSTAT,** Production data for primary livestock, primary crops, and crops by area harvested, 1998, for Africa. Available online at: apps.fao.org, 1999.
2. **London, L.,** The "DOP" system, alcohol abuse and social control amongst farm workers in South Africa: A public health challenge, *Soc. Sci. Med.*, 48, 1999.
3. **FAOSTAT,** Pesticide Use Data, 1997, for Africa. Available online at: apps.fao.org, 1999.
4. **London, L. and Myers, J.,** General patterns of agrichemical usage in the southern region of South Africa., *S. Afr. J. Sci.*, 91, 509, 1995.
5. **London, L. and Myers, J.,** Agrichemical usage patterns and workplace exposure in the major farming sectors in the southern region of South Africa, *S. Afr. J. Sci.*, 91, 515, 1995.
6. **George, A. O.,** Contact leucoderma from paraquat dichloride?, *Contact Dermatitis*, 20, 225, 1989.
7. **Bailie, R. and London, L.,** Enhanced surveillance for pesticide poisoning in the Western Cape–an elusive target., *S. Afr. Med. J.*, 88, 1105, 1998.
8. **Vandekar, M.,** Observations on the toxicity of carbaryl, folithion and 3-isopropylphenyl N-methylcarbamate in a village-scale trial in Southern Nigeria., *Bull. WHO*, 33, 107, 1965.
9. **Verhagen, A.,** Contact dermatitis in Kenya, *Trans. St. John's Hosp. Dermatol. Soc.*, 60, 86, 1974.
10. **Mwanthi, M. A. and Kimani, V. N.,** Patterns of agrochemical handling and community response in Central Kenya, *J. Environ. Health*, 55, 11, 1993.
11. **Moretto, A.,** Indoor spraying with the pyrethroid insecticide lambda-cyhalothrin: effects on spraymen and inhabitants of sprayed houses, *Bull. WHO*, 69, 591, 1991.
12. **Herz, K.,** Agricultural research in selected countries of sub-Saharan Africa. Available online at: www.fao.org, 1996.

9.7 AUSTRALIA, NEW ZELAND, AND PAPUA NEW GUINEA

Livestock play a dominant role in the agriculture of Australia, as indicated by a 1993-94 commonwealth report:

> About 60% of Australia's land area is used for agriculture — most is used for animal grazing; only about 10% is under crops or improved pasture, a small proportion of which is irrigated. Australia is a major producer and exporter of agricultural products which are mainly sold at world market prices and are produced at relatively low costs. In 1993–1994 the gross value of agricultural commodities produced was $23.646 billion, made up of crops (approximately 49%), and livestock sales (29%).*

Other major agricultural products include grain, sugar crops, oil crops, fruits, and vegetables. The livestock sector (dairy, sheep, and farming) is also important in

* Commonwealth of Australia Environmental Economics Unit; Agricultural Chemicals; 1996: http://www.environment.gov.au/epcg/eeu/subs/subs13.htm.

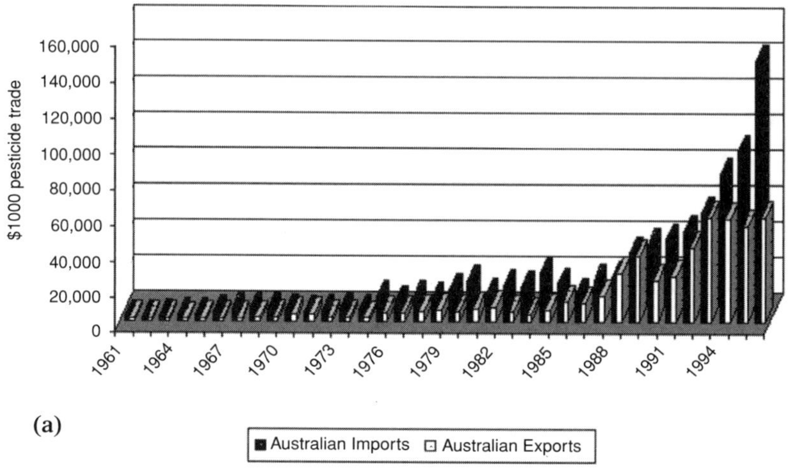

(a)

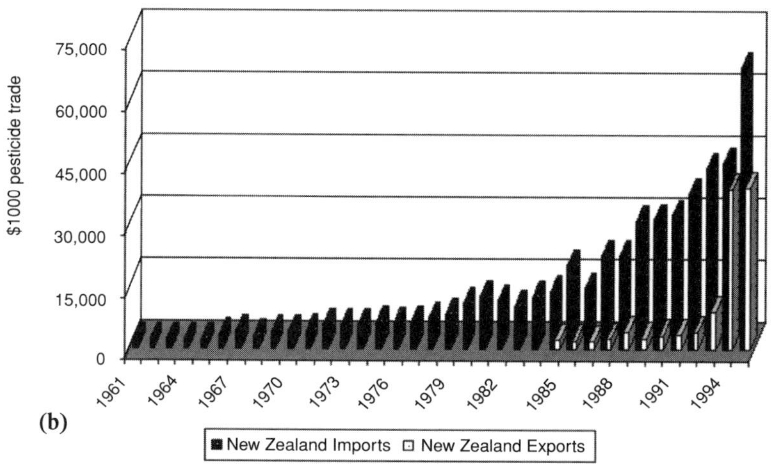

(b)

FIGURE 9.56 Pesticide trade data for Australia **(a)** and New Zealand **(b)**.

New Zealand*, but New Zealand also produces timber, hops, kiwis, avocados, peaches, apples, grapes and other fresh fruit. Fruit (bananas and other tropical fruit), root crops (sweet potato, cassava, taro, and yams), and oil crops (coconut and oil palm fruit) dominate agricultural production in New Guinea, but the country also produces grain, vegetables, and pyrethrum flowers.[1]

A principal use of pesticides in livestock production is control of ectoparasites. Compounds used for this purpose include organophosphates, carbamates, organochlorines, and synthetic pyrethroids. Apart from direct exposure to these compounds, the residue of organochlorine pesticides in lanolin derived from the Australian sheep industry has given rise to unique concerns.[2] Use data indicate that

* New Zealand Ministry of Agriculture and Fisheries web site: http://www.maf.govt.nz/Dairy/.

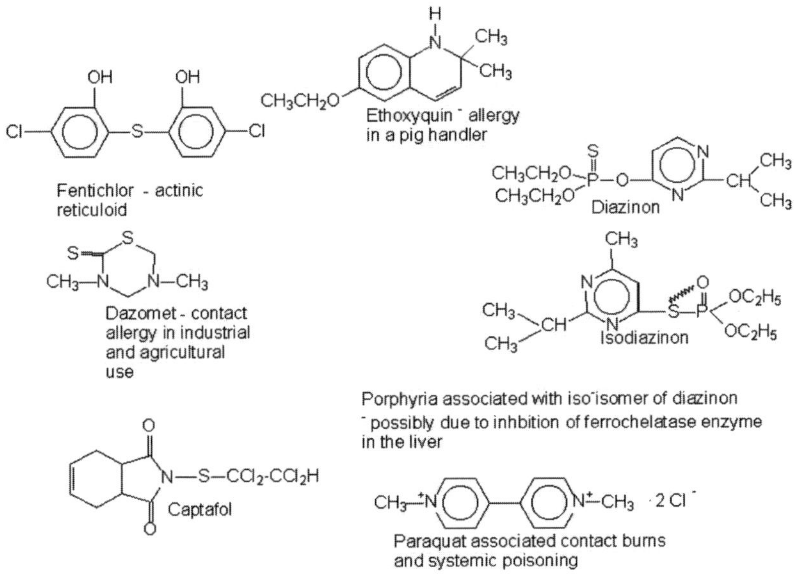

FIGURE 9.57 Pesticide-related skin conditions reported in Australia, New Zealand, and Papua New Guinea.

fungicides account for almost 80% of total pesticide consumption. Trade data indicate pesticide use in the three countries appears to be increasing, based on the annual dollar value of imports between 1961 and 1997 (Figures 9.56a and b). Figure 9.57 summarizes experience with pesticide-associated skin disease in the region.

9.7.1 Case Reports and Case Series

Cases associated with agricultural production included a report describing persistent allergy in a farmer handling chicken feed, proved by provocation testing to be due to the antioxidant additive, ethoxyquin;[3] two cases of allergy to the thiodiazine compound dazomet, used as an industrial slimicide, and as an agricultural fumigant;[4] actinic reticuloid associated with the sulfide compound fenticlor;[5] and two cases of the systemic dermatosis porphyria cutanea tarda in workers handling the organophosphate compound diazinon,[6] possibly attributable to the contaminant isodiazinon, shown by experimental work to inhibit ferrochelatase, a liver enzyme in the porphyrin pathway.[7] In addition to the cases associated with pesticides, Australian agricultural workers have experienced dermatitis from a number of imported and native Australian plants.[8-11]

Reports from New Guinea included 5 cases of severe dermatitis due to paraquat accompanied by fatal systemic poisoning,[12, 13] irritant dermatitis in a 54-year-old pest control worker due to dichlorvos accompanied by systemic organophosphate poisoning,[14] and cases of allergic dermatitis associated with thiocarbamates.[15],[16]

9.7.2 Prevalence, Incidence, Surveillance, and Outbreak Data

Pesticide illness surveillance data have not been routinely reported from Australia. The national poison service published information in 1969 regarding 5001 cases of poisoning, including 49 deaths. Pesticides accounted for 284 cases (4%), but the report did not describe the precise nature of the cases.[17]

In 1979, Stoke[18] reported a survey of 133 workers from 14 timber treatment plants and found that 23% gave questionnaire responses suggesting occupational contact dermatitis probably related to their employment. He focused attention on captafol, recently introduced as resplacement for pentachlorophenol as an antisapstain agent. Walk-through surveys conducted in the plant identified poor workplace hygiene as a possible contributing factor in many of the cases. Patch testing was not carried out; the author presumed that the majority of the cases represented irritant dermatitis.

A 1997 questionnaire survey[19] determined that 21% of pig farm workers handling in-feed antibiotics had a history of contact dermatitis in the 2 years prior to the survey; 15% had had dermatitis within 12 months of the survey; and 7% had symptoms when the questionnaire was administered. As determined from questionnaire responses, the only antibiotic associated with the history of dermatitis was tylosin (a macrolide antibiotic obtained from cultures of *Streptomyces fradiae*, effective against many microorganisms in animals but not in humans). The study also determined that few feed handlers took precautions to minimize skin contact with the antibiotics.

The prevalence of allergic contact dermatitis in 46 farmers was also evaluated in a 1998 study in New Zealand.[20] Results included 28 subjects with positive patch tests, including 23 considered to be relevant to prior exposures, and 20 related to their employment. Pesticide allergens included captafol, maneb, and copper sulfate. Non-pesticidal positive responses included IPPD and 4-phenylene diamine.

REFERENCES

1. **FAOSTAT,** Production data for primary livestock, primary crops, and crops by area harvested, 1998, for Australia, New Guinea and Oceania, New Zealand. Available online at: apps.fao.org, 1999.
2. **Rosanove, R.,** Dangers of the application of lanolin [letter], *Med. J. Aust.*, 146, 232, 1987.
3. **Rubel, D. and Freeman, S.,** Allergic contact dermatitis to ethoxyquin in a farmer handling chicken feeds, *Aust. J. Dermatol.*, 39, 89, 1998.
4. **Black, H.,** Dazomet and chloropicrin, *Contact Dermatitis Newsletter*, 14, 410, 1973.
5. **Burry, J.,** Fenticlor, actinic reticuloid, and antihistamines, *Br. Med. J.*, 2, 556, 1974.
6. **Collins, A., Nichol, A., and Elsbury, S.,** Porphyria cutanea tarda and agricultural pesticides, *Aust. J. Derm.*, 23, 70, 1982.
7. **Nichol, A., Elsbury, S., Angel, L., and Elder, G.,** The site of inhibition of porphyrin biosynthesis by an isomer of diazinon in rats, *Biochem. Pharmacol.*, 32, 2653, 1983.
8. **McFadyen, R.,** Parthenium weed and human health in Queensland, *Aust. Fam. Physician.*, 24, 1455, 1995.
9. **Southcott, R. and Haegi, L.,** Plant hair dermatitis, *Med. J. Aust.*, 156, 623, 1992.

10. **Tilsley, D.,** Australian blackwood dermatitis, *Contact Dermatitis*, 23, 40, 1990.
11. **Lothian, N.,** Grevillea species and hybrids causing contact dermatitis, *Aust. J. Dermatol.*, 30, 111, 1989.
12. **Wohlfahrt, D.,** Paraquat poisoning in Papua New Guinea, *Papua New Guinea Med. J.*, 24, 164, 1982.
13. **Davies, R.,** Skin absorption of paraquat, *Med. J. Aust.*, 2, 222, 1982.
14. **Bisby, J. A. and Simpson, G. R.,** An unusual presentation of systemic organophosphate poisoning, *Med. J. Aust.*, 2, 394, 1975.
15. **Burry, J. N.,** Contact dermatitis from agricultural fungicide in South Australia, *Contact Dermatitis*, 2, 289, 1976.
16. **Oakley, A. M.,** Contact allergy to fungicide [letter], *N. Z. Med. J.*, 101, 180, 1988.
17. **Travers, D. B.,** Poisoning in Australia, 1967, *Med. J. Aust.*, 1, 472, 1969.
18. **Stoke, J. C.,** Captafol dermatitis in the timber industry, *Contact Dermatitis*, 5, 284, 1979.
19. **Pirkis, J., O'Regan, K., and Bailie, R.,** Contact dermatitis and in-feed exposure to antibiotics among pig feed handlers, *Aust. J. Rural. Health.*, 5, 76, 1997.
20. **Rademaker, M.,** Occupational contact dermatitis among New Zealand farmers, *Aust. J. Dermatol.*, 39, 164, 1998.

9.8 LATIN AMERICA

Although most pesticide use occurs in developed countries, greater human exposure occurs in developing countries as a result of patterns of farming, type of equipment used for application and maintenance schedules, location of housing, storage of pesticides in living areas, and cultural factors. As many as 99% of all deaths from pesticide poisoning may occur in the developing countries. In Latin America, underreporting of pesticide poisoning is a major obstacle to assessment of the problem. However, surveillance data demonstrate that poisoning is a major public health problem, at least in Mexico, Colombia, Brazil, and throughout Central America, among other countries. As much as 12-13% of rural agricultural workers self-report an episode of pesticide poisoning every year. Existing surveillance systems could be used to identify cohorts for study of chronic effects of poisoning. In addition, new methods for estimating the true incidence of pesticide poisoning merit investigation. These methods are necessary in order to measure the impact of public health interventions, which should be a primary focus of applied research.[1]

9.8.1 Pesticide Use and Agricultural Data

The diversity of agriculture in both Central and South America is reflected in FAO data showing that principial crops for the region including grains, root crops, legumes, oil crops, nuts, vegetables, forage crops, coffee, spices, pyrethrum daisies, rubber, cotton, and other fiber crops.[2] Average pesticide use reported to FAO for South America between 1993 and 1997 included 97,000 metric tons of herbicides, 44,000 metric tons of herbicides, and 47,000 metrict tons of insecticides.[3] Over the period for which FAO trade data are available, there has been a marked apparent increase in the use of imported pesticides in both regions, rising from a value of approximately $30 million in 1961 in South America to more than $1 billion in 1997 (Figures 9.58a and b).

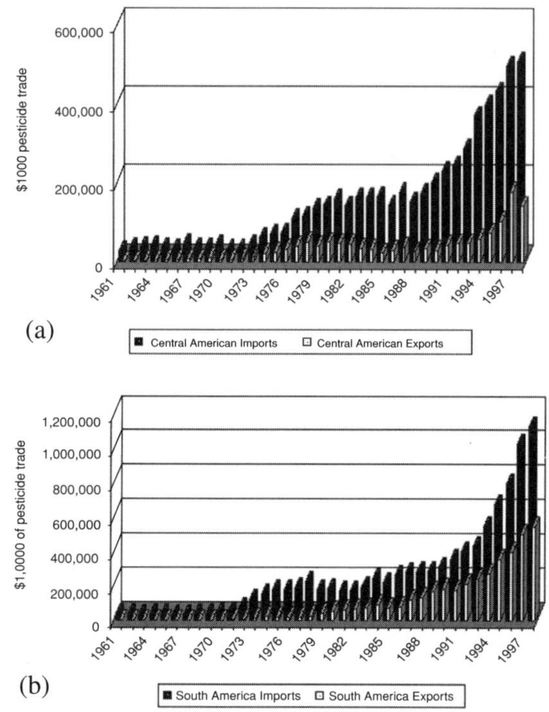

FIGURE 9.58 Pesticide trade data for Central America (**a**) and South America (**b**).

Central American imports rose during the same time period from \$30 million to approximately \$650 million.. Subcategories of chemicals corresponding to the use data in the figure are shown for Costa Rica in Table 9.17. Figure 9.59 summarizes experience with pesticide-associated skin disease in the region.

The public health response to this growing quantity of pesticide use has recently included a growing emphasis on illness surveillance. Routine surveillance reports are now published from Costa Rica[4] and Nicaragua (Figure 9.60)[5] and regional work by the Pan American Health Organization has also produced intermittent surveillance efforts in Mexico,[6] Panama,[7] El Salvador,[8] Brazil,[9] and Uruguay.[10] Most of the reporting has understandably focused on systemic poisoning and only scattered data on skin effects of pesticides in Latin America are available.

9.8.1.1 Case Reports and Case Series

Typical cases associated with banana production in Panama are discussed in Chapter 2. These range from contact irritation and burns due to paraquat to fungicide related allergic contact dermatitis (imazalil, thiabendazole, and chlorothalonil), urticaria, and toxic epidermal necrolysis (TEN). In the experience of the consulting dermatologic clinic for Seguro Social (social security) in the Chiriquí region, the northernmost section of the country, adjacent to the Costa Rican frontier, paraquat burns (Figure 9.61) were the most frequently encountered pesticide skin problems, accounting for a total of 15 cases between 1986 and 1996.[11] In addition, several

TABLE 9.17
Metric Tons of Pesticide Use in Costa Rica

Compound/use category	1995	1996
Insecticides	1575	5159
Chlorinated hydrocarbons	0	1626
Organophosphates	771	1701
Carbamates insecticides	621	111
Pyrethroids	15	8
Botanical and biological	36	1
Other insecticides	132	1712
Herbicides	2746	5047
Phenoxy hormone products	382	331
Triazine	30	66
Amides	233	29
Carbamates herbicides	1	2
Dinitroanilines	2	274
Urea derivates	0	17
Other herbicides	1928	2062
Sulfonyl ureas	29	1
Bipiridils	141	2261
Uracil		4
Fungicides and bactericides and seed treatment	1008	632
Inorganics	215	19
Dithiocarbamates	262	392
Benzimidazoles	27	30
Triazoles, diazoles	29	29
Diazines, morpholines	54	25
Other fungicides	421	137
Plant growth regulators	8	3
Rodenticides	4	8
Anticoagulants	4	8

cases were evaluated that had apparent cumulative irritant mechanisms, possibly related to pesticide residues. Evaluation in these cases did not identify any source of acute irritation and no cause of allergic contact dermatitis was identified from antigens series used in patch testing. Examples include:

TABLE 9.18
Patch Testing on Panamanian Banana Workers with Chronic Pigmented Dermatitis and Controls With Fungal Skin Conditions

	Positive patch reactions	
Substance	Cases[a] (n = 39)	Controls (n = 41)
Chlorothalonil (0.001% acetone)	34	0
PPD	0	3
Formaldehyde	0	2

[a] Statistcally significant (χ^2 corrected for continuty = 58,6, p <0,001).

TABLE 9.19
1995–1998 Surveillance Data from Costa Rica — Cases of Skin Reaction

Pesticide	1995	1996	1997	1998	Total
2,4-D	2	2			4
Diquat			1		1
Herbicides, unspecified	6				6
Fluazifop-butyl				1	1
Metsulfuron methyl	1				1
Glyphosate	6	2	12	6	26
Picloram		1		1	2
Paraquat	57	55	51	31	194
Total Herbicides					**235**
Cipropanazol			1		1
Clorotalonil	3	1	1	2	7
Streptomycin			1		1
Triforine	6				6
Mancozeb	1	1			2
Total Fungicides					**17**
Clorpirifos	2		1		3
Cadusafos (Ebefos)	2	1			3
Coumafos		1			1
Diazinon	6	1	3	1	11
Etoprofos		2		2	4
Fenamifos	1	2	4	4	11
Malathion	2		2	2	6
Metamidophos	3	3		4	10
Triclorfon				1	1
Monocrotophos			1		1
Methyl parathion	1			1	2
Terbufos	14	12	4	4	34
Total organophosphates					**87**
Carbamate, unspecified	1				1
Carbofuran		19	13	11	43
Methomyl	2	4	1	5	12
Oxamyl	2				2
Propoxur			2	1	3
Total Carbamates					**61**
Cypermethrin	2				2
Deltamethrin			1	5	6
Lambdacyhalothrin			2	2	4
Total pyrethroids					**12**
Endosulfan	3	2		1	6
Fenbutatin oxide			1		1
Insecticide mixtures	19				19
Sulfluramid	1				1
Other insecticides	22			1	23
Total Insecticides					**209**
Other and unspecified	25	27	35	24	111
TOTAL	**232**	**137**	**137**	**110**	**616**

Source: Control Estatal de la Actividad Bananera, Direction Protection al Ambiente Humano, M.S.

A female, who has worked several years at a banana packing plant, sought medical care for hand dermatitis of 3 years' duration, characterized by dryness and fissuring and cracking skin. Diagnostic studies including standard and patch testing were negative.

TABLE 9.20
Costa Rican Pesticide Cases by Gender

Gender	1995	1996	1997	1998	Total
Male	200	133	128	101	562
Female	32	4	9	9	54
Total	**232**	**137**	**137**	**110**	**616**

A 43-year-old male, working as a "seleccionador," developed a bilateral dermatitis in the arms and forearms, characterized by dryness and fissuring and cracking skin. Diagnostic studies including standard and pesticide series were negative.

9.8.1.2 Surveillance, Outbreak Investigations, Prevalence Studies, and Case Control Investigations

Surveillance data from Costa Rica* for 616 pesticide skin effects cases reported for the years 1995, 1996, and 1997 are shown in Tables 9.19-9.21. Five hundred sixty two (91.2%) were men and 54 (8.8%) were women; 542 (88.0%) resulted from occupational exposure, 48 (7.8%) from accidental non-occupational exposure, 9 (1.5%) from attempted suicide, and 17 (2.8%) from unspecified circumstances of exposure. The compounds were similar to those reported from Panama by Penagos, including cases associated with the fungicide chlorothalonil, glyphosate, and numerous cases associated with bipyridyl compounds. The surveillance data also contain 86 cases associated with organophosphates (etoprofos, fenamiphos, cadusafos, malathion, methyl parathion, chlorpyrifos, diazinon, monocrotophos, terbufos), 61 cases associated with carbamates (carbofuran, oxamil, propoxur, and methomyl), and 12 associated with pyrethroids (deltamethrin and lambdacyhalothrin), a case associated with the novel insecticide sulfuramid, and 19 cases associated with pesticide mixtures.

TABLE 9.21
Costa Rican pesticide dermatitis cases by activity

Circumstances of exposure	1995	1996	1997	1998	Total
Unspecified	5	2	3	7	17
Occupational	217	123	117	85	542
Accidental, non-occupational	10	12	14	12	48
Attempted suicide			3	6	9
TOTAL	**232**	**137**	**137**	**110**	**616**

Source: Control Estatal de la Actividad Bananera, Direction Protection al Ambiente Humano, M.S.

* Source: Control Estatal de la Actividad Bananera, Dirección Protección al Ambiente Humano, M.S.

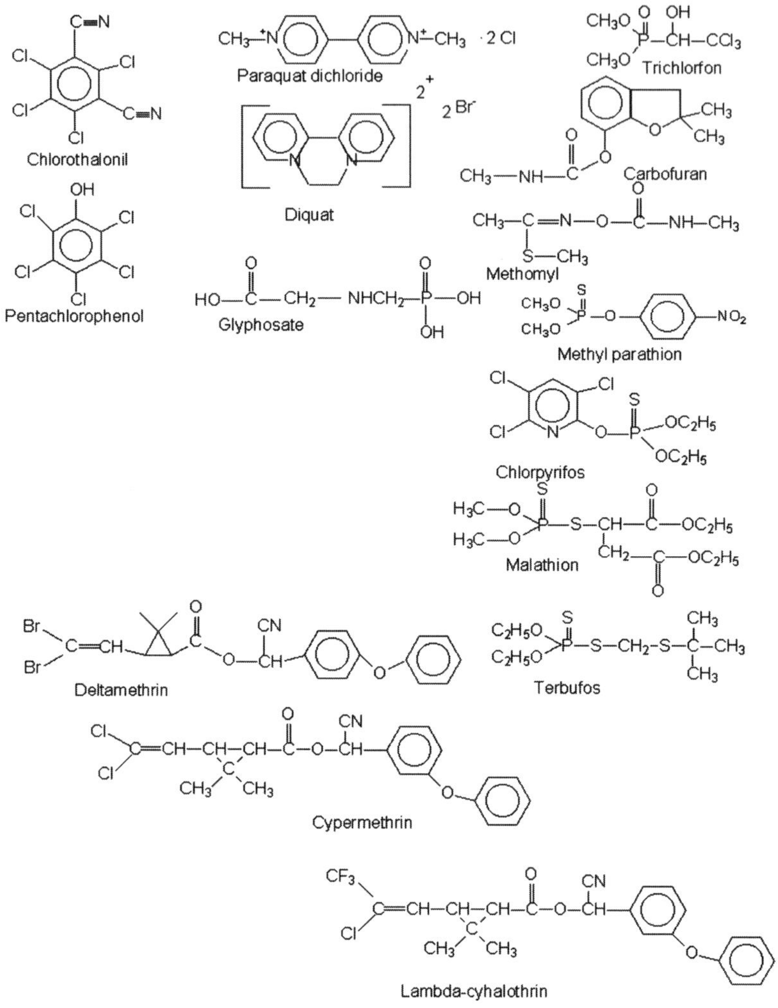

FIGURE 9.59 Pesticide-related skin conditions in Latin America.

In 1997, the surveillance program in Nicaragua,which does not routinely publish information on skin conditions associated with pesticides, described an outbreak of dermatitis and eye irritation associated with environmental exposure to metam sodium.[12] The problems described closely resemble those associated with metam sodium in Germany and California.

> In September 1997, residents of the Los Angeles neighborhood in the Esteli municipality complained about fumigations conducted by the tobacco farm (ASP-NIC) with a product called Fumisol (metam sodium) in the pre-planting preparation of seven strips of land. As a result, six residents presented with

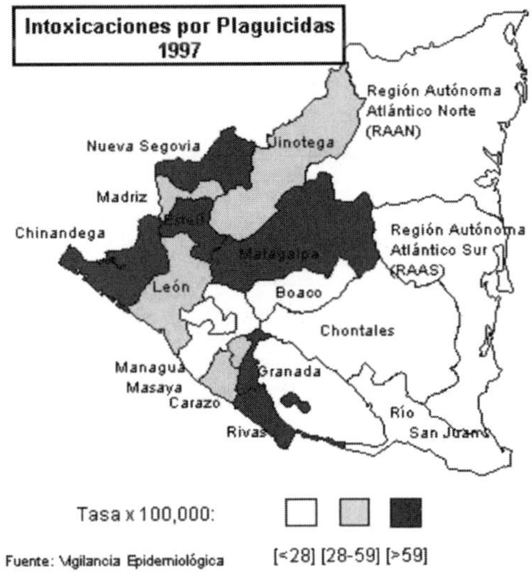

FIGURE 9.60 Surveillance information from Nicaragua, rates/100,000 population for 1997. (Courtesy of the Republic of Nicaragua, Ministry of Health, Directorate of Hygiene.)

skin and eye symptoms requiring medical attention. The neighboring population complained of skin effects.

The local pesticide commission visited the site on various occations and concluded the outbreak was due to incorrect application of the product — the failure to immediately cover the fumigated site with water — which permitted the accumulation of metam's gaseous byproducts in the environment and the subsequent windborne dispersion that affected the neighboring population.

The commission developed nine recommendations including: institution of quality checks on the physical and chemical integrity of the product; no applications within 200 m of the ; community residences tarping of the periphery of the fumigated area; water treatment after each application; strict supervision of the applications; direction not to apply high odor products; training of applicators in use and safe handling of pesticides.

In Panama, Penagos[13] conducted a case control study, evaluating the frequency of patch test reactions to a battery of fungicides (captafol, chlorothalonil, benomyl, propioconazole, thiabendazole, imazalil, ziram, maneb, dichlorophen, $Al(OH)^3$), herbicides (dalapon, ametrine, paraquat), and other pesticides (lindane, hexachlorophene, coal tar) in 39 patients at the outpatient clinic of occupational dermatology, Changuinola Hospital, with a unique, chronic pigmented dermatitis of the face, neck, trunk, and upper extremities (Figure 9.62). Skin biopsy showed deterioration of the basal layer of the epidermis and pigment deposits in the epidermis (Figure 9.63). The comparison group consisted of 41 banana plantation workers treated at the same clinic for other skin conditions (predominantly onychomycosis, tinea versicolor, and dermatomycosis). The pigmented dermatitis was strongly associated with positive patch tests to the fungicide chlorothalonil. Positive reactions in controls were confined to standard

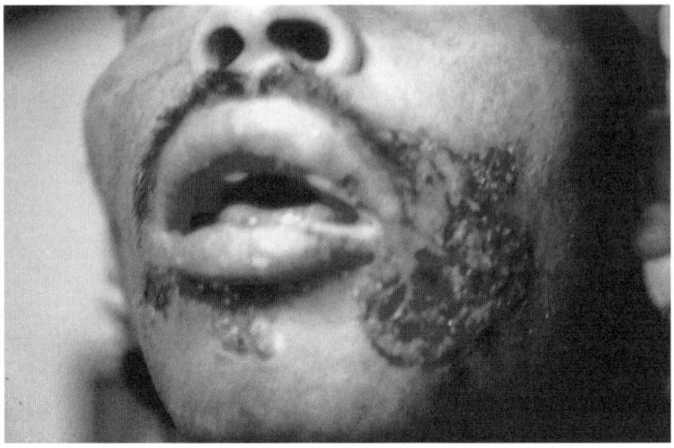

FIGURE 9.61 Facial irritation secondary to paraquat in Panamanian banana plantation worker. (Photo courtesy of H. Penagos; originally printed in O'Malley, M.A., *State-of-the-Art Reviews in Occupational Medicine*, Hanley and Belfus.)

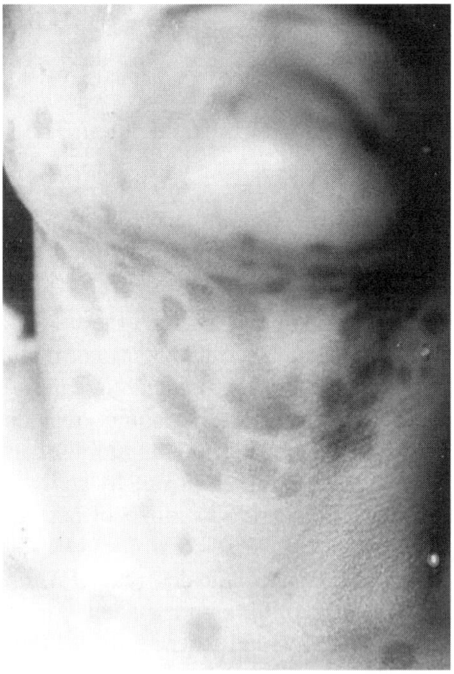

FIGURE 9.62 Chlorothalonil-associated pigmentary disorder.

series antigens—PPD in three patients and formaldehyde in two. The probable causal relationship between pigmented dermatitis and allergy to chlorothalonil suggested by the strong statistical association was strengthened by the occurrence of the pigmented lesions at the site of the patch test reaction 6 months after testing.

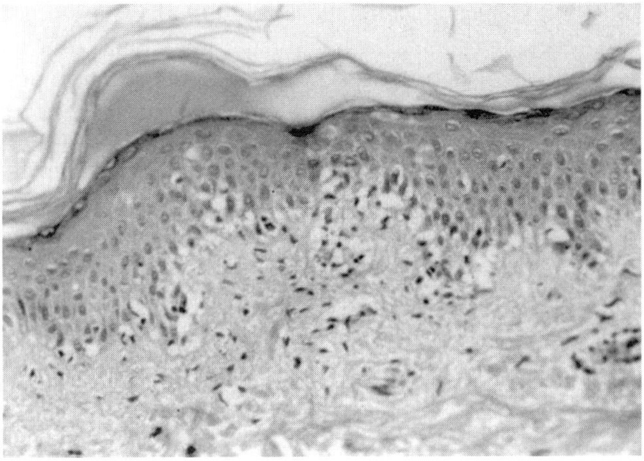

FIGURE 9.63 Biopsy from pigmented skin lesion showing deterioration of dermal-epidermal junction and melanin deposits in the dermis.

Cannavo and Mirande describe below, in one of the few published reports from Argentina, an outbreak of chloracne occurring in the 1979s and 1980s in an Argentine plant, associated with the production of pentachlorophenol, accompanied by secondary cases in family members of the workers.

In 1979 an epidemic of chloracne associated with production of pentachlorophenol occurred in the province of Buenos Aires, affecting 20 workers and and 10 relatives of

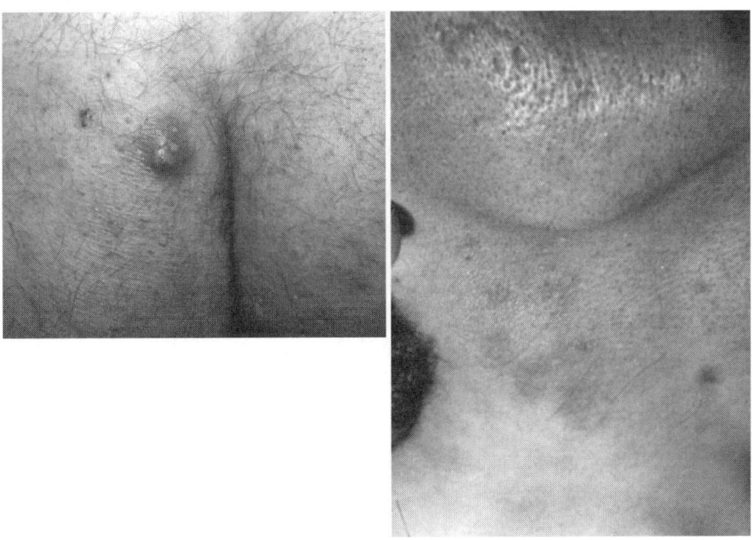

FIGURE 9.64 Gluteal acne and facial scars associated with pentachlorophenol exposure, Argentina, 1979.

those exposed at work. Clinical manifestations in the workers included lesions on the face, flanks, penis, testes, and perineal region (Figure 9.64). Some patients had generalized lesions. Scars and poikilodermatous skin changes were also seen in some patients.

Systemic manifestations included pulmonary failure and hemoptysis, gastrointestinal disorders such as diarrhea, vomiting, and gastritis, and neurologic symptoms including headache, irritability, and dizziness. Other reported symptoms were muscular fatigue, decrease of sexual interest, backache, and facial hypertrichosis.

One of the affected workers had seven children brought in for evaluation and treatment of numerous scaly papule and follicular hyperkeratoses their skin. As confirmed by the Italian dermatologist Gianotti, visiting Argentina, the clinical manifestations seen in the children were similar to those seen in children affected by the toxic cloud of trichlorophenol that occurred in 1976 at Seveso.

Histopathology changes included cyclic dilation of the hair bulb of the philosebaceous follicle and epidermal cyst formation. In the children, we observed the presence of giant cell granulomas in the inferior reticular dermis.

Pesticide patch testing was not performed because of concern about the systemic effects of pentachlorophenol and its dioxin contaminants. Public health measures resulted in closure of the factory for 95 days, with exposure controls imposed upon reopening.

In a cross-sectional survey of potato farm workers in northern Ecuador, Cole[14] examined the relationship between pesticide exposure and skin disorders. From a farm population census, all pesticide applicators and one half of exposed field workers were selected and controls were age-matched from urban occupations. Individual exposure measures included overall years and current hours working with pesticides. Farm-level measures included numbers of applications over the last 6 months and an application practices score (range 0-4). Each participant underwent a clinical skin examination. The thiocarbamate compounds maneb and zineb accounted for most of the fungicide applications and dermatitis cases wiith a history of maneb contact were patch tested with the compound. Applicators had a 7% prevalence of conjunctivitis relative to 0% in other groups examined. Dermatitis (usually on the hands are forearms) was present in 68% of those exposed, including 55% of the applicators, compared to 31% of the controls. Pigmentary disorders were found in 25% of those exposed compared to 10% of controls, a difference that approached statistical significance. Of the 117 dermatitis cases, 5% reacted to maneb on patch tests. In logistic regression analysis, significant predictors (p- <0.1) of dermatitis included years using fungicides (OR = 1.12 per year) and poor application practices (OR = 1.42 per score unit). Because the patch testing was conducted with a single compound, rather than with a pesticide battery, the true prevalence of allergic reactions to pesticides could not be evaluated, but exposures to fungicides appeared to be a significant risk factor for dermatitis in the population studied.

REFERENCES

1. **McConnell, R,** Corey, G., Henao, S., Nieto, O., Rosenstock, L., Trapé, A., and Wesseling, C., Pesticide Epidemiology in Latin America: opportunities for North-South Collaboration, 1993, Pan American Center for Human Ecology and Health, Pan American Health Organization, World Health Organization.
2. **FAOSTAT**, Production data for primary livertock, primary crops, and crops by area harvested, 1998, for Mexico, Central America, and South America. Available online at: apps.fao.org, 1999.
3. **FAOSTAT,** Pesticide Use Data, 1993-1997, for Central America and South America, Available online at: apps.fao.org, 1999.
4. **Castro, R, Morera, N., and Jarquin, C.,** *International Sinposium on Pesticides in the Developing World,* San Jose, Costa Rica, 1998.
5. **Ministerio de Salud, D. D.H., Republica de Nicaragua,** Boletin epidemiologico e informativo No 10. Available online at: http://salud. ops.org. ni/opsnic/p-salud/boletin/ 10/bol-10-0.htm, 1999.
6. **Servicios Coordinados de Salud Pública En el Estado de Baja California,** *Curso de Diagnóstico, Tratamiento y Registro de Intoxicaciones por Plaguicidas,* Instituto Mexicano del Seguro Social, Agencia de Protección del Medio Ambiente de California, Departamcnto de Regulación de Plaguicidas Sacramento, California., Mexicali, Baja California, 1995.
7. **Esquivel, M. and Coacioppo, S.,** *International Sinposium on Pesticides in the Developing World,* San Jose, Costa Rica, 1998.
8. **Garay, J.,** *International Sinposium on Pesticides in the Developing World,* San Jose, Costa Rica, 1998.
9. **Oliveira, L.,** *International Sinposium on Pesticides in the Developing World,* San Jose, Costa Rica, 1998.
10. **Burger, M.,** *International Sinposium on Pesticides in the Developing World,* San Jose, Costa Rica, 1998.
11. **Penagos, H.,** Contact dermatistis among banana plantation workers in Panama, *Epiderniology,* 6, S129, 1995.
12. **Ministerio de Salud, D. D.H., Republica de Nicaragua,** Boletin epidemiologico e informativo No 13. Available online at: http://salud. ops.org. ni/opsnic/p-salud/boletin/ 13/bol-1398.htm, 1999.
13. **Penagos, H., Jimenez, V., Fallas, V., O'Mnlley, M., and Maibach, H. I.,** Chlorothalonil, a possible cause of erythema dyschromicum perstans (ashy dermatitis), *Contact Dermatitis,* 35, 214, 1996.
14. **Cole, D, C., Carpio, F., Math, J. J., and Leon, N.**, Dermatitis in Ecuadorean farm workers, *Contact Dermatitis,* 37, 1, 1997.

9.9 CONCLUSIONS

Skin conditions related to pesticides are reported intermittently from many areas of the world. This is due to the lack of routinely available pesticide iIlness surveillance data in both developed and developing countries. Limited systematic evaluations, such as that conducted by Rodriguez and O'Malley (Section 9.2.7) and by Lisi (Section 9.3.2.5) indicate that the prevalence of positive patch test reactions to pesticides in agricultural workers with skin problems may be as high as 25%. The study reported by Cole (Section 9.7) demonstrated that under adverse working conditions, the overall prevalence of dermatitis probably attributable to pesticides may range as high as 68%.

Review of international experience with pesticides demonstrates a remarkable coherence of experience with the effects of individual compounds. This is illustrated by the stereotypical occurrence of outbreaks of chloracne following process accidents in the manufacture of trichlorophenol, and the nearly simultaneous occurrence of outbreaks of irritant dermatitis among applicators following the introduction of propargite in Japan (Section 9.4.2) and in California (Section 9.2.4). Likewise, although its manufacturer has made valid points about the irritant effects of chlorothalonil and denied vigorously that it is an allergen, international experience with the compound in the U.S. (Section 9.2.3), Europe (Sections 9.3.2.2 and 9.3.2.3), and Latin America (Section 9.7) have clearly demonstrated its allergenicity and work remains to find means of evaluating the prevalence of pesticide related skin conditions in different populations. The use of standardized questionnaires (see appendix II) has great promise in this regard, but its practical utility in identifying specific problems related to pesticides remains to be fully demonstrated.

Additional work also remains in detailing already existing information contained in the functioning surveillance databases in the U.S., Latin America, and elsewhere.

Appendix IA

Abbreviated field interview/questionnaire and examination form

Name ___________________ Crew Leader: ________________ Crew number ____

Questions regarding your work history (Preguntas tocante su trabajo:)
Do you use a long sleeve shirt when you are working? Yes No (Usa camisa con mangas largas cuando esta trabajando? Si No)
Do you use gloves when you are working? Yes No (Usa guantes cuando esta trabajando? Si No)

Questions dealing with skin problems (Preguntas tocante problemas de la piel):
As a child did you have a problem with rashes or eczema? Yes No (De niño/niña tuva problemas con salpullido o erupciones de la piel? Si No)
During the last two weeks have you had a rash (redness, itching or burning of the skin? (If yes: On what day or date did it begin? ___) Durante las dos ultimas semanas tuvo salpullido (ronchas or comezón, ardor de la piel? Si No)
Did you inform your supervisor? Yes No (Avisó a su mayordomo? Si No)
When was that? What date? (Cuando fue? Que fecha?) ___________
Did you go to the doctor? Yes No (If yes, when did you go? __________) (Fue al doctor? Si No (Que fecha? __________)
Did you receive medicine? Yes No (If yes: Cream? Injection? Pills?) Se le dio medicina? Si No (Cremas? Inyecciones? Pastillas?)
Did you lose time from work? Yes No (indicate how many days _____) Perdió dias de trabajo? Si No (Cuantos dias _____)

Exam: 1 = slight; 2 = moderate; 3 = severe blank = normal; crossed out = not examined

Periorbital:	erythema ____	blistering _____	post inflammatory
Jaw:	erythema ____	blistering _____	post inflammatory
Upper Neck:	erythema ____	blistering _____	post inflammatory
V of neck:	erythema ____	blistering _____	post inflammatory
Chest:	erythema ____	blistering _____	post inflammatory
Abdomen:	erythema ____	blistering _____	post inflammatory

L Shoulder:	erythema ____	blistering _____	post inflammatory
Upper arm:	erythema ____	blistering _____	post inflammatory
Antecubital:	erythema ____	blistering _____	post inflammatory
Forearm:	erythema ____	blistering _____	post inflammatory
Wrist:	erythema ____	blistering _____	post inflammatory
Hand:	erythema ____	blistering _____	post inflammatory

R Shoulder:	erythema ____	blistering _____	post inflammatory
Upper arm:	erythema ____	blistering _____	post inflammatory
Antecubital:	erythema ____	blistering _____	post inflammatory
Forearm:	erythema ____	blistering _____	post inflammatory
Wrist:	erythema ____	blistering _____	post inflammatory
Hand:	erythema ____	blistering _____	post inflammatory

The abbreviated questionnaire and examination form shown above has proved useful for investigating acute illness outbreaks among migrant workers in situations where access to study subjects is often limited.[1,2] In field situations, a complete dermatologic examination is often not possible. It may, nevertheless, be possible to examine the torso, upper extremities, head, and neck. Questions on workplace exposures were not included, because the workers did not have specific knowledge of materials applied to the fields in which they worked. However, the questionnaire was used in conjunction with an extensive administrative records and environmental sampling.

REFERENCES

1. **O'Malley, M.,** Surveillance of occupational skin disease, *Case Studies in Occupational Epidemiology*, Steenland, K., Ed., Oxford University Press, New York, 1993, viii.
2. **O'Malley, M., Smith, C., Krieger, R., and Margetich, S.,** Dermatitis among stone fruit harvesters, Tulare County, 1988, *Am. J. Contact Dermatitis*, 1, 100, 1990.

Appendix IB

Contact dermatitis and non-specific skin symptoms, The Tuohilampi Questionnaire

Paivikki Susitaival

Irritant and allergic contact dermatitis are among the most common occupational diseases in occupations where the skin is in contact with water, detergents, chemicals, plants or exposed to mechanical irritation. In most cases, hands or forearms are involved. Some studies have shown that the morphology of eczema (dyshidrotic, nummular, etc.) can be variable, and does not necessarily tell anything about the etiology of hand eczema.[1-5]

Eczema is the most common chronic dermatosis in hands. Others include psoriasis, palmoplantar pustulosis, lichen planus, and genetic hyperkeratoses. There are also many less chronic dermatoses like urticaria and scabies that can affect hands and forearms. In clinical surveys by Coenraads et al.[6, 7] eczema has constituted from 54% (general population, males) to 72% (construction workers, males) of all reported skin diseases in hands. Meding examined respondents from the general population in Sweden who had reported having hand eczema in a questionnaire, and considered 89% of the self-diagnoses to be correct[8]. In Agrup's questionnaire survey[9] of all skin changes in hands except common warts, eczema was found to constitute 53%, psoriasis 8%, senile lentigo 9% and vitiligo 5% of all cases with positive answers.

Variations of two methods in diagnosing hand eczema have been used in questionnaire studies: a self-report ("self-diagnosis") of hand dermatosis or eczema ("Have you had..."), or a diagnosis based on a symptom list with study specific criteria (symptom-based diagnosis). Some studies have included the forearms in hand eczema questions ("hands and/or forearms").[6, 7, 10] Forearms are often in contact with occupational and environmental allergens and irritants even when the hands are protected with gloves. Some studies have included only dermatoses of at least 3 weeks duration.[6, 7, 11] For the diagnosis of eczema, symptoms of at least 3 weeks duration or repeating symptoms[12] have been included.

Many studies have clinically validated questionnaire answers in epidemiological studies of skin diseases. There are difficulties in interpreting and comparing these data. The clinical criteria for dermatosis cases are usually not stated in the publications, and also the conceptualization of different terms (eczema, allergic eczema, skin change, skin condition) by the respondent is not clear. Eczema as a word may be familiar to laymen in many languages, but not, e.g., in Finnish.

Smit et al.[12] (The Netherlands) evaluated questions detecting hand eczema in populations. They used a symptom-based diagnosis. Skin symptoms of hands and fingers were grouped in five groups (redness and edema, redness and fissures, vesicles, scaling and fissuring, and itching and fissuring). The criteria of eczema were at least one positive symptom group more than once or at least for 3 weeks within the past 12 months. This symptom-based diagnosis and a self-reported diagnosis (the respondent's own opinion of having suffered from hand dermatitis in the past 12 months) were compared with the medical diagnosis (clinician's opinion). The symptom-based diagnosis gave a much higher prevalence (48%) than the self-reported (17%) or medical diagnoses (18%). According to the study, the self-reported diagnosis was a good indicator of the presence of hand dermatitis. 68% of those reporting hand dermatitis and 8% of those not reporting hand dermatitis had a hand eczema from the dermatologist's point of view (sensitivity 65%, specificity 93%).

The symptom-based diagnosis, on the other hand, could find all hand eczema cases, but included also many non-cases as cases, since only 38% of questionnaire hand eczema diagnoses were confirmed by the clinician (sensitivity 100%, specificity 64%). The results by Smit et al. are in accordance with other studies validating the self-diagnosis of hand eczema.[9, 13, 14] In Agrup's study, 94% of those reporting skin changes in hands were clinically found to have them.

In a survey asking about any "rash or other disorders of the skin, hair or nails" 14% of those reporting were found to have no dermatosis, and in 13% of those reporting "no," a moderate to severe dermatosis was present.[13] Berg et al.[15] (Sweden) evaluated clinically a questionnaire asking skin symptoms (morphology and subjective symptoms) during the last 2 years and the area of involvement. They came to a conclusion that the questionnaire gave reasonably adequate answers regarding clear-cut symptoms over a period of time (sensitivity 85%, specificity 99%). The clinical diagnosis confirmed the self-reported answers in atopic eczema (in 92%), seborrheic dermatitis (in 86%), and hand dermatosis (in 83%). The agreement in other diagnoses was poorer. Reporting dermatosis involvement in hands had both a good sensitivity (85%) and specificity (95%) in Berg's study.[15] Meding et al.[16] surveyed and examined skin complaints (questions not specified in the paper) in a paper mill and in a dairy; 90% of those with reported skin symptoms and 2 of 15 of those with no skin complaints were diagnosed as having a skin disease. Thus, the sensitivity of the questionnaire was 69%, and the specificity was 97%.

As a part of a questionnaire study of Finnish farmers,[17] a random sample of those reporting no dermatosis and another random sample of those reporting a hand dermatosis in the questionnaire were invited to visit the municipal healthcare center for a "health check," without informing them about the aim of the examination. Before the examination, the farmers were given a form to fill out that asked: Do you have any skin disease now? After this their skin was immediately examined by a dermatologist. All the farmers answering affirmatively to the question of having a skin disease now were diagnosed as having one. Of those farmers answering "no skin disease now" in the pre-examination questionnaire, 22% (27/122) were diagnosed as having a dermatosis including mostly interdigital macerations in feet (16/27) and 3 (2.5%) hand eczema cases. Thus, a self-diagnosis of having "a skin disease now" had a high specificity (100%) but a lower sensitivity (60%).

Constitutional risk factors for hand eczema include atopic (childhood) eczema, especially hand eczema[10, 18-22] and also at least in some populations respiratory atopy.[10, 12, 21-23] Also a so-called "atopic skin diathesis" meaning a dry and itchy skin, itch when sweating, and intolerance to wool, has in many studies been connected with an increased risk of hand eczema.[20, 24, 25] Also nickel allergy or symptoms of metal allergy have significantly raised the risk of hand eczema in many epidemiological studies.[21, 24, 26-29] Symptoms of metal allergy probably act as a marker of skin sensitivity.

A person with (hand) eczema can usually tell the specific exacerbating factors if there are any. This is well known to dermatologists in clinic, but there is also scientific proof to the fact that the patient's suspicion of the aggravating factor and the skin test results are often in concordance.[30]

This questionnaire includes several series of questions for different purposes. For a specific study, the questions or question series relevant for the study can be chosen. There is more information in the questionnaire about the data given by the individual questions.

QUESTIONS

A. Prevalence of self-reported hand dermatosis

D.1. Have you ever had a dermatosis (e.g., rash, eczema) on your fingers, finger webs, palms, backs of hands, wrists, or forearms?

0 no *(proceed to question D.18)*
1 yes

In some languages the word "eczema" can be used. The "cutting point" of what is "eczema or dermatosis" and what is not depends most likely on the population, both culture and occupation.

D.2. Has this hand or forearm dermatosis appeared (only one answer)

only once and continued for less than 2 weeks □ 3 (proceed to question D.17)
only once but continued for at least 2 weeks □ 4
several times or has been persistent □ 5

D.3. **Which parts of hands or forearms did the dermatosis usually involve?** *(one or more answers, please shade the areas of your usual dermatosis involvement in the picture)*

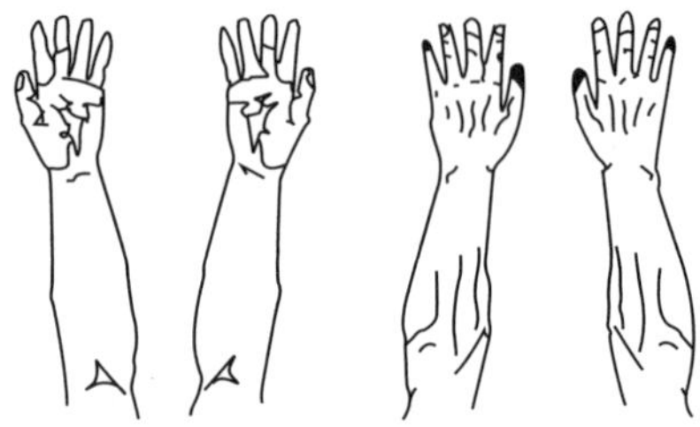

fingers or finger webs□ 2
backs of hands □ 2
palms □ 2
wrists □ 2
forearms □ 2
hollows (fronts) of the elbows □ 2
elbows □ 2

D.4 **When was the last time you had this hand or forearm dermatosis** *(only one answer)*

I have it at this moment □ 3
not at this moment but within the past
3 months □ 4
3-12 months ago □ 5
over a year ago. □ 6
Which year? _____ *(give the best estimate you can)*

With questions D.1. and D.3. we get a self-report of the point-, 3-month, and 12-month prevalence of hand dermatoses. Also the life-time prevalence is asked.

D.5 **Has this hand or forearm dermatosis ...** *(only one answer)*

2 appeared only *once* and continued for *less than two weeks (proceed to question D.18)*
3 appeared only *once* and continued for *at least two weeks*
4 appeared *several times* or has been been *persistent*

Question D.4 is designed to distinguish between those with acute episodes of hand dermatosis from recurring and more persistent cases.

D.6 **When did you have this hand dermatosis the first time?** *(Only one answer, give the best estimate you can. If you cannot estimate the year, still circle the number of the appropriate answer)*

in childhood (under 6 years of age) □ 3 in (year) _______
at school age (6 to 18 years of age) □ 4 in (year) _______
in adult life (over 18 years of age) □ 5 in (year) _______

Question D.6 gives data of the history of hand dermatosis as far as the respondent has information or remembers the facts. Those known to have had hand dermatosis already as children belong to a high risk group which is an important factor for occupational hand eczema studies.

D.7 **Since the hand dermatosis first appeared, how many whole years** *(meaning 12 consecutive months)* **have you been free of it?** ***If you have not been free of hand dermatosis for one whole year, give number of full months*** *(meaning 4 consecutive weeks).*

________ years *(give the best estimate you can)*
or ________ months
0 no time at all (hand dermatosis has persisted since it started)

Questions D.6 and D.7 are designed to delineate the length of time during which the respondent has suffered from current or previous hand dermatosis. Hand eczema is seldom a continuous one-course disease. Typically it comes and goes in periods of weeks to months and continues for months or years.

B. Symptoms of Hand Dermatosis

D.8. **What symptoms have you had in your hand or forearm dermatosis?** *(one or more answers)*

redness	□ 2	
dry skin with scaling	□ 2	
hardening of skin with scaling	□ 2	
itching	□ 2	
pain or tenderness	□ 2	
burning, prickling, or stinging	□ 2	
rough skin, rash	□ 2	
tiny water blisters (vesicles) or oozing	□ 2	
fissures, cracks, or crusting	□ 2	
swelling or dropsy (edema)	□ 2	
wheals, welts, or urticaria	□ 2	
papules	□ 2	
pustules	□ 2	
pimples	□ 2	
sensory loss (e.g., fingertips)	□ 2	
other	□ 2	**what?**____________
no specific symptoms	□ 1	

Question D.8 is important when making hand eczema diagnosis with the self-report of hand dermatosis. There are not yet generally accepted criteria for this diagnosis and therefore the criteria formed with these tools should be validated in clinical studies.

C. Familiarity with diagnosis or cause

The next question endeavors to find out if the respondent is aware of the nature of his/her skin problem(s). It is likely that many of those with a long-standing skin disease like psoriasis or atopic eczema may be well informed about their dermatosis. Also there are commonly known skin diseases like acne. In environmental studies it is useful to obtain this data.

D.9 Do you know the diagnosis or the cause of your hand or forearm dermatosis? *(one or more answers)*

Yes, I know the diagnosis or cause which is:

	opinion	
	own	**doctor's**
sensitive or itchy skin	□ 3	□ 4
dry skin	□ 3	□ 4
childhood eczema or atopic eczema	□ 3	□ 4
psoriasis	□ 3	□ 4
allergy	□ 3	□ 4
other skin disease	□ 3	□ 4
stress	□ 3	□ 4
other cause,	□ 3	□ 4
what?______________________		

I don't know the diagnosis or cause because:

	I did not consult a doctor	I did consult a doctor
diagnosis or cause not found	□ 5	□ 6
I don't know or remember the cause or diagnosis	□ 1	

D. Exacerbating factors

D.10 Have you noticed if contact with certain materials, chemicals, or other things aggravates your hand dermatosis?

no	□ 1	*(proceed to question D.11)*
yes	□ 2	
I don't know	□ 0	*(proceed to question D.11)*

D.11 Which are the most important of the aggravating factors? *(your opinion, one or more answers)*

soaps, liquid soaps, shampoos, e.g., products used for personal hygiene	□ 2
detergents and other products used in household cleaning and laundry	□ 2
protective gloves	□ 2
machine maintenance (e.g., cars) handling oils, building, and decorating	□ 2

gardening, handling plants, soil, vegetables, berries, fruits etc. □ 2
handling food stuffs □ 2
handling animals □ 2
emotional state, stress □ 2
menstrual periods or other hormonal factors □ 2
other ____ □ 2

These two questions (D.10, D.11) can be useful in getting a rough estimate of the role of specific causes or allergies in, e.g., occupational groups. The data are probably more valid if the question is open with no coding and thus no hints to the respondent. Open questions, on the other hand, are more time consuming in data processing. This problem is eliminated in telephone surveys where the interviewer codes the open answer to a pre-coded list. Coding should be kept concise with the most relevant factors individually coded and the obvious factors grouped together (or else, what?). Here, as an example, coding of general factors only. The list should include work-specific factors when studying occupational groups. Also, additional questions on work are needed in such studies.

E. Medical consultations and skin tests

D.12 Have you seen a doctor about your hand or forearm dermatosis?
(consultations in adult age)

no □ 1 *(proceed to question D.15)*
yes □ 2
when was the last time? *(year)* _____ *(give the best estimate you can)*

D.13 Have you been skin tested for allergy because of your dermatosis?

no □ 1 *(proceed to question D.15)*
yes, □ 2 **when was the last time?** *(year) 19___ (give the best estimate you can)*
I do not know □ 0 *(proceed to question D.15)*

Answers to questions D.12 and D.13 will give information with some idea of the grade of the dermatosis, or more likely, information on the dermatosis being a problem to the respondent, assuming that consulting a doctor means that there is a problem. Skin tests are (in Finland) made only by specialists, and they are made when environmental causes or aggravators are suspected. Both assumptions can also be biased. Consulting a doctor for a problem not threatening your life depends on the culture, surroundings, and many personal matters, but the bigger the problem gets the more liable one is to seek professional help. Indications for testing should be the same but unfortunately they vary also from country and doctor to another.

D.14 Were you ever found to be allergic to anything in the skin tests?

no	□ 1	*(proceed to question D.15)*
yes	□ 2	
I do not remember	□ 0	*(proceed to question D.15)*

D.15 Which allergies have been found in your skin tests? *(write in)*

____

I don't remember □ 0

Answers to questions D.14 and D.15 must be interpreted with caution. There were different opinions in the work group designing the questions about the understanding of the test results by the patients. This will, of course, vary from one population to another depending on, e.g., occupation and education, but also on the quality of information given to the patient. It is important to validate the answers to these questions in the future.

F. CONSEQUENCES AND LIFE-IMPACT OF DERMATOSIS

D.16 Has the hand or forearm dermatosis affected the everyday work in your occupation or job *(including housewife's work)* **in any specific way? Which of the following are true for you?** *(one or more answers)*

in no specific way	□ 1
hands must be protected with gloves	□ 2
I have been on sick leave or otherwise off work because of the dermatosis	□ 2
How many weeks or months during the past 12 months? __ weeks,__months	□ 2
I have changed jobs because of the dermatosis	□ 2
I have changed occupation or profession because of the dermatosis	□ 2
The attitude of the employer(s) is negative toward me because of the dermatosis	□ 2
The attitudes of the fellow workers are negative toward me because of the dermatosis	□ 2
Dermatosis has influenced my choice of profession or occupation	□ 2
My income has diminished because of dermatosis	□ 2
I have lost my job because of dermatosis	□ 2
I am retired because of dermatosis	□ 2
I am not working at present (unemployed, retired, etc., for reasons not connected with dermatosis)	□ 1

D.17. During the past 12 months, has the hand or forearm dermatosis affected your work at home, spare time, social life, or hobbies in any specific way? Which of the following are true for you? *(one or more answers)*

no other inconveniences or effects	□	1
I have to use moisturizers or skin creams daily	□	2
dermatosis or protective gloves make managing daily routines difficult	□	2
I have cut down household duties because of dermatosis	□	2
I have cut down hobbies or activities because of dermatosis	□	2
washing is difficult because of the dermatosis	□	2
itching and pain are disturbing and cause sleeplessness		
problems in socializing with other people because of dermatosis	□	2
problems in intimate relationships because of dermatosis	□	2
treatment of and protective measures are costly		
other____	□	2

Appendix II

Summary of Pesticide Patch Test Concentrations Reported in the Public Domain Literature

Source material: Material for patch testing with pesticide active ingredients for the California Nursery Study [O'Malley, 1998 #1195; O'Malley, 1995 #806] was obtained from Chem Service, Inc (PO Box 559, West Chester, PA 19381-0599, U.S. Phone: 1-610-692-3026; Fax: 1-610-692-8729 (WWW.CHEMSERVICE.COM) and the same source was also used more recently by Guo (Guo et al. 1996).

Current sources for the standard test series antigens, antimicrobials, and fungicides (thiocarbamate and phthalimido compounds) include:

Glaxo Pharmaceuticals: http://www.truetest.com

Chem-Diagnostics, Sweden: http://www.dormer.com/html/chem_diagnostics.html

Hermal: http://www.hermal.com. Select product range, then click on link for TroLab®

A concentration of 1% in petrolatum may be used as a default for compounds without a previously established test concentration that are not recognized from animal testing to be severe irritants. A summary of test concentration information from published sources is given in the table below.

Material Insecticides	CAS #	Concentration %	Vehicle	Ref. and commentary
Organophosphates				
Acephate	30560-19-1	1	butanol	O'Malley et al., 1995
Azamethiphos		2	aqueous	Iliev and Elsner, 1997
Azinphos methyl	86-50-0	1	Vaseline	Jung et al., 1987; Lisi, 1992a
Bromophos	2104-96-3	0.1	Vaseline	Jung et al., 1987
Butonate	126-22-7	5	aqueous	Jung et al., 1987
Chlorpyrifos	2921-88-2	1	butanol	O'Malley et al., 1995
Diazinon	333-41-5	1	butanol	O'Malley et al., 1995
		1	petrolatum	Lisi, 1992a
Demephion	8065-62-1	0.5	aqueous	Jung et al., 1987, 1989
Dichlorvos (DDVP)	62-73-7	0.02	aqueous	Ueda et al., 1994
		1	petrolatum	Lisi, 1986; Lisi et al., 1986
		0.5	petrolatum	(Sharma and Kaur, 1990) 20 negative controls
		0.1,1	petrolatum	Mathias, 1983
		0.05	aqueous	Lisi, 1992a
		0.05-0.1	aqueous	Jung et al., 1987
Dichlorvos 50		1	aquaeous	Stoermer, 1985
Dimethoate	60-51-5	1	petrolatum	Lisi, 1992a; Schena and Barba, 1992 20 negative controls, Lisi, 1986; Lisi et al., 1986; Sharma and Kaur, 1990

Material Insecticides	CAS #	Concentration %	Vehicle	Ref. and commentary
		1	alcohol	Haenen et al., 1996
		1	Oil	Jung et al., 1989
		1-2	aqueous	Jung et al., 1987
Malathion	121-75-5	1	butanol	21 negative controls, O'Malley et al., 1995
		0.5	petrolatum	Guo et al., 1996; Jung et al., 1987; Lisi, 1992a; Sharma and Kaur 1990 Adams, 1990b; Cronin, 1980
		10	alcohol	Milby and Epstein, 1964
Methidathion	950-37-8	1-2	Vaseline	Jung et al., 1987
Mevinphos	7786-34-7	0.25	aqueous	
Naled	300-76-5	0.01	petrolatum	Guo et al., 1996
Naled technical		0.5	aqueous	Jung et al., 1987; + reactions noted in 4/6
		1	petrolatum	Lisi, 1992a
		60%	xylene	Edmundson and Davies, 1967
Omethoate		1	aqueous	Haenen et al., 1996
Oxydemeton-methyl	301-12-2	1	Vaseline	Jung et al., 1987; Sharma and Kaur, 1990
Parathion ethyl	56-38-2	1	alcohol	Jung et al., 1987; Lisi 1986; Lisi, 1992a; Lisi et al., 1986
		1	petrolatum	van Ketel, 1976
		0.01	aqueous	Guo et al., 1996
Parathion methyl	298-00-0	1	petrolatum	Lisi, 1992a; Sharma and Kaur, 1990 irritant reactions in 2/20 controls; photopatch test, Jung et al., 1987; Lisi, 1986; Lisi et al., 1986
Phorate	298-02-2	1	petrolatum	Lisi, 1986; 1992a; Lisi, et al., 1986
Pirimiphos	29232-93-7	1	Vaseline	Jung et al., 1987
Quinalphos	13593-03-8	1	petrolatum	Sharma and Kaur, 1990
Trichlorfon	52-68-6	1	aqueous	Jung et al., 1987
N-methyl-carbamates				
Methomyl	16752-77-5	0.5	petrolatum	Ueda et al., 1994
		0.2 (a.i.); 1% formulated product	aqueous	Bruynzeel, 1991, 26 Controls
Carbaryl	63-25-2	1	petrolatum	Lisi, 1992a; Sharma and Kaur, 1990
Carbaryl 50%		0.25	Vaseline	Jung et al., 1989
Carbofuran	1563-66-2	1	petrolatum	Sharma and Kaur, 1990
Methiocarb	2032-65-7	0.5		Willems et al., 1997
Propoxur (Baygon EC 20)		1	aqueous	Stoermer, 1985
Organotin compounds				
Cyhexatin (plictran)		0.3	petrolatum	Ueda et al., 1994
Fenbutatin oxide	13356-08-6	0.1	butanol	O'Malley et al., 1995
Tributyltin oxide(TBTO)		0.01	aqueous	3/22 Controls had irritant at this concentration, Gammeltoft, 1978
Triphenyl tin fluoride		0.01,0.1,1,1	acetone	Andersen and Petri; 1982
Organochlorines				
Aldrin	309-00-2	1	petrolatum	Sharma and Kaur, 1990
DDT	50-29-3	1	petrolatum	Sharma and Kaur, 1990; Fisher, 1986
DDT		10%,1%	acetone	Cases of pemphigus negative on testing with 1:10 and 1:100 dilutions Tsankov et al., 1998
DDT		1	petrolatum	Adams, 1990b
DDT technical		1	acetone, Vaseline	Jung et al., 1987
DDT lotion		5	acetone	Citing U.S. case reported by Leider; Behrbohm, 1962
DDT technical		undiluted	none	Behrbohm, 1962
Purified DDT		undiluted	none	Behrbohm, 1962

Material Insecticides	CAS #	Concen- tration %	Vehicle	Ref. and commentary
Endosulfan	115-29-7	0.5	petrolatum	Lisi 1986; Lisi et al., 1986
		1	petrolatum	Lisi, 1992a
Fentichlor	97-24-5	0.1,0.5,1.0	aqueous	Phototesting - persistent light reaction Norris et al., 1988
Lindane (Hexachlorocyclohexane)	58-89-9	1	petrolatum	Adams, 1990b; Cronin, 1980; Lisi, 1992a; Lisi et al., 1986; Sharma and Kaur, 1990
(Hexachlorocyclohexane technical)		not specified	acetone or Vaseline	Jung et al., 1987
(Hexachlorocyclohexane technical)		15-20	acetone	Behrbohm, 1962
Dienochlor		1	butanol	O'Malley et al., 1995
		1	petrolatum	10 negative controls; van Joost et al., 1983
Pyrethroid				
Allethrin		1	petrolatum	Lisi, 1992b
Cypermethrin		1	petrolatum	Lisi, 1992b; Sharma and Kaur, 1990
Deltamethrin		1	petrolatum	Lisi, 1992b
Fenothrin		1	petrolatum	Lisi, 1992b
Fenvalerate		1	petrolatum	Sharma and Kaur, 1990
Permethrin		1	petrolatum	Lisi, 1992b
		1	butanol	O'Malley et al., 1995
Fluvalinate		1	butanol	O'Malley et al., 1995
Resmethrin		1	petrolatum	Lisi, 1992b
Pyrethrum		2	petrolatum	Dannaker et al., 1993; Jung et al., 1987; Lisi 1992a; Piraccini et al., 1991
Miscellaneous insecticides				
Arsenic trioxide		5	starch powder	
Benzyl benzoate	120-51-4	1	petrolatum	Stransky et al., 1996
Bacillus thuringiensis	68038-71-1			Adams
Chlorobenzene	108-90-7	5		Adams
DEET (diethyl toluamide)	134-62-3	5%	ethanol/ petrolatum	Lamberg and Mulrennan, 1969
		0.1 cc undiluted Ready-to-use product in open test; prick test	NA	Maibach and Johnson, 1975; Wantke et al., 1996
Imidacloprid		0.5	petrolatum	Willems et al., 1997
Nicotine		1	aqueous	Adams, 1990b
		5	aqueous	Lisi, 1992a
Propargite	2312-35-8	0.05		Irritant reactions in controls; Nishioka, et al., 1970
Oxythioquinox		0.1	petrolatum	Ueda et al., 1994
Piperonyl butoxide		—	—	Jung et al., 1987
Tetrachloroacetophenone		0.01	petrolatum	Lisi, 1992a
Insecticide precursors				
Ethoxymethylenemalono-nitrile		0.01-1	petrolatum	Wakelin et al., 1998
Malononitrile		0.01-1	petrolatum	Wakelin et al., 1998
TCAP (Tetrachloroacetophenone)		0.01-2	petrolatum	van Joost and Wiemer, 1991
Rodenticides				
Warfarin		0.05	petrolatum	Cronin, 1980
		0.05	Vaseline	Jung et al., 1987
Antimicrobials and antibiotics				
Isothiazolines				
Methylchloroisothiazolinone-methylisothiazolinone (Kathon CG)		0.01	aqueous	Zemtsov, 1991

Material Insecticides	CAS #	Concen-tration %	Vehicle	Ref. and commentary
Benzisothiazoline	2634-33-5	1	petrolatum	Andersen and Rycroft, 1991; Fisher, 1986
Mercury compounds				
Mercuric ammonium		1	petrolatum	Guo et al., 1996
Thimerosal	54-64-8	0.1	petrolatum	Guo et al., 1996
		0.05	aqueous	Pambor and Timmel, 1989
Mercury		0.01	petrolatum	Sharma and Kaur, 1990
Mercuric chloride	33631-63-9	0.03	petrolatum	Pambor and Timmel, 1989
Phenylmercuric acetate	62-38-4	0.01	petrolatum	Guo et al., 1996
		0.1	aqueous	Jung et al., 1987
		0.02	aqueous	Jung et al., 1989
		0.2-1.0	petrolatum	Bonnetblanc and Delrous, 1996
Aldehydes				
Formaldehyde (formalin 40% solution)	50-00-0	2	aqueous	Fisher, 1986
Glutaraldehyde	111-30-8	1	aqueous	Andersen and Rycroft, 1991
Metaldehyde	9002-91-9	1		Lisi et al., 1986
Phenols				
Dowicide 32 (4-chloro-2-phenyl phenol; 6-chloro-2- phenyl phenol)	607-12-5; 85-97-2	1	aqueous	Adams, 1972
Dowicide 6 (tetrachlorophenol)	25167-83-3	1	petrolatum	Fisher, 1986
Dowicide B sodium 2,3,5-trichlorophenate	136-32-3	1	petrolatum	Fisher, 1986
Dowicide E 2-bromo-4-phenyl-phenol	92-03-5	1	petrolatum	Fisher, 1986
Dowicide F sodium tetrachlorophenate	25567-55-9	1	petrolatum	Fisher, 1986
o-Phenylphenol Dowicide 1	90-43-7	1	petrolatum	Andersen and Rycroft, 1991
o-Benzyl-p-chlorophenol (dichlorophene)	120-32-1	1	aqueous	Fisher, 1986
Pentachlorophenol Dowicide 7	87-86-5	1	alcohol	Fisher, 1986
p-chloro-*o*-cresol	1570-64-5	0.1	alcohol	Cronin Fisher, 1986
Organotin compounds				
Triphenyltin (Fentin) Hydroxide	76-87-9	1		Lisi, 1986; Lisi et al., 1986
		0.5	petrolatum	Lisi, 1992a
Tributyltin oxide	56-35-9	0.01	water	Cronin, 1980; Gammeltoft, 1978; 3 irritant reactons at this concentration among 22 controls — probably preferable to use 0.001% concentration — or use petrolatum suspension
Tributyltinhydroxide	1067-97-6	0.01		Cronin, 1980
Miscellaneous				
Alantalactone		0.1	petrolatum	Guo et al., 1996
Benzalkonium chloride		0.1	aqueous	Guo et al., 1996
Gentamicin sulfate		20	petrolatum	Guo et al., 1996
Benzoyl peroxide	94-36-0	1.0	petrolatum	Guo et al., 1996
Chlormequat				Jung et al., 1987
Carboxin		not specified	not specified	Jung et al., 1987
Chlorhexidine digluconate		0.5	aqueous	Guo et al., 1996
1-chloro-2,4-dinitrobenzene		0.00025	Acetone	Adams et al., 1971
Tylosin tartrate	1401-69-0	5	petrolatum	Guo et al., 1996
Chlorphenesin	104-29-0	1	petrolatum	Burns, 1986; Goh, 1987; Wakelin and White, 1997 Adams, 2
2-phenoxyethanol (Euxyl K 400)	122-99-6			

Material Insecticides	CAS #	Concen-tration %	Vehicle	Ref. and commentary
Hexahydro, 1,3,5,tris (2-hydroxyethyl) triazine (Grotan)		0.2	aqueous	Keczkes and Brown, 1976
Sodium hypochlorite	10022-70-5	1	petrolatum	Hostynek et al., 1990
		1	aqueous	Urticarial reaction in open test; Hostynek et al., 1989
Tetrachlorsalicylanilide	1154-59-2	0.1	alcohol	Wilkinson, 1961
Piscicide				
Rotenone		5%	talcum	[Lisi, 1992 #833; Fisher, 1986 #1965]
		1%	talcum	[Adams, 1990 #1992]
Fungicides				
Phthalimido compounds				
Captafol, Difolatan	2939-80-2	0.05- 0.1, 0.1	petrolatum	Lisi, 1986; Lisi et al., 1986
		1	petrolatum	Camarasa, 1975; Groundwater, 1977; Sharma and Kaur, 1990; 2 irritant reactions in 20 controls
		0.1	petrolatum	Guo et al., 1996; Peluso et al., 1991; Piraccini et al., 1991
		1,5,10	aqueous	Brown, 1984
Captan	133-06-2	0.25	petrolatum	Jung et al., 1987 Adams,2
Captan		0.1	petrolatum	Dannaker et al., 1993; Piraccini et al., 1991, concentrations used in material from Troh Lab/Pharmascience
		1	petrolatum	Rudner, 1977
Folpet	133-07-3	0.05		Adams, 2
		0.1	petrolatum	Guo et al., 1996; Lisi 1992a; Peluso et al., 1991; Piraccini et al., 1991
Ditalimfos (plondrel)	5131-24-8	0.01	petrolatum	Jung et al., 1987; van Ketel 1975; van Ketel 1977
Phaltan	133-07-3	0.01	petrolatum	Adams, 1990 #1992]
Carbamates and carbamate-like compounds				
Benomyl	17804-35-2	0.1	petrolatum or water	Dannaker et al., 1993; Jung et al., 1987; Kuhne et al., 1985; Piraccini et al., 1991 Lisi, 1986
Benomyl	17804-35-2	1.0	petrolatum	No reaction in 10 controls; Ueda et al., 1994; van Joost et al., 1983; van Ketel, 1976
		undiluted and diluted 1:5	olive oil	Savitt, 1972
Thiocarbamates				
Mancozeb	8018-01-7	1	petrolatum	Lisi, 1992a
Mancozeb/pyrifenox		2	aqueous	Iliev and Elsner, 1997
Maneb	12427-38-2	1	petrolatum	Lisi, 1992a; Peluso et al., 1991; Piraccini et al., 1991; Sharma and Kaur, 1990; Adams and Manchester, 1982; Nater et al., 1979
		1	petrolatum	Dannaker et al., 1993
Maneb technical		0.5	vaseline	Jung et al., 1989; Jung et al., 1987
		1	aqueous	Jung et al., 1987
Zineb	12122-67-7	0.002%	not specified	low concentration selected as 1% of end use concentration - testing positive in 2 patients; negative in 3 controls; Kleibl and Rackova, 1980
		1	petrolatum	Adams, 1990a; Dannaker et al., 1993; Jung et al., 1987; Lisi, 1992a; Piraccini et al., 1991
Metiram — ammonia complex of zinc EBDC	9006-42-2	1	petrolatum	Koch, 1996

Material Insecticides	CAS #	Concentration %	Vehicle	Ref. and commentary
Thiram (TMTD)	137-26-8	1	petrolatum	Lisi, 1992a; Rudzki and Napiorkowska, 1980; Sharma and Kaur, 1990; Adams, 2
		2	Vaseline	Jung et al., 1987
		1	petrolatum	Cronin, 1980
Disulfiram (TETD)	European standard		petrolatum	
Tetmosol (sulfiram)	95-05-6	1	petrolatum	Cronin
Propineb	12071-83-9 9016-72-2	0.01-1	aqueous	Jung et al., 1989
Ziram	137-30-4	1	petrolatum	Dannaker et al., 1993; Guo et al., 1996; Jung et al., 1987; Lisi, 1992a
		0.1	petrolatum	Piraccini et al., 1991
Nitro and nitrile chlorinated benzenes				
Chlorothalonil	1897-45-6	0.001	petrolatum or acetone or butanol	Flannigan et al., 1986; O'Malley et al., 1995
		0.005-0.0001	aqueous	Liden, 1990; 10 controls negative at 0.0001%
		0.02	aqueous	Ueda et al., 1994
		0.01	aqueous	Immediate anaphylactoid reaction observed on open test, Dannaker et al., 1993
		0.01	Pet	Lisi, 1992a
		0.01	acetone	No reaction in 10 individuals; Bach and Pedersen, 1980; no reactions in unspecified number of controls; Johnsson et al., 1983
		0.002	aqueous	Patch and photopatch - stronger reaction on; Matsushita et al., 1996
		0.01	petrolatum	Borderline toxic reactions at 48-72 h in 4/17 control subjects; Bruynzeel and van Ketel, 1986
Tetrafluoro-terephthalonitrile	1835-49-0	0.001-0.1	petrolatum	Carmichael et al., 1989
PCNB (pentachloronitro-benzene)	82-68-8	0.5		Cronin
		10	aqueous	Jung et al., 1987
		1	petrolatum	Sharma and Kaur, 1990
		1	butanol	O'Malley et al., 1995
Dichloronitroaniline		1	butanol	O'Malley et al., 1995
Dinobuton	973-21-7	1	petrolatum	Adams, 2 Cronin
		30-40	not specified	Wahlberg, 1974
Dinocap	39300-45-3	0.5	petrolatum	Sharma and Kaur, 1990
		1	petrolatum	Lisi, 1992a
DNCB		0.01		Cronin
		0.25	aqueuos	Malten, 1974
Rodannitrobenzene		1	petrolatum	Fregert, 1967
Dichlornitroaniline		1	Butanol	O'Malley et al., 1995
Trichlorodinitrobenzene		0.01	aqueous	Jung et al., 1987
Miscellaneous fungicides				
Antisapstain		1		Adams, 2
Benzimidazole carbamate		1	petrolatum	Sharma and Kaur, 1990
Carbendazim		0.5	petrolatum	Bruynzeel et al., 1995
CuSO^4	7758-98-7 77	1	petrolatum	Lisi, 1986; Lisi 1992a; b; Lisi et al., 1986; 1987
Copper fungicide (uns)		1	petrolatum	Sharma and Kaur, 1990
Dichlofluanid	1085-98-9	10%	petrolatum	Hansson and Wallengren, 1995; testing on controls not described
Dichlorophen		0.5		Adams, 2

Material Insecticides	CAS #	Concentration %	Vehicle	Ref. and commentary
Dithianon	3347-22-6	1	petrolatum	Calnan, 1969; Jung et al., 1987; Koch, 1996; Lisi 1992a; van Joost et al., 1983
Dodemorph	1593-77-7	0.25	petrolatum	van Ketel, 1975
Fluazinam	79622-59-6	0.5	petrolatum	Bruynzeel et al., 1995
Iprodione	36734-19-7	1		default
Ketoconazole	65277-42-1	1	ethanol	Valsecchi et al., 1993
Methyl-2-benzimidazole		1	petrolatum	Sharma and Kaur, 1990
Pyrifenox (as part of Rondo-M)	88283-41-4	2	aqueous	Iliev and Elsner, 1997
Triforine (Saprol)		0.02	aqueous	58 male subjects and 47 female in cross sectional study; Ueda et al., 1994. No control subjects identified — irritant threshold 0.5% comparable to dichlorvos 1% in guinea pig studies; strong allergic reaction on guinea pig test
Streptocycline		3	petrolatum	Sharma and Kaur, 1990
Streptomycin		1		Adams,2 citing van Ketel, 1974 - citation not located in bibliography
Sulfur	7704-34-9	1	butanol	O'Malley et al., 1995
Inorganic sulfur	63705-05-5	5	petrolatum	Sharma and Kaur, 1990
		5-20	alcohol, Vaseline	Schneider, 1978
		1,5	petrolatum	Wilkinson, 1975
Triadimefon	43121-43-3	1	petrolatum	UC Davis Unpublished study
Vinclozolin	50471-44-8	1	butanol	O'Malley et al., 1995
Dichlone	117-80-6	NS	not specified	Adams, 2
Animal feed additives				
Olaquindox	23696-28-8	0.01-5.0	petrolatum	Schauder et al., 1996
Ethoxyquin	91-53-2	0.5	petrolatum	Savini et al., 1989
Fumigants				
Chloropicrin		0.25	petrolatum	Adams, 2
Dazomet		0.1	petrolatum	Jung et al., 1987; Vilaplana et al., 1993, Black, 1973
		0.1-0.25	petrolatum	Lisi et al., 1986
		0.025-0.0025	Vaseline	Warin, 1992; negative tests in 20 controls
Metam sodium		0.03	aqueous/ Vaseline	Jung et al., 1987
		0.1	aqueous	Schubert, 1978
		0.5-1	petrolatum	Lisi et al., 1986
MITC		0.1	aqueous	Schubert, 1978
1,2 dichloropropane		10%	petrolatum	Baruffini et al., 1989; material used as a solvent in painting, metalworking
		1%	petrolatum	2 patients, no controls tested; Kleibl and Rackova, 1980
D-D mixture (dichloropropene and dichloropropane)	26952-23-8	0.05	petrolatum	Bousema et al., 1991; 20 negative controls; patient was positive at concentrations ranging from 0.005-2%
		0.02-2.0	petrolatum	van Joost and de Jong, 1988
Herbicides				
Phenoxy compounds				
2,4-D	94-75-7	1	petrolatum	Sharma and Kaur, 1990
		1-2	aqueous	Jung et al., 1987 Adams, 2
2,4-D/2,4,5-T		0.4	diesel oil	Jung, 1975 #1237
2,4-D/2,4,5-T		0.1-2	olive oil	
p-chloro-*o*-cresol		1	aqueous	Fregert, 1968
MCPA		5-10	aqueous	Jung et al., 1987
		1	aqueous	Jung et al., 1989

Material Insecticides	CAS #	Concen- tration %	Vehicle	Ref. and commentary
2,4,5-T		—	—	Tested as part of a mixture with 2,4-D; Jung et al., 1987
Dichlorprop		1	aqueous	Jung et al., 1989
Dicamba		0.5	aqueous	Jung et al., 1989
Acetanilides				
Alachlor (Lasso)	15972-60-8	0.1	petrolatum	Vilaplana et al., 1993; Adams, 2 citing Iden and Schroeter, 1977
		1	petrolatum	Lisi et al., 1986
Allidochlor (Randox)	93-71-0	0.01-0.1	petrolatum	Cronin, 1980
Butachlor		1	aqueous	Sharma and Kaur, 1990
Propachlor (Ramrod)		1	?	Schubert, 1979
		0.05	aqueous	Jung et al., 1987 citing Korossy 1978 meeting presentation
Propachlor		0.01-0.5	aqueous	Jung et al., 1987; 1989
Triazines				
Atrazine	1912-24-9	0.1		Jung et al., 1987; Vilaplana et al., 1993
		1	petrolatum	Sharma and Kaur, 1990
Amitrole		1		English et al., 1986
		NS	not specifid	Jung et al., 1987
Simazine		5-10	aqueous	Jung et al., 1987
Ametryn		0.1	aqueous	Jung et al., 1987
Carbamates				
Carbyne (Barban)	101-27-9	10	mineral oil	Brancaccio and Chamales 1977
Molinate	2212-67-1	1	petrolatum	Lisi et al., 1986
Thiobencarb		1	aqueous	Sharma and Kaur, 1990; 2 irritant reactions
Chloropropham		0.5	Vaseline	Jung et al., 1987
Bipyridyls				
Diquat		0.1	petrolatum	Lisi et al., 1986; Vilaplana et al., 1993
Paraquat		0.1	petrolatum	Lisi et al., 1986; Sharma and Kaur 1990; Vilaplana et al., 1993
Nitrobenzene				
4,6-DiNitroOrthoCresol	534-52-1	0.5-1	aqueous	Lisi et al., 1986
Dinitro-*o*-cresol		0.5		Lisi, 1986; Lisi et al., 1986
		50	aqueous	Jung et al., 1987
Dinoterbe				Sabouraud et al., 1997
Nitralen	4726-14-1	1	?	Nishioka et al., 1983
Nitrofen		0.5	olive oil	Jung et al., 1987
Chloranil	118-75-2	1		Adams, 2
Chloridazon	1698-60-8	0.1		Lisi et al., 1986
Miscellaneous				
Anilofos		1	aqueous	Sharma and Kaur, 1990
Cacodylic acid		0.1-1%	aqueous	Bourrain et al., 1998
Cymoxanil	57966-95-7			
Dichlofopmethyl		1	aqueous	Sharma and Kaur, 1990
Fluchloralin		1	aqueous	Sharma and Kaur, 1990
Glyphosate	1071-83-6	10	aqueous	Maibach, 1986
Glyphosate end use formulation with unspecified concentration of active ingredient		1-5	aqueous	Recorded reaction to preservative ingredient, Hindson and Diffey 1984a; 1984b
Hexythiazox		1	petrolatum	Willems et al., 1997
Oxadiazon		1	aqueous	Sharma and Kaur, 1990
Pendimethalin		1	aqueous	Sharma and Kaur, 1990
Phenmedipham	13684-63-4	2		Adams, 2
Propanil	709-98-8	1	petrolatum	Lisi et al., 1986
		0.1	petrolatum	Vilaplana et al., 1993
Trichlorobenzyl chloride	1344-32-7	1		Adams,1
Choline chloride (chlormequat chloride) (PGR)	999-81-5	1		Adams,2
Dodecyl gallate		0.25	petrolatum	Guo et al., 1996
Chloropropham	101-21-3	0.5	Vaseline	Jung et al., 1987

Material Insecticides	CAS #	Concentration %	Vehicle	Ref. and commentary
Dalapon	75-99-0	50	aqueous, Vaseline	Jung et al., 1987
Ethephon	16672-87-0	2	aqueous	Jung et al., 1987
Methabenzthiazinon		1	petrolatum	Sharma and Kaur, 1990
Metoxuron		1	petrolatum	Sharma and Kaur, 1990
Isoproturon		1	petrolatum	Sharma and Kaur, 1990; 2 irritant reactions
Nitrofurazone		1	petrolatum	Guo et al., 1996
Fenazox	495-48-7	unknown	not specified	Jung et al., 1987

REFERENCES

1. Abrams, K., Hogan, D. J. and Maibach, H. I.: Pesticide-related dermatoses in agricultural workers. *Occup. Med.,* 1991, **6,** 463-92.
2. Adams, R. M., Photoallergic contact dermatitis to chloro-2-phenylphenol, *Arch. Dermatol.,* 1972, **106,** 711-4.
3. Adams, R. M., *Occupational Skin Disease*, W.B. Saunders, Philadelphia, 1990a, pp. xvii, 706 , 2 of plates.
4. Adams, R. M., Pesticides and other agricultural chemicals, in *Occupational Skin Disease,* W.B. Saunders, Philadelphia, 1990b, pp. 361-378.
5. Adams, R. M., Zimmerman, M. C., Bartlett, J. B., and Preston, J. F., 1-chloro-2,4-dinitrobenzene as an algicide. Report of four cases of contact dermatitis. *Arch. Dermatol.,* 1971, **103,** 191-3.
6. Andersen, K. E. and Petri, M., Occupational irritant contact folliculitis associated with triphenyl tin fluoride (TPTF) exposure, *Contact Dermatitis* 1982, **8,** 173-7.
7. Andersen, K. E. and Rycroft, R. J., Recommended patch test concentrations for preservatives, biocides and antimicrobials, *Contact Dermatitis* 1991, **25,** 1-18.
8. Bach, B. and Pedersen, N., Contact dermatitis from a wood preservative containing tetrachloroisophthalonitrile, *Contact Dermatitis* 1980, **6,** 142.
9. Baruffini, A., Cirla, A., Pisati, G., Ratti, R., and Zedda, S., Allergic contact dermatitis from 1,2-dichloropropane, *Contact Dermatitis* 1989, **20,** 379.
10. Behrbohm, Uber allergische Krahnkheiten durch DDT und HCH, *Allerg. Asthma,* 1962, **8,** 237-48.
11. Bonnetblanc, J. M. and Delrous, J. L., Connubial dermatitis from phenylmercuric nitrate, *Contact Dermatitis,* 1996, **34,** 367.
12. Bourrain, J. L., Morin, C., Beani, J. C., and Amblard, P., Airborne contact dermatitis from cacodylic acid, *Contact Dermatitis,* 1998, **38,** 364-5.
13. Bousema, M. T., Wiemer, G. R., and van Joost, T., A classic case of sensitization to DD-95, *Contact Dermatitis,* 1991, **24,** 132-3.
14. Brancaccio, R. R. and Chamales, M. H., Contact dermatitis and depigmentation produced by the herbicide Carbyne, *Contact Dermatitis,* 1977, **3,** 108-9.
15. Brown, R., Contact sensitivity to Difolatan (Captafol), *Contact Dermatitis,* 1984, 10, 181-182.
16. Bruynzeel, D. P., Contact sensitivity to lannate, *Contact Dermatitis,* 1991, 25**,** 60-61.
17. Bruynzeel, D. P., Tafelkruijer, J., and Wilks, M. F., Contact dermatitis due to a new fungicide used in the tulip bulb industry [published erratum appears in *Contact Dermatitis* 1996 Jul; 35(1):70], *Contact Dermatitis,* 1995, 33**,** 8-11.

18. Bruynzeel, D. P. and van Ketel, W. G., Contact dermatitis due to chlorothalonil in floriculture, *Contact Dermatitis,* 1986, **14,** 67-8.
19. Burns, D. A., Allergic contact sensitivity to chlorphenesin, *Contact Dermatitis,* 1986, 14, 246.
20. Calnan, C., Dithianone sensitivity, *Contact Dermatitis Newsletter* , 1969, **6,** 119.
21. Camarasa, G., Difolatan dermatitis, *Contact Dermatitis,* 1975, **1,** 127.
22. Carmichael, A. J., Foulds, I. S., and Sadhra, S., Allergic contact dermatitis from tetrafluoroterephthalonitrile and TFX diamine, *Contact Dermatitis,* 1989, **20,** 233-234.
23. Cronin, E., Pesticides, in *Contact Dermatitis,* Churchill Livingstone, Edinburgh, 1980, 393-414.
24. Dannaker, C. J., Maibach, H. I., and O'Malley, M., Contact urticaria and anaphylaxis to the fungicide chlorothalonil, *Cutis,* 1993, **52,** 312-5.
25. Edmundson, W. F. and Davies, J. E., Occupational dermatitis from naled. A clinical report, *Arch. Environ. Health.,* 1967, **15,** 89-91.
26. English, J. S., Rycroft, R. J., and Calnan, C. D., Allergic contact dermatitis from aminotriazole, *Contact Dermatitis,* 1986, **14,** 255-6.
27. Fisher, A., Appendix [Recommended patch test concentrations], Lea and Febiger, Philadelphia, 1986, 851-917.
28. Flannigan, S. A., Tucker, S. B., and Calderon, V. Jr., Irritant dermatitis from tetrachloroisophthalonitrile, *Contact Dermatitis,* 1986, **14,** 258-9.
29. Fregert, S., Allergic contact dermatitis from the pesticide rodannitrobenzene, *Contact Dermatitis Newsletter,* 1967, **2,** 4.
30. Fregert, S., Allergic contact dermatitis from *p*-chloro-*o*-cresol in a pesticide, *Contact Dermatitis Newsletter,* 1968, **3,** 46.
31. Gammeltoft, M., Tributyltinoxide is not allergenic, *Contact Dermatitis,* 1978, **4,** 238-9.
32. Goh, C. L., Dermatitis from chlorphenesin in a deodorant, *Contact Dermatitis,* 1987, **16,** 287.
33. Groundwater, J. R., Difolatan dermatitis in a welder; non-agricultural exposure, *Contact Dermatitis,* 1977, **3,** 104.
34. Guo, Y. L., Wang, B. J., Lee, C. C., and Wang, J. D., Prevalence of dermatoses and skin sensitisation associated with use of pesticides in fruit farmers of southern Taiwan, *Occup. Environ. Med.,* 1996, **53,** 427-31.
35. Haenen, C., de Moor, A., and Dooms-Goossens, A., Contact dermatitis caused by the insecticides omethoate and dimethoate, *Contact Dermatitis,* 1996, **35,** 54-5.
36. Hansson, C. and Wallengren, J., Allergic contact dermatitis from dichlofluanid, *Contact Dermatitis.* 1995, **32,** 116-7.
37. Hindson, C. and Diffey, B., Phototoxicity of glyphosate in a weedkiller, *Contact Dermatitis,* 1984a, **10,** 51-2.
38. Hindson, T. C. and Diffey, B. L., Phototoxicity of a weedkiller: A correction, *Contact Dermatitis,* 1984b, **11,** 260.
39. Hostynek, J. J., Patrick, E., Younger, B., and Maibach, H. I., Hypochlorite sensitivity in man, *Contact Dermatitis,* 1989, **20,** 32-7.
40. Hostynek, J. J., Wilhelm, K. P., Cua, A. B., and Maibach, H. I., Irritation factors of sodium hypochlorite solutions in human skin, *Contact Dermatitis,* 1990, **23,** 316-24.
41. Iden, D. L. and Schroeter, A. L., Allergic contact dermatitis to herbicides [letter], *Arch. Dermatol.,* 1977, **113,** 983.
42. Iliev, D. and Elsner, P., Allergic contact dermatitis from the fungicide Rondo-M and the insecticide Alfacron, *Contact Dermatitis,* 1997, **36,** 51.
43. Johnsson, M., Buhagen, M., Leira, H. L., and Solvang, S., Fungicide-induced contact dermatitis, *Contact Dermatitis,* 1983, **9,** 285-8.

44. Jung, H. D., Honemann, W., Kloth, C., Lubbe, D., Pambor, M., Quednow, C., Ratz, K. H., Rothe, A., and Tarnick, M., Contact eczema caused by pesticides in East Germany, *Dermatol. Monatsschr.,* 1989, **175,** 203-14.
45. Jung, H. D., Rothe, A., and Heise, H., Epicutaneous testing with plant protective and pest control agents (pesticides), *Derm. Beruf. Umwelt.,* 1987, **35,** 43-51.
46. Keczkes, K. and Brown, P., Hexahydro, 1,3,5,tris (2-hydroxyethyl) triazine, a new bacteriocidal agent as a cause of allergic contact dermatitis, *Contact Dermatitis,* 1976, **2,** 92-98.
47. Kleibl, K. and Rackova, M., Cutaneous allergic reactions to dithiocarbamates, *Contact Dermatitis,* 1980, **6,** 348-9.
48. Koch, P., Occupational allergic contact dermatitis and airborne contact dermatitis from 5 fungicides in a vineyard worker. Cross-reactions between fungicides of the dithiocarbamate group?, *Contact Dermatitis,* 1996, **34,** 324-9.
49. Kuhne, G., Heise, H., Plottke, B., and Puskeiler, T., Dermatitis after Benlate contact, *Z. Gesamte Hyg.,* 1985, **31,** 710- 1.
50. Lee, S., Chin, Y., Chang, W. and Kim, J., A study on hypersensitivity of Korean farmers to various agrochemicals. Determination of concentration for patch test of fruit-tree agrochemicals and hypersensitivity of orange orchard farmers in Che-ju Do, Korea. *Seoul Med. J.,* 1981, **22,** 137-42.
51. Liden, C., Facial dermatitis caused by chlorothalonil in a paint, *Contact Dermatitis,* 1990, **22,** 206-211.
52. Lisi, P., Pesticide contact dermatitis, *Ann. Ital. Dermatol. Clin. Sper.,* 1986, **40,** 67-81.
53. Lisi, P., Pesticides in occupational contact dermatitis, *Clin. Dermatol.,* 1992a, **10,** 175-184.
54. Lisi, P., Sensitization risk of pyrethroid insecticides, *Contact Dermatitis,* 1992b, **26,** 349-350.
55. Lisi, P., Caraffini, S., and Assalve, D., A test series for pesticide dermatitis, *Contact Dermatitis,* 1986, **15,** 266-9.
56. Lisi, P., Caraffini, S. and Assalve, D., Irritation and sensitization potential of pesticides, *Contact Dermatitis,* 1987, **17,** 212-8.
57. Maibach, H. I., Irritation, sensitization, photoirritation and photosensitization assays with a glyphosate herbicide, *Contact Dermatitis,* 1986, **15,** 152-6.
58. Maibach, H. I. and Johnson, H. L., Contact urticaria syndrome. Contact urticaria to diethyltoluamide (immediate-type hypersensitivity), *Arch. Dermatol.,* 1975, **111,** 726-30.
59. Malten, K., DNCB in cooling water, *Contact Dermatitis Newsletter,* 1974, **15.**
60. Mathias, C. G., Persistent contact dermatitis from the insecticide dichlorvos, *Contact Dermatitis,* 1983, **9,** 217-8.
61. Matsushita, S., Kanekura, T., Saruwatari, K., and Kanzaki, T., Photoallergic contact dermatitis due to Daconil, *Contact Dermatitis,* 1996, **35,** 115-6.
62. Milby, T. and Epstein, W., Allergic contact sensitivity to malathion, *Arch. Environ. Health,* 1964, **9,** 434- 437.
63. Nater, J. P., Terpstra, H., and Bleumink, E., Allergic contact sensitization to the fungicide Maneb, *Contact Dermatitis,* 1979, **5,** 24-6.
64. Nishioka, K., Asagami, C., Kurata, M., and Fujita, H., Sensitivity to the weed killer DNA-nitralin and cross-sensitivity to dinitrochlorobenzene, *Arch. Dermatol.,* 1983, **119,** 304-6.
65. Nishioka, K., Kozuka, T., and Tashiro, M., Agricultural miticide (BPPS) dermatitis, *Skin Res.,* 1970, **12,** 15-18.
66. Norris, P. G., Hawk, J. L., and White, I. R., Photoallergic contact dermatitis from fentichlor, *Contact Dermatitis,* 1988, **18,** 318-320.

67. O'Malley, M., Rodriguez, P., and Maibach, H. I., Pesticide patch testing: California nursery workers and controls, *Contact Dermatitis,* 1995, **32,** 61-63.
68. Pambor, M. and Timmel, A., Mercury dermatitis, *Contact Dermatitis,* 1989, **20,** 157.
69. Peluso, A. M., Tardio, M., Adamo, F., and Venturo, N., Multiple sensitization due to bis-dithiocarbamate and thiophthalimide pesticides, *Contact Dermatitis,* 1991, **25,** 327.
70. Piraccini, B. M., Cameli, N., Peluso, A. M., and Tardio, M., A case of allergic contact dermatitis due to the pesticide maneb, *Contact Dermatitis,* 1991, **24,** 381-2.
71. Rudner, E., North American Group Results, *Contact Dermatitis,* 1977, **3,** 208-209.
72. Rudzki, E. and Napiorkowska, T., Dermatitis caused by the Polish fungicide Sadoplon 75, *Contact Dermatitis,* 1980, **6,** 300-1.
73. Sabouraud, S., Testud, F., Rogerie, M. J., Descotes, J., and Evreux, J. C., Occupational depigmentation from dinoterbe, *Contact Dermatitis,* 1997, **36,** 227.
74. Savini, C., Morelli, R., Piancastelli, E., and Restani, S., Contact dermatitis due to ethoxyquin, *Contact Dermatitis,* 1989, **21,** 342-3.
75. Savitt, L., Contact dermatitis due to benomyl insecticide, *Arch. Dermatol.,* 1972, **105.**
76. Schauder, S., Schroder, W., and Geier, J., Olaquindox-induced airborne photoallergic contact dermatitis followed by transient or persistent light reactions in 15 pig breeders, *Contact Dermatitis,* 1996, **35,** 344-54.
77. Schena, D. and Barba, A., Erythema multiforme-like contact dermatitis from dimethoate, *Contact Dermatitis,* 1992, **27,** 116-7.
78. Schneider, H., Schwefelallergie [Sulfur allergy], *Hautartz,* 1978, **29,** 340-342.
79. Schubert, H., Contact dermatitis to sodium N-methyl dithiocarbamate, *Contact Dermatitis,* 1978, **6,** 370-1.
80. Schubert, H., Allergic contact dermatitis due to propachlor (author's transl.), *Dermatol. Monatsschr.,* 1979, **165,** 495-8.
81. Sharma, V. K. and Kaur, S., Contact sensitization by pesticides in farmers, *Contact Dermatitis,* 1990, **23,** 77-80.
82. Stoermer, D., Occupational contact dermatitis caused by dichlorvos pesticide, *Dermatol. Monatsschr.,* 1985, **171,** 245-9.
83. Stransky, L., Vasileva, S., and Mateev, G., Contact bullous pemphigoid?, *Contact Dermatitis,* 1996, **35,** 182.
84. Tsankov, N., Kazandjieva, J., and Gantcheva, M.: Contact pemphigus induced by dihydrodiphenyltrichlorethane, *Eur. J. Dermatol.,* 1998, **8,** 442-3.
85. Ueda, A., Aoyama, K., Manda, F., Ueda, T., and Kawahara, Y., Delayed-type allergenicity of triforine (Saprol), *Contact Dermatitis,* 1994, **31,** 140-5.
86. Valsecchi, R., Pansera, B., di Landrom, A., and Cainelli, T., Contact dermatitis from ketoconazole, *Contact Dermatitis,* 1993, **29,** 162.
87. van Joost, T. and de Jong, G., Sensitization to DD soil fumigant during manufacture, *Contact Dermatitis,* 1988, **18,** 307-8.
88. van Joost, T., Naafs, B. and van Ketel, W. G., Sensitization to benomyl and related pesticides, *Contact Dermatitis,* 1983, **9,** 153-4.
89. van Joost, T. and Wiemer, G. R., Contact dermatitis from tetrachloroacetophenone (TCAP) in an insecticide plant, *Contact Dermatitis,* 1991, **25,** 66-7.
90. van Ketel, W. G., Allergic dermatitis from a new pesticide, *Contact Dermatitis,* 1975, **1,** 297-300.
91. van Ketel, W. G., Sensitivity to the pesticide benomyl, *Contact Dermatitis,* 1976, **2,** 290-1.
92. van Ketel, W. G., Active sensitization by *O,O,* diethyl-phtalimido-phosphothioate (Plondrel), *Contact Dermatitis,* 1977, **3,** 51.

93. Vilaplana, J., Azon, A., Romaguera, C., and Lecha, M., Phototoxic contact dermatitis with toxic hepatitis due to the percutaneous absorption of paraquat, *Contact Dermatitis,* 1993, **29,** 163-4.
94. Wahlberg, J., Yellow staining of hair and nails and contact sensitivity to dinobuton, *Contact Dermatitis Newsletter,* 1974, **16,** 481.
95. Wakelin, S. H., Price, A. E., Basketter, D. A., and Rycroft, R. J., Allergic contact dermatitis from ethoxymethylenemalononitrile in an agrochemical chemist, *Contact Dermatitis,* 1998, **38,** 237.
96. Wakelin, S. H. and White, I. R., Dermatitis from chlorphenesin in a facial cosmetic, *Contact Dermatitis,* 1997, **37,** 138-9.
97. Wantke, F., Focke, M., Hemmer, W., Gotz, M. and Jarisch, R., Generalized urticaria induced by a diethyltoluamide- containing insect repellent in a child, *Contact Dermatitis,* 1996, **35,** 186-7.
98. Warin, A. P., Allergic contact dermatitis from dazomet, *Contact Dermatitis,* 1992, **26,** 135-136.
99. Wilkinson, D., Photodermatitis due to tetrachlorsalicylanilide, *Br. J. Derm.,* 1961, **73,** 213-219.
100. Wilkinson, D., Sulphur sensitivity, *Contact Dermatitis,* 1975, **1,** 58.
101. Willems, P. W., Geursen-Reitsma, A. M. and van Joost, T., Allergic contact dermatitis due to methiocarb (Mesurol), *Contact Dermatitis,* 1997, **36,** 270.
102. Zemtsov: A case of allergy to Kathon CG in the United States, *Contact Dermatitis,* 1991, **25,** 135.

Appendix III

Pesticides Included in the California Handler Database

Class/Compound	Number of applicator cases 1982–1995
Organophosphates	
Chlorpyrifos	10
Malathion	5
Diazinon	4
Acephate	3
Naled	2
DDVP	1
Dimethoate	1
Ethion	1
Methamidophos	1
Oxydemeton-methyl	1
Parathion	1
Phorate	1
Phosmet	1
Total	**32**
Carbamates	
Propoxur	3
Bendiocarb	2
Carbaryl	2
Methomyl	2
Carbamate Total	9
Pyrethrin/pyrethroids	
Pyrethrins/piperonyl butoxide	10
Permethrin	3
Bifenthrin	2
Allethrin	1
Cypermethrin	1
Fenvalerate	1
Flucythrinate	1
Total	**19**
Organochlorines	
Dicofol	3
Dienochlor	3
Lindane	2
Total	**8**
Miscellaneous insecticides	
Propargite	55
Cyhexatin	7
Sodium arsenite	2
Abamectin	2
Boric acid	1
Cryolite	1
D-limonene	1
Oxythioquinox	1
Total	**70**
Carbamate/phthalimido fungicides	
Benomyl	5
Captafol	1
Captan	6

Total	**12**
Thiocarbamates	
Ziram	2
Mancozeb	1
Maneb	1
Thiram	1
Total	**5**
Copper compounds	
Copper sulfate	4
Copper hydroxide	3
Copper	2
Copper naphthenate	2
Total	**11**
Miscellaneous fungicides	
Sulfur	28
Dinoseb	9
Chlorothalonil	8
Pentachlorophenol	5
Amitrole	3
DCNA	2
Metalaxyl	2
Sulfur dioxide	2
Anilazine	2
Streptomycin	2
Tributyltin oxide	1
Dodemorph acetate	1
Ethylene oxide	1
Fenarimol	1
Lime sulfur	1
Piperalin	1
Triadimefon	1
Tridimefon	1
Iprodione	1
Myclobutanil	1
Total	**73**
Fumigants	
Methyl Bromide	59
Metam sodium	34
1,3-dichloropropene	11
D-d mixture	5
Ethylene dibromide	4
Sulfuryl fluoride	4
Dazomet	1
Aluminum phosphide	1
Bipyridyl herbicides	
Diquat	15
Paraquat	17
Total	**32**
Herbicides	
Trifluralin	7
Benfluralin	1
Ethalfluralin	1
Glyphosate	68
Fluazifop-butyl	5
Oryzalin	5
Weed oil	3
Pendimethalin	2
Oxyfluorfen	2
Prometon	2
Triclopyr	2
Bensulide	1
EPTC	1
Bromoxynil	1
MCPA	1
Molinate	1
Alachlor	1
Diethatyl ethyl	1

DSMA/MSMA	1
Endothall	1
Norflurazon	1
Oxadiazon	1
Picloram	1
Sethoxydim	1
Simazine	1
Total	**112**
Antimicrobials	
Sodium hypochlorite	131
Quaternary ammonia	107
Creosote	37
Phenolic disinfectants	29
Gluteraldehyde	19
Cyanuric acid	16
Isothiazolin compounds	13
Calcium hypochlorite	8
Trichloromelamine	4
Hydrogen chloride	4
Octhilinone	2
1br-3cl-5,5(ch3)2-hydanto	1
chlorine	1
Ethyl alcohol	1
Halogenated hydantoins	1
Iodine-complex	1
Isopropyl alcohol	1
Sodium chlorite	1
Sodium hydroxide	1
Acrolein	1
Formaldehyde	1
Phenylmercury acetate	1
Total	**381**
Miscellaneous compounds and adjuvants	
Gibberellic acid	4
Polybutenes	1
Adjuvant	2
Petroleum hydrocarbons	1
Petroleum distillates	3
Petroleum oil	3
Pine oil	9
Rotenone	1
Ethephon	1
Total	**25**

Appendix IV

Cross-Sectional Data on Skin Diseases in the U.S. Population from the NHANES Study

TABLE 1
Prevalence of Significant Skin Pathology of all Types and the Most Frequently Occurring Types of Conditions among Persons 1–74 Years, by Sex, Showing Rates, Number of Persons Affected, and Selected Standard Errors: United States, 1971–1974[1]

Condition	Both sexes	Male	Female	Both sexes	Male	Female
		(Rate per 1,000 population)			(Number in thousands)	
Persons with one or more significant skin conditions	312.4	339.8	286.6	60,601	31,988	28,613
Significant skin conditions, all types	439.7	499.4	383.4	85,287	47,018	38,269
Diseases of sebaceous glands	84.8	87.9	81.9	16,465	8,279	8,186
Acne vulgaris	68.1	70.5	65.9	13,217	6,638	6,579
Cystic acne	1.9	3.3	0.6	375	312	63
Acne scars	1.7	2.0	1.3	321	192	129
Xerosis	6.5	5.3	7.7	1,260	497	763
Dermatophytoses	81.1	131.4	33.7	15,733	12,372	3,361
Tinea pedis	38.7	68.4	10.7	7,509	6,438	1,071
Tinea unguium	21.8	30.3	13.9	4,232	2,856	1,376
Tinea versicolor	8.4	10.9	6.0	1,623	1,022	601
Tinea cruris	6.7	12.9	0.9	1,301	1,214	87
Tumors[a]	56.5	59.6	53.7	10,968	5,606	5,362
Malignant	5.9	6.4	5.3	1,136	602	534
Basal-cell epithelioma	1	4.1	4.7	3.5	796	44,5351
Benign	38.2	35.8	40.5	7,412	3,372	4,040

TABLE 1 (continued)
Prevalence of Significant Skin Pathology of all Types and the Most Frequently Occurring Types of Conditions among Persons 1–74 Years, by Sex, Showing Rates, Number of Persons Affected, and Selected Standard Errors: United States, 1971–1974[1]

Condition	Both Sexes	Male	Female	Both Sexes	Male	Female
		(Rate per 1,000 population)			(Number in thousands)	
Lipoma	2.1	2.6	1.6	401	244	157
Precancerous and not specified	12.4	17.4	7.9	2,420	1,632	788
Actinic keratosis	9.6	13.9	5.5	1,858	1,306	552
Seborrheic dermatitis	28.5	26.7	30.1	5,520	2,514	3,006
Seborrheic dermatitis, type not indicated	28.2	26.4	29.9	5,476	2,490	2,986
Atopic dermatitis, eczema	18.4	19.5	17.4	3,575	1,837	1,738
Atopic dermatitis	6.9	8.2	5.6	1,332	771	561
Lichen simplex chronicus	4.5	4.7	4A	882	441	441
Hand eczema	1.6	1.1	2.1	311	105	206
Nummular eczema	13	1.0	2.4	332	92	240
Dyshidrotic eczema	2.1	3.1	1.2	405	288	117
Contact dermatitis	13.6	13.8	13.4	2,641	1,303	1,338
Ichthyosis, keratosis	9.5	9.3	9.6	1,835	880	955
Verruca vulgaris	8.5	10.3	7.2	1,684	967	717
Folliculitis	8.0	12.3	4.0	1,553	1,154	399
Psoriasis	5.5	5.9	5.1	1,070	556	514

TABLE 1 (continued)
Prevalence of Significant Skin Pathology of all Types and the Most Frequently Occurring Types of Conditions among Persons 1–74 Years, by Sex, Showing Rates, Number of Persons Affected, and Selected Standard Errors: United States, 1971–1974[1]

Condition	Both Sexes	Male	Female	Both Sexes	Male	Female
	(Rate per 1,000 population)			(Number in thousands)		
Seborrheic keratosis	5.2	4.6	5.8	1,010	431	579
Vitiligo	4.9	3.6	6.2	957	341	616
Urticaria (hives, etc.)	4.8	3"11	5,8	933	354	578
Herpes simplex	4.2	4.0	4.5	824	376	448
All other skin conditions	106.2	106.7	105.0	20,519	10,048	10,472
	Standard error of rate					
Persons with one or more significant skin conditions	20.05	21.06	20.13			

[a]Excludes seborrheic keratosis which is listed with benign tumors.

TABLE 2

Prevalence of Significant Skin Pathology of all Types and the Most Frequently Occurring Types of Conditions among Persons 1–74 Years, by Age, Showing Rates and Standard Errors: U.S., 1971–1974

Condition	1–5	6–11	12–17	18–24	25–34	35–44	45–54	55–64	65–74
				(Rate per 1,000 Population)					
Persons with one or more significant skin conditions	142.3	174.2	362.0	365.1	318.0	328.2	356.8	361.0	409.5
Significant skin conditions, all types	166.4	195.4	468.5	499.9	452.9	475.2	529.7	563.9	654.8
Diseases of sebaceous glands	6.3	21.4	249.9	191.1	99.6	44.1	21.5	32.0	25.5
Acne vulgaris	0.4	11.9	231.9	172.2	83.6	25.4	8.7	2.5	0.7
Cystic acne	—	—	2.0	5.4	5.2	2.6	—	—	0.1
Acne scars	—	—	4.5	5.2	0.9	1.7	02	—	—
Xerosis	2.8	6.7	3.9	3.7	4.0	6.8	5.4	15.3	14.9
Dermatophytoses	3.4	8.4	33.2	61.7	87.7	121.5	155.7	150.8	126.8
Tinea pedis	0.4	4.0	14.1	24.9	49.1	62.0	80.9	57.9	59.6
Tinea unguium	—	0.9	1.9	7.0	13.4	24.9	46.2	68.5	54.7
Tinea versicolor	1.1	1.5	10.2	17.8	9.9	13.0	7.2	7.3	2.3
Tinea cruris	—	—	4.7	6.7	10.2	11.7	13.1	8.9	1.3
Tumors	21.7	18.9	31.3	43.0	39.6	46.8	84.9	100.2	184.1
Malignant	—	0.5	0.1	0.1	3.2	9.5	18.9	36.4	
Basal-cell epithelioma	—	—	—	0.1	—	3.2	9.1	13.3	19.9
Benign[a]	21.7	18.9	30.7	40.7	38.9	38.2	52.9	44.9	68.1
Lipoma	—	0.8	—	0.9	0.6	1.6	5.3	4.4	7.7
Precancerous and not specified	0.1	2.2	0.6	5.4	22.5	36.4	79.6		
Actinic keratosis	—	—	—	1.5	15.9	32.8	65.1		
Seborrheic dermatitis	9.8	6.7	18.8	38.7	41.9	41.6	31.9	28.7	36.4
Seborrheic dermatitis, type not indicated	9.6	6.7	18.5	38.7	41.9	41.1	31.9	27.4	36.3
Atopic dermatitis, eczema	24.9	20.0	17.0	15.4	28.0	14.7	13.9	12.8	18.3
Atopic dermatitis	19.3	13.8	10.7	5.1	6.0	1.4	2.4	1.0	1.7
Lichen simplex chronicus	—	1.7	4.1	1.2	5.8	5.5	5.2	9.4	10.3

TABLE 2 (continued)
Prevalence of Significant Skin Pathology of all Types and the Most Frequently Occurring Types of Conditions among Persons 1–74 Years, by Age, Showing Rates and Standard Errors: U.S., 1971–1974

Condition	1–5	6–11	12–17	18–24	25–34	35–44	45–54	55–64	3.2
				(Rate per 1,000 Population)					
Hand eczema	—	—	0.2	0.3	4.9	3.7	1.5	1.9	0.9
Nummular eczema	1.9	1.3	0.6	4.9	2.5	0.1	0.7	0.6	3.2
Dyshidrotic eczema	—	1.9	0.4	2.6	5.9	1.3	4.0	—	0.4
Contact dermatitis	13.9	4.9	8.6	14.2	9.6	17.0	16.7	23.6	20.4
Ichthyosis, kerstosis	8.6	15.5	16.7	8.8	11.5	7.7	2.7	7.0	1.5
Verruce vulgaris	4.7	13.1	15.5	12.4	6.3	6.2	8.9	2.4	3.6
Folliculitis	1.2	3.2	5.3	18.9	15.4	12.8	5.5	2.1	1.1
Psoriasis	3.1	1.9	2.5	1.5	6.2	4.9	11.4	10.9	9.5
Seborrheic keratosis	0.2	—	—	0.4	0.4	4.5	9.7	18.3	24.4
Vitiligo	0.6	5.5	1.0	3.6	3.0	3.4	5.9	12.6	13.6
Urticaria (hives, etc.)	3.1	1.8	3.4	4.3	4.6	3.6	8.8	7.6	7.4
Herpes simplex	3.4	5.3	5.1	3.2	5.2	4.8	2.8	3.3	4.9
All other skin conditions	61.5	68.8	60.2	82.7	93.9	141.6	149.4	151.6	177.3
				Standard error of rate					
Persons with one or more significant skin conditions	13.03	18.45	28.57	27.50	24.90	25.28	23.08	23.27	24.91

[a] Seborrheic keratosis is listed with benign tumors.

TABLE 3
Prelevance of Significant Skin Pathology of All Types and the Most Frequently Occuring Types of Conditions Among Males 1–74 Years, by Age — U.S., 1971–1974

Condition	1-5	6-1	12-17	18-24	25-34	35-44	45-54	55-64	65-74
				(Rate per 1,000 males)					
Persons with one or more significant skin conditions	135.9	182.4	363.6	393.4	358.4	335.8	438.0	424.4	467.8
Significant skin conditions, all types	155.8	209.7	477.2	564.7	532.8	508.4	677.4	717.9	782.4
Disease of sebaceous glands	4.9	19.5	248.0	206.6	102.6	38.2	30.0	31.7	25.7
Acne vulgaris	—	6.7	232.5	182.8	85.1	17.7	15.3	5.3	1.6
Cystic acne			2.4	9.3	10.8	3.5	—	—	0.2
Acne scars			3.5	8.9		—	3.1	1.2	
Xerosis	3.1	9.1	3.9	2.7	1.5	5.5	4.6	9.6	11.8
Dermatophytoses	3.6	12.4	54.O	101.8	152.9	198.4	268.5	247.5	186.5
Tinea pedis	0.8	7.1	2&0	46.7	94.0	106.4	153.4	98.8	95.9
Tinea unguium		1.2	3.7	10.0	21.4	34.2	64.7	103.2	71.3
Tinea versicolor	1.0	2.5	10.9	24.7	11.5	17.7	9.4	10.2	4.9
Tinea cruris	—	—	9.2	13.8	19.3	21.8	24.6	18.5	2.9
Tumors	14.5	22.1	25.5	44.9	41.4	41.1	92.3	128.4	222.7
Malignant	—	—	1.0		—	3.5	11.3	20.6	44.0
Basal cell epithelioma	—	—	—	—	—	3.5	10.7	15.5	7.1
Benign[a]	14.5	22.1	24.3	42.1	40.1	30.8	48.0	50.2	65.5
Lipoma	—	—	—	1.8	1.3	0.9	5.7	8,1	10.9
Pre-cancerous and not specified			0.2	2.8	1.3	6.8	33.0	57.6	113.2
Actinic keratosis	—	—	—	—	—	1.8	25.4	53.2	96.3
Seborrheic dermatitis	7.1	5.5	13.9	37.6	40.5	28.84	0.1	32.0	39.3
Seborrheic dermatitis, type not indicated	7.1	5.5	13.9	37.6	40.5	28.8	40.1	29.3	39.3
Atopic dermatitis, eczema	21.2	29.3	19.4	10.7	28.9	10.1	17.4	14.2	23.3
Atopic dermatitis	17.0	19.7	15~5	5.6	5.7	—	3.6	—	1.6
Lichen simplex chronicus		3.0	2.5	1.1	6.9	1.1	4.7	11.9	18.0
Hand eczema		—	—		3.1	4.3	—	2.3	
Nummular eczema		2.6			2.9	—	1.2	—	1.6
Dyshidrotic eczema		2.5		3.5	8.1	2.4	7.7		0.2
Contact dermatitis	11.2	5.7	6.9	19.5	9.6	14.2	14.1	33.1	18.0
Ichthyosis, kerstosis	10.7	18.4	12.7	10.3	10.4	6.8	3.1	3.5	2.0

TABLE 3 (continued)
Prelevance of Significant Skin Pathology of All Tipes and the Most Frequently Occuring Types of Conditions Among Males 1–74 Years, by Age — U.S., 1971–1974

Condition	1-5	6-1	12-17	18-24	25-34	35-44	45-54	55-64	65-74
				(Rate per 1,000 males)					
Verruca vulgaris	5.2	12.8	14.8	21.5	7.4	8.5	8.7	2.6	4.5
Folliculitis	2.0	3.5	7.5	29.9	25.4	22.2	6.3	1.2	2
Psoriasis	5.8	1.2	—		9.9	3.1	10.4	15.2	14.4
Seborrheic keratosis						4.3	10.4	15.3	24.4
Vitiligo	0.2	4.3	1.1	3.1	2.1		3.5	9.5	ISA
Urticaria (hives, etc.)	4.4	1.7	1.3	4.0	2.0	0.8	10.5	4.8	7.1
Herpes simplex	2.7	4.8	7.4	0.5	7.6	3.1	4.4	0.6	1.8
All other skin conditions	62.3	68.5	64.7	74.3	92.1	128.8	157.7	1783	193.8

[a] Includes seborrheic keratosis.

TABLE 4

Prelavence of Significant Skin Pathology of All Types and the Most Frequently Occuring Types of Conditions Among Females 1–74 Years, by Age, U.S., 1971–1974

Condition	1–5	6–1	12–17	18–24	25–34	35–44	45–54	55–64	65–74
				(Rate per 1000 females)					
Persons with one or more significant skin conditions	148.9	166.0	360.4	338.5	281,2	321.1	282.3	306.1	365.5
Significant skin conditions, all types	177.4	180.7	459.3	439.2	380.0	444.8	394.0	426.5	558.5
Disease of sebaceous glands	7.7	23.5	251.8	177.1	96.7	49.5	13.7	32.2	25.3
Acne vulgaris	0.7	17.3	231.3	162.2	82.2	32.4	2.7	—	—
Cystic acne	—	—	1.6	1.7	—	1.8	—	—	—
Acne scars			5.5	1.9	1.8	0.4	0.7	—	—
Xerosis	2.5	4.1	3.9	4.5	6.4	8.0	6.2	20.7	17.5
Dermatophytoses	3.2	4.1	11.4	22.3	28.3	51.1	52.2	64.6	81.8
Tinea pedis	—	0.8	1.5	4.5	8.3	21.3	14.5	21.1	32.2
Tinea unguium	—	0.7	4.2	6.1	16.3	29.2	37.7	42.2	
Tinea versicolor	1.0	0.6	9.4	11.4	8.5	8.5	5.1	4.7	0.4
Tinea cruris	—	—	—	—	1.7	2.4	2.5	0.4	—
Tumors	29.2	15.7	37.4	41.1	38.1	52.1	78.3	75.2	154.8
Malignant	—	—	—	0.2	0.2	3.0	8.0	17.5	30.5
Basal cell epithelioma	—	—	—	0.2	—	3.0	7.6	11.4	14.6
Benign[a]	29.2	15.7	37.4	39.4	37.9	45.2	57.5	40.3	70.2
Lipoma	—	1.7				2.4	4.9	1.0	5.2
Pre-cancerous and not specified	—	—	—	1.5	—	3.9	12.8	17.4	54.1
Actinic keratosis	—	—	—	—	—	1.2	7.3	14.6	41.6
Seborrheic dermatitis	12.6	7.9	23.8	39.8	43.2	53.3	24.3	25.7	34.2
Seborrheic dermatitis, type not indicated	12.3	7.8	23.3	39.8	43.2	52.4	24.3	25.7	34.1
Atopic dermatitis, eczema	28.7	10.6	14.4	19.7	27.2	18.9	10.8	11.7	14.6
Atopic dermatitis	21.4	7.7	5.5	4.6	6.3	2.6	1.3	1.9	1.6
Lichen simplex chronicus	—	0.4	5.7	1.2	4.7	9.6	5.7	7.1	4.4
Hand eczema	—	—	0.3	0.7	6.6	3.2	2.9	1.6	1.6
Nummular eczema	4.0	—	1.2	9.6	2.1	0.2	0.2	1.1	4.4
Dyshidrotic eczema	—	1.3	0.7	1.7	3.9	0.3	0.7	—	0.4
Contact dermatitis	16.8	4.1	10.4	9.2	9.4	19.5	19.1	15.3	22.3
Ichthyosis, keratosis	6.4	12.5	20.8	7.3	12.6	8.2	2.3	10.1	1.1
Verruca vulgaris	4.1	13.4	16.4	4.0	5.1	4.1	9.2	1.1	2.1
Folliculitis		0.4	2.8	3.1	8.7	6.1	4.1	4.7	2.8

TABLE 4 (continued)
Prelavence of Significant Skin Pathology of All Types and the Most Frequently Occuring Types of Conditions Among Females 1–74 Years, by Age, U.S., 1971–1974

Condition	1–5	6–1	12–17	18–24	25–34	35–44	45–54	55–64	65–74
				(Rate per 1000 females)					
Psoriasis	0.2	2.8	5.1	3.1	2.6	6.5	12.4	7.1	5.8
Seborrheic keratosis	0.5			0.8	0.9	4.7	9.1	21.1	24.5
Vitiligo	1.1	6.7	0.8	4.0	3.9	6.5	8.2	15.4	11.5
Urticaria (hives, etc.)	1.8	1.9	5.6	4.6	6.9	6.2	7.3	10.1	7.7
Herpes simplex	4.1	5.9	2.8	5.6	Z9	6.5	1.2	1.5	7.1
All other skin conditions	60.6	68.8	55.5	91.9	96.1	153.6	141.2	132.6	165.7

[a] Includes seborrheic keratosis.

TABLE 5
Range, Mean, Median, and Quartiles in the Distribution of Dermatologist Examiners, by the Age-Sex Adjusted Proportion of Examinees with Significant Skin Pathology, Skin Conditions of Concern (Complaints) and their Ratios: Health and Nutrition Examination Survey, 1971–1974

Measure and sex	Significant skin pathology	Skin condition of concern to examinee	Ratio of 100 x skin complaint/significant skin pathology
	(Rate per 1,000 population)		
Both sexes			
Range	0-903.9	0-707.8	10-428.9
Mean	312.4	113.6	36.4
Median	265.4	88.6	43.5
P25	136.7	27.3	8.5
P75	461.1	198.0	84.8
Males			
Range	0-882.5	0-805.6	10-617.6
Mean	339.8	127.7	37.6
Median	279.7	118.8	43.5
P25	149.2	30.0	9.9
P75	508.3	261.5	99.0
Females			
Range	0-900.2	0-827.4	10-4424.6
Mean	286.6	100.3	35.0
Median	241.0	90.6	47.5
P25	122.8	25.8	84.0
P75	429.5	204.9	92.8

Note: Excludes examiner finding of no significant pathology but with examinee having conditions of concern (24.2%).

Source: Johnson, M., Skin conditions and related need for medical care among persons 1-74 years, United States, 1971-1974, Hyattsville, MD, 1978, U.S. Department of Health, Education, and Welfare, Publication No. (PHS) 79-1660.

REFERENCES

1. Johnson, M., Skin conditions and related need for medical care among persons 1-74 years, United States, 1971-1974., Hyattsville, Md., 1978, U.S. Department of Health, Education, and Welfare, DHEW Publication No. (PHS) 79-1660

EXAMINER VARIABILITY

The Data Collection Unit of the Committee on Planning for the National Academy of Dermatology (NAD), under chairman Dr. Marie Louise T. Johnson, was responsible for planning the content of the dermatology examination and for recruiting and training the 101 dermatologist examiners employed at the 65 examination locations in the Health and Nutrition Examination Survey of 1971-1974.

Advance training in the dermatology examination protocol was given each dermatologist before the survey examinations were started. In addition, members of the Data Collection Unit periodically reviewed the methods used and the recording of findings by the dermatologists during the survey. After completion of the survey, Dr. Johnson thoroughly reviewed the 20,637 recorded examinations to ensure consistency with the standard examination protocol, including the selection of the significant diagnoses, and resolved the diagnoses of skin cancer and dermatophytoses on the basis of subsequent findings from the biopsy and laboratory fungal determinations, respectively.

The number of examinees per dermatologist ranged from 4 to 638, with 27% examining fewer than 100 persons, 28% 100-199 persons, 26% 200-299 persons, and 20% 300 persons or more.

Variation would be expected among the examiners in the proportion of their examinees found to have significant skin pathology and in the proportion with skin complaints because of differences in the age-sex distribution in the groups examined, regional differences, and the extent or type of environmental exposure that may have affected the prevalence of various types of skin pathology and other factors. To assess the extent of examiner variability and the effect that it may have had on the findings from this dermatology examination, the effects of age-sex differences in the groups examined by each dermatologist have been controlled through a direct adjustment method. The age-sex-specific rates for each examiner were applied against the number of persons in that age-sex group for the total population and the age-sex-adjusted rate recomputed.

Among persons examined, the age-sex-adjusted proportions were found to have a significant skin pathology range from 0 to 90.4 % per examiner, with one-fourth of the dermatologists finding 13.7% or less of their examinees to have such pathology, one-half finding between 13.7 and 46.1%, and the remaining one-fourth finding 46.1% or more with significant skin pathology. The range in such findings among examiners is negligibly greater among females (0 to 90.0%) than among males (0 to 88.2%).

Examiner variability in the proportion of their examinees with skin complaints, when the effect of age-sex differences among the groups are removed by direct adjustment, is somewhat less than that for significant pathology because of the lower rates in general for the complaints. The range in the age-sex-adjusted rates of skin complaints is from 0 to 70.8% per examiner, with one-fourth of the examiners recording 2.7% or less, one-half of the examiners between 2.7 and 19.8%, and the remainder 19.8% or more of their examinees with skin complaints.

The ratio of skin complaints to significant skin pathology per examiner ranges from 0 to 4.29, with one fourth of the examiners showing 0.08 or less, one half from 0.08 to 0.85, and the remainder 0.85 or more.

Index

D

G

H

I

J

K

L

M

N

O

P

Q

R

S

T

U